Medicinal Plants of the Asteraceae Family

Hari Prasad Devkota • Tariq Aftab
Editors

Medicinal Plants of the Asteraceae Family

Traditional Uses, Phytochemistry and Pharmacological Activities

Springer

Editors
Hari Prasad Devkota
Headquarters for Admissions and Education
Kumamoto University
Kumamoto, Japan

Tariq Aftab
Department of Botany
Aligarh Muslim University
Aligarh, Uttar Pradesh, India

ISBN 978-981-19-6079-6 ISBN 978-981-19-6080-2 (eBook)
https://doi.org/10.1007/978-981-19-6080-2

This Springer imprint is published by the registered company Springer Nature Singapore Pte Ltd.
The registered company address is: 152 Beach Road, #21-01/04 Gateway East, Singapore 189721, Singapore

Preface

Medicinal plants have long been used for the primary healthcare of humans. Many of these plant species are also used as components of regular foods, functional foods, beverages, and cosmetics. Recent years have seen a growing interest in the natural components in these medicinal plants that are being widely used in healthy food industries, pharmaceutical, cosmetic, and aroma industries.

Asteraceae is one of the largest families of flowering plants comprising more than 1600 plant genera and 32,000 plant species. Many plants belonging to this family have a long history of being used as medicinal plants for the treatment of various diseases. Many of them are also used in the preparation of foods, beverages, and also in pharmaceutical and cosmetic industries. Plants such as *Artemisia annua* have played an important role in the discovery of novel drugs.

The purpose of this book is to highlight the various aspects related to traditional uses as food and medicine, phytochemistry, and pharmacological activities of widely used medicinal plants belonging to the Asteraceae family. It includes a total of 13 chapters with one chapter providing an overview of medicinal plants of the Asteraceae family and 12 chapters dealing with individual medicinal plants. Comprehensive compilation of the knowledge related to the traditional uses and modern scientific evidence of these plants will help to understand their relation and impact on human health. Proper understanding of current scientific knowledge will not only help in the commercialization of products based on these plants but also to find the research gaps that should be fulfilled in future for optimal use. It will also help in increasing the awareness of the plant species regarding their conservation, cultivation, and sustainable utilization.

We would like to thank all the authors for their contribution in preparing the chapters. We are also thankful to Springer for this opportunity.

Kumamoto, Japan — Hari Prasad Devkota
Aligarh, Uttar Pradesh, India — Tariq Aftab

Contents

1 **An Overview of Medicinal Plants of the Asteraceae Family and Their Role in Human Health** 1
Hari Prasad Devkota

2 ***Artemisia annua* L.: Traditional Uses, Phytochemistry, and Pharmacological Activities** 17
Kaiser Iqbal Wani, Andleeb Zehra, Sadaf Choudhary, M. Naeem, M. Masroor A. Khan, Riyazuddeen Khan, and Tariq Aftab

3 ***Artemisia indica* Willd.: Ethnobotany, Phytochemistry, Pharmacological Attributes, and Safety Profile** 43
Rupa Sanyal, Sharmila Pandey, Samapika Nandy, Saikat Dewanjee, Abdel Rahman Al-Tawaha, Ercan Bursal, Protha Biswas, Manoj Kumar, Radha, Abilash Valsala Gopalakrishnan, Md. Habibur Rahman, Mahipal S. Shekhawat, Devendra Kumar Pandey, Tabarak Malik, and Abhijit Dey

4 ***Arnica montana* L.: Traditional Uses, Bioactive Chemical Constituents, and Pharmacological Activities** 61
Nisha Gyawali, Asmita Rayamajhi, Deepa Karki, Tamlal Pokhrel, and Achyut Adhikari

5 ***Aster tataricus* L.f.: Ethnomedicinal Uses, Phytochemistry, and Pharmacological Activities** 77
Deepak Timalsina and Hari Prasad Devkota

6 *Atractylodes lancea* (Thunb.) DC.: Ethnobotany, Phytochemistry, Pharmacological Attributes, and Safety Profile ... 89
Rupa Sanyal, Akash Deep Paul, Tuyelee Das, Saikat Dewanjee, Abdel Rahman Al-Tawaha, Ercan Bursal, Protha Biswas, Manoj Kumar, Radha, Samapika Nandy, Abilash Valsala Gopalakrishnan, Md. Habibur Rahman, Mahipal S. Shekhawat, Devendra Kumar Pandey, Tabarak Malik, and Abhijit Dey

7 *Carthamus tinctorius* L.: Traditional Uses, Phytochemistry, and Pharmacological Activities ... 103
Gopal Lamichhane, Hari Prasad Devkota, Kusum Sai, and Prakash Poudel

8 *Chrysanthemum morifolium* Ramat.: A Medicinal Plant with Diverse Traditional Uses, Bioactive Constituents, and Pharmacological Activities ... 125
Jitendra Pandey, Tonking Bastola, Bhawana Dhakal, Amrit Poudel, and Hari Prasad Devkota

9 *Crassocephalum crepidioides* (Benth.) S. Moore: Traditional Uses, Chemical Constituents, and Biological Activities ... 145
Sila Gurung, Prakash Poudel, Namuna Adhikari, Gopal Lamichhane, and Rashmi Thapa

10 *Dolomiaea costus* (Falc.) Kasana and A. K. Pandey: Ethnomedicinal Uses, Bioactive Chemical Constituents, and Pharmacological Activities ... 153
Abhay Prakash Mishra, Manisha Nigam, Hari Prasad Devkota, and Motlalepula Gilbert Matsabisa

11 *Eclipta prostrata* (L.) L.: Traditional Use, Phytochemistry, and Pharmacology ... 173
Rupa Sanyal, Saheli Nandi, Sujata Mandal, Saikat Dewanjee, Abdel Rahman Al-Tawaha, Ercan Bursal, Protha Biswas, Manoj Kumar, Radha, Samapika Nandy, Abilash Valsala Gopalakrishnan, Md. Habibur Rahman, Mahipal S. Shekhawat, Devendra Kumar Pandey, Tabarak Malik, and Abhijit Dey

12 *Helianthus annuus* L.: Traditional Uses, Phytochemistry, and Pharmacological Activities ... 197
Arpit Singh, Shivaraman Pandey, Ravi Prakash Srivastava, Hari Prasad Devkota, Lav Singh, and Gauri Saxena

13 *Silybum marianum* (L.) Gaertn.: Traditional Uses, Phytochemistry, and Pharmacological Activities ... 213
Deepa Khatri, Sumit Bahadur Baruwal Chhetri, and Hari Prasad Devkota

About the Editors

Hari Prasad Devkota currently works as Assistant Professor at Headquarters for Admissions and Education, and Graduate School of Pharmaceutical Sciences, Kumamoto University, Kumamoto, Japan. He completed his PhD in Pharmaceutical Sciences from Graduate School of Pharmaceutical Sciences, Kumamoto University, in 2013. His main research interests are plant-derived bioactive natural products, functional foods, and ethnopharmacology. To date, he has authored more than 150 articles including original research articles and review articles, 3 books, and more than 20 book chapters. Dr. Devkota is currently involved as an editorial board member of various international journals related to medicinal chemistry, pharmacology, and ethnobotany. Dr. Devkota also received Kumamoto University Education Award in 2017 and 2019 for his activities related to multidisciplinary education and science communication.

Tariq Aftab received his PhD in the Department of Botany at Aligarh Muslim University, India, and is currently an Assistant Professor there. He is the recipient of a prestigious Leibniz-DAAD fellowship from Germany, Raman Fellowship from the Government of India, and Young Scientist Awards from the State Government of Uttar Pradesh (India) and Government of India. After completing his doctorate, he has worked as Research Fellow at National Bureau of Plant Genetic Resources, New Delhi, and as Post-doctorate Fellow at Jamia Hamdard, New Delhi, India. Dr. Aftab also worked as Visiting Scientist at Leibniz Institute of Plant Genetics and Crop Plant Research (IPK), Gatersleben, Germany, and in the Department of Plant Biology, Michigan State University, USA. He is a member of various scientific associations from India and abroad.

He has edited 10 books with international publishers, including Elsevier Inc., Springer Nature, and CRC Press (Taylor & Francis Group), coauthored several book chapters, and published over 60 research papers in peer-reviewed international journals. His research interests include physiological, proteomic, and molecular studies on medicinal and aromatic plants.

Chapter 1
An Overview of Medicinal Plants of the Asteraceae Family and Their Role in Human Health

Hari Prasad Devkota

Abstract The Asteraceae family, commonly known as sunflower family, is one of the largest families of flowering plants comprising more than 1600 plant genera and 32,000 plant species. Members of this family have long history of being used as medicinal plants, ornamental flowers, vegetables, food ingredients, pigments, and dyes along with many other purposes. Many members of this family are widely studied for their bioactive chemical constituents, and phenolic acids, flavonoids, terpenoids, volatile components, etc. are reported as major compounds. Pharmacological studies have revelated their potent antioxidant, anti-inflammatory, antibacterial, anticancer, and antiparasitic activities, among others. Extensive chemical and pharmacological analysis of the less explored species can lead to discovery and development of novel drug molecules, functional food ingredients, and cosmetic products.

Keywords Sunflower family · Flavonoids · Phenolic acids · Pharmacological activities · *Artemisia* · Chrysanthemum

1.1 Introduction

Medicinal plants have been an important source of the primary healthcare for prevention and treatment of diseases. Many of these plant species are also used as components of foods, nutraceuticals, functional foods, beverages, cosmetics, dyes, and many other purposes (Khanal et al. 2021). Medicinal plants are one of the important sources of modern drug discovery and development and more than 30% of the drugs currently marketed are derived from natural products (Newman and Cragg 2016; Atanasov et al. 2021).

H. P. Devkota (✉)
Headquarters for Admissions and Education, Kumamoto University, Kumamoto, Japan

Graduate School of Pharmaceutical Sciences, Kumamoto University, Kumamoto, Japan

Pharmacy Program, Gandaki University, Pokhara, Nepal
e-mail: devkotah@kumamoto-u.ac.jp

H. P. Devkota, T. Aftab (eds.), *Medicinal Plants of the Asteraceae Family*,
https://doi.org/10.1007/978-981-19-6080-2_1

The Asteraceae family (synonym: Compositae), commonly known as sunflower family, is one of the largest families of flowering plants comprising more than 1600 plant genera and 32,000 plant species (POWO 2022). Plants of the Asteraceae family can be found from sea level to high mountains, and their main characteristic is the presence of capitula containing many florets (Willis 2017). Many of this family have long history of being used as medicinal plants for the treatment of various diseases as some of them are reported to be cultivated from 3000 years ago (Rolnik and Olas 2021). Apart from being medicinally important, they also have great importance as ornamental flowers, vegetables, food ingredients, pigments, and dyes. Some of these plants are also potent insecticides. Many members of this family are widely studied for their bioactive chemical constituents and pharmacological activities. Some of the plants have shown promising results as potential agents in the development of modern drugs.

1.2 Traditional Uses of the Plants of the Asteraceae Family as Medicine, Food, Cosmetic, and Other Purposes

Asteraceae family plants have long history of use for various purposes. Having beautiful flowers, many plant species of this family are cultivated for ornamental purposes. Plants such as marigold (*Tagetes erecta* L.), French marigold (*Tagetes petula* L.), sunflower (*Helianthus annuus* L.), pot marigold (*Calendula officinalis* L.), and corn flower (*Centaurea cyanus* L.) are cultivated all over the world for their beautiful flowers (Fig. 1.1). Many of these plants are also used in cultural and religious ceremonies and functions.

Hundreds of plant species of the Asteraceae family are used as medicinal plants and as important ingredients in traditional medicines. Many of such species are in the form of herbal tea prepared from flowers or other plant parts or as one of the ingredients in the traditional medicine formulations. Echinacea (*Echinacea purpurea* (L.) Moench), chamomile (*Matricaria chamomilla* L.), Tatarian aster (*Aster tataricus* L.f.), dandelion (*Taraxacum officinale* F.H.Wigg.), and plants of *Cirsium* and *Artemisia* genera are some of the common examples among many (Fig. 1.2). Their traditional uses vary from one species to another species and also depend upon the ethnic knowledge of these species in various communities. Some common uses are for the treatment of fever, cold, allergy, inflammation, etc. (Watanabe et al. 2021). Many plants of *Artemisia* genus are used to treat cough, cold, gastrointestinal disorders, bacterial and parasitic diseases, etc. (Nigam et al. 2019). The young leaves are also used to treat cuts and wounds.

Dandelion is widely used as a diuretic agent (Rolnik and Olas 2021). Blessed milk thistle (*Silybum marianum* (L.) Gaertn.) and false daisy (*Eclipta prostrata* (L.) L.) are well known as potent hepatoprotective agents (Ball and Kowdley 2005; Timalsina and Devkota 2021). Various species of *Arctium* genus are commonly used as medicinal species and have anti-inflammatory and diuretic properties (Wang et al.

Fig. 1.1 Photographs of some common ornamental flowers of Asteraceae family

2019). They are also used to treat skin diseases, wounds, rheumatic pain, etc. Various plants of this family are rich source of essential oils and have string antimicrobial activities. Lakshamn et al. reported a review of some common Asteraceae plants used in dermatological diseases (Lakshman et al. 2014). Many of these plants are widely used as antiparasitic agents and also have shown similar activities in lab experiments (Panda and Luyten 2018).

Many plants of Asteraceae family are also used as vegetables and components in foods. Flower buds of Artichole (*Cynara cardunculus* var. *scolymus* L.), tubers of Jerusalem artichoke (*Helianthus tuberosus* L.), young leaves and roots of greater burdock (*Arctium lappa* L.), roots of elecampane (*Inula helenium* L.), leaves and roots of chicory (*Cichorium intybus* L.) and young leaves of lettuce (*Lactuca sativa* L.), garland chrysanthemum (*Xanthophthalmum coronarium* (L.) P.D.Sell) (Fig. 1.3), and Okinawan spinach (*Gynura bicolor* (Roxb. ex Willd.) DC.) are some of the examples of plants commonly used as vegetables (Watanabe et al. 2021; Rolnik and Olas 2021). Many of these vegetables are rich in vitamins, minerals, polyphenols, fibers, and other bioactive compounds (Rolnik and Olas 2021). Plants such as paracress (*Acmella oleracea* (L.) R.K. Jansen) are used as spices in many south Asian countries (Khanal et al. 2021). Flowers of chamomile,

Fig. 1.2 Some common medicinal plants of Asteraceae family

chrysanthemum (*Chrysanthemum morifolium* Ramat.), Indian chrysanthemum (*Chrysanthemum indicum* L.), etc. are used as herbal tea (Fig. 1.4).

Some of the plants of Asteraceae family are also source of natural pigments used in cosmetic and food and as textile dyes from ancient times till today. One of the most prominent examples is safflower (*Carthamus tinctorius*) rich in quinochalcone *C*-glucosides. Carthamin in petals is responsible for its red color and

Fig. 1.3 Photographs of artichoke and garland chrysanthemum

Fig. 1.4 Photographs of *Chrysanthemum* species

hydroxysafflower yellow A, safflower yellow A and B, safflomin A and C, and tinctormine are responsible for yellow coloration (Kazuma et al. 2000; Shinozaki et al. 2016; Menegaes and Nunes 2020).

Sunflower and safflower seeds are also rich sources of vegetable oils. Sunflower oil is the most widely produced and consumed vegetable oil in the world (Pilorgé

Fig. 1.5 Photographs of stevia and chemical structure of stevioside

2020). Similarly, safflower oil has been widely used in Asian countries (Menegaes and Nunes 2020).

Stevia (*Stevia rebaudiana* (Bertoni) Bertoni) is widely used as natural sweetener in various food products and beverages. Many diterpene glycosides such as stevioside (Fig. 1.5) and rebaudioside A are responsible for its sweetness (Ceunen and Geuns 2013; Ahmad et al. 2020).

Dalmatian pyrethrum (*Tanacetum cinerariifolium* (Trevir.) Sch. Bip.) and pyrethrum (*Tanacetum coccineum* (Willd.) Grierson) are traditionally used as insect repellants. Isolated compounds such as pyrethrin I, pyrethrin II, and cinerin I (Fig. 1.6) are widely used in insect repellant and insecticide products (Fradin 2013; Watanabe et al. 2021).

1.3 Bioactive Chemical Constituents and Pharmacological Activities of the Plants of Asteraceae Family

Medicinal plants of the Asteraceae family are widely studied for the chemical constituents and biological/pharmacological activities. Being used as foods and traditional medicines, they have received attentions from researchers to explore their active chemical constituents for the discovery and development of new drugs, functions foods, cosmetic products, and also scientific evidences for their traditional uses.

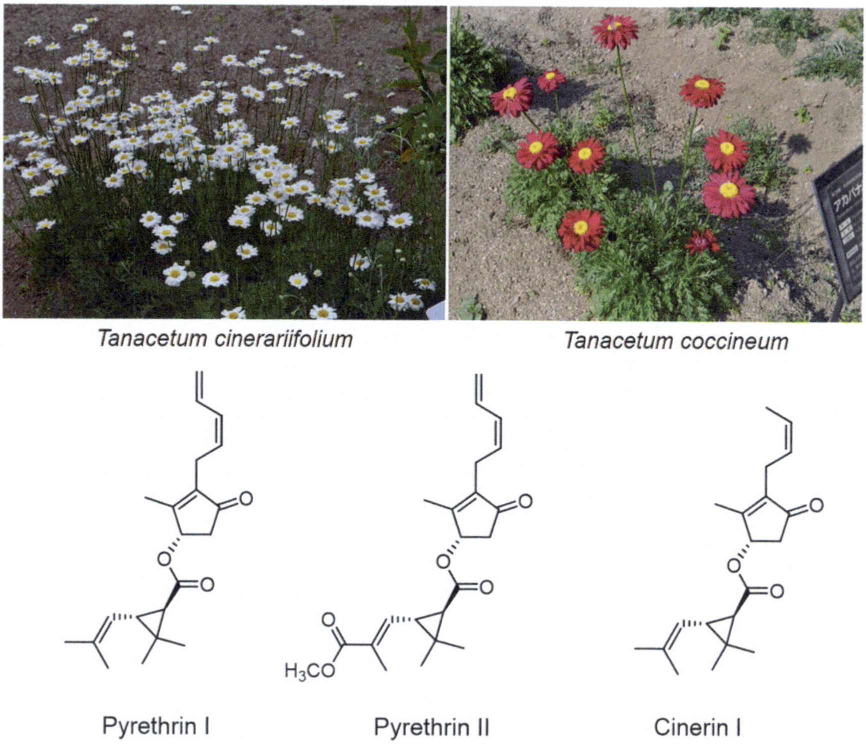

Fig. 1.6 Photographs of *Tanacetum* species and chemical structures of pyrethrin derivatives

Phenolic acid derivatives, flavonoids, tersesquiterpene lactones and other terpenoids, and polyacetylenes are some of the most commonly reported compounds from Asteraceae plants. The nature of these compounds varies depending upon plant species, plant parts used, extraction methodologies, etc. Seeds of some plants such as sunflower and safflower are used to obtain edible oils which are rich unsaturated fatty acids such as linoleic acid and linoleic acids. For example, sunflower oil is reported to contain unsaturated and saturated fatty acids in the approximate ratio of 85:15, and oleic acid and linoleic acids are the main unsaturated fatty acids (Akkaya 2018). Tubers of Jerusalem artichoke *Helianthus tuberosus* and roots of *Inula helenium* and *Cichorium intybus* are rich in inulin and nondigestible polysaccharide receiving great attention in recent years as food functional ingredient (Flamm et al. 2001; Mensink et al. 2015; Shoaib et al. 2016).

Phenolic acids such as gallic acid, caffeic acid, rosmarinic acid, and mono- or di-caffeoyl quinic acids are widely reported (Cheng et al. 2005; He et al. 2021; Barral-Martinez et al. 2021; Devkota et al. 2022; Nalewajko-Sieliwoniuk et al. 2019). Structures of some of these phenolic compounds are presented in Fig. 1.7. Phenolic acids are widely reported as active antioxidant, enzyme inhibitory, antibacterial, anti-inflammatory, and cytotoxic components (Cheng et al. 2005;

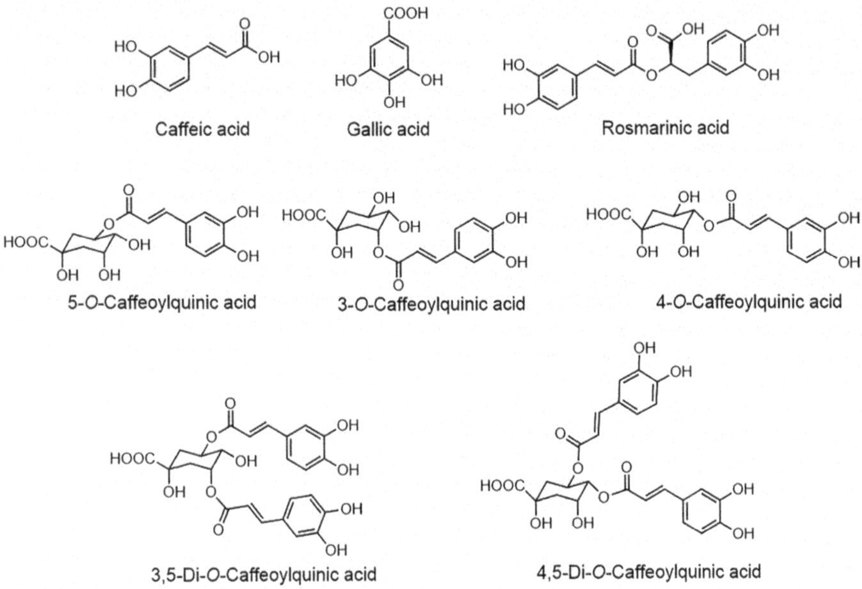

Fig. 1.7 Chemical structures of common phenolic acid derivatives

Chen et al. 2014; Mekinic et al. 2014; Kashif et al. 2015; Barral-Martinez et al. 2021; Devkota et al. 2022).

Hundreds of flavonoids have been reported from plants of the Asteraceae family. Commonly reported flavonoids include apigenin, luteolin, kaempferol, quercetin, and 6-hydroxylated flavonoids such as scutellarein, hispidulin eupatorine, sinesetin, and their methylated, acylated, and glycosylated derivatives (Fig. 1.8) (El-karemy et al. 1987; Bohm and Stuessy 2001; Emerenciano et al. 2001; Chicaro et al. 2004; Wang et al. 2013; Nigam et al. 2019). Similar to phenolic acid derivatives, flavonoids are also reported as potent antioxidant, anti-inflammatory, and anticancer compounds.

Sesquiterpene lactones and furanosesquiterpenes are another class of characteristic compounds in Asteraceae which are mainly involved in plant protection (Stefani et al. 2006; Hristozov et al. 2007; Sakamoto et al. 2010; Salapovic et al. 2013; Chadwick et al. 2013; Sokovic et al. 2017). They are reported to possess various biological activities such as anti-inflammatory and anti-tumor/anticancer activities (Stefani et al. 2006; Chadwick et al. 2013; Sokovic et al. 2017). However, many such compounds, especially those with an activated exocyclic methylene group are reported as potent allergens responsible for allergic dermatitis (Salapovic et al. 2013; Chadwick et al. 2013). Structures of some common sesquiterpene lactones are presented in Fig. 1.9.

One of the most important sesquiterpene lactone derivative is artemisinin, isolated from sweet wormwood (*Artemisia annua* L.), used worldwide as a potent antimalarial agent (Fig. 1.10). Many other natural and synthetic derivatives of

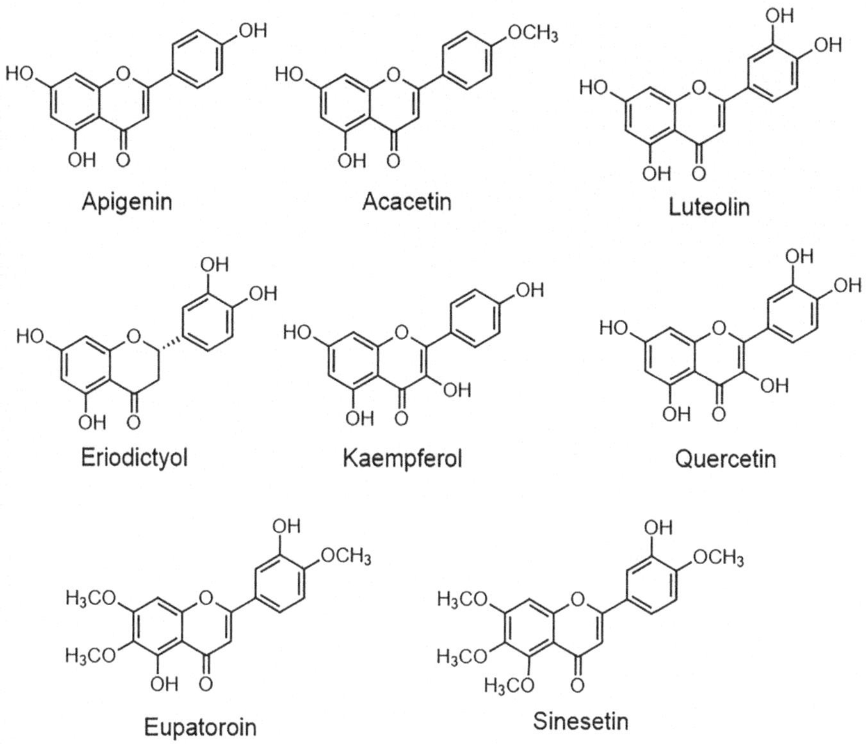

Fig. 1.8 Chemical structures of common flavonoids

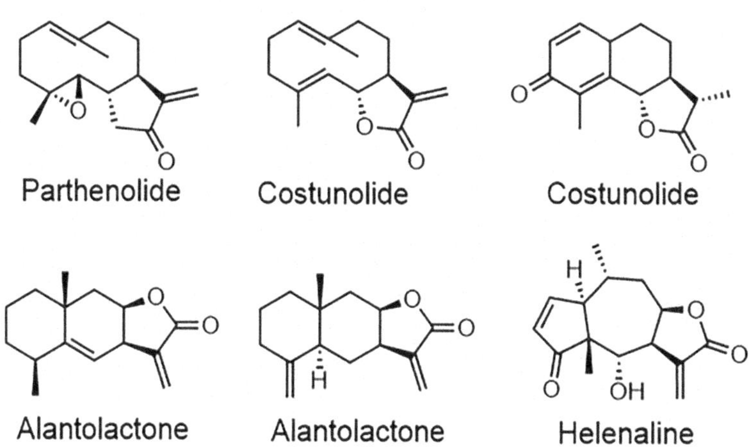

Fig. 1.9 Chemical structures of common sesquiterpene lactones

Fig. 1.10 Photograph of *Artemisia annua* and chemical structure of artemisinin

artemisinin are being studied and evaluated as potential antiparasitic, antimicrobial, and anticancer agents (Ansari et al. 2013; Lanteri et al. 2014; Corsello and Garg 2015; Tiwari and Chaudhary 2020; Augustin et al. 2020; Ma et al. 2021).

Many other classes of compounds from Asteraceae plants are being evaluated for diverse pharmacological activities such as antidiabetic, anti-obesity, antihypertensive, cardiovascular protective, and anticancer activities, among others (de Souza et al. 2013; Cifuentes et al. 2016; Atawodi et al. 2017; Dib et al. 2017; Spínola and Castilho 2017; Michel et al. 2020). Few studies have also reported promising results. For example, Awale et al. (2006) evaluated the antitumor activity of arctigenin, arctiin (Fig. 1.11), and other lignans, isolated from the seeds of *Arctium lappa*, in pancreatic cancer (PANC-1) cell lines and in animal model. Arctigenin showed preferential cytotoxicity against PANC-1 cell lines and also suppressed tumor growth in nude mice through inhibiting the tolerance of cancer cells to nutrient starvation.

Asteraceae family plants are also rich in essential oils having strong antimicrobial and antioxidant activities (Abad et al. 2012; Abu-Darwish et al. 2015; Bandeira Reidel et al. 2018; Salehi et al. 2018; Cazella et al. 2019; Nigam et al. 2019; Youssef et al. 2020; Valarezo et al. 2021). Many of these essential oils are also used as fragrances. Structures of some of the main components in these essential oils are represented in Fig. 1.12.

1.4 Conclusion and Future Prospective

Asteraceae is one of the largest families of flowering plants, and many members of this family are used as medicinal plants from ancient times. It also includes various plants with agricultural and food values and economic importance. There are still many plants species which have not been studies in detail for their chemical constituents and pharmacological activities. Extensive chemical and pharmacological analysis of the less explored species can lead to discovery and development of novel drug molecules, functional food ingredients and cosmetic products. However,

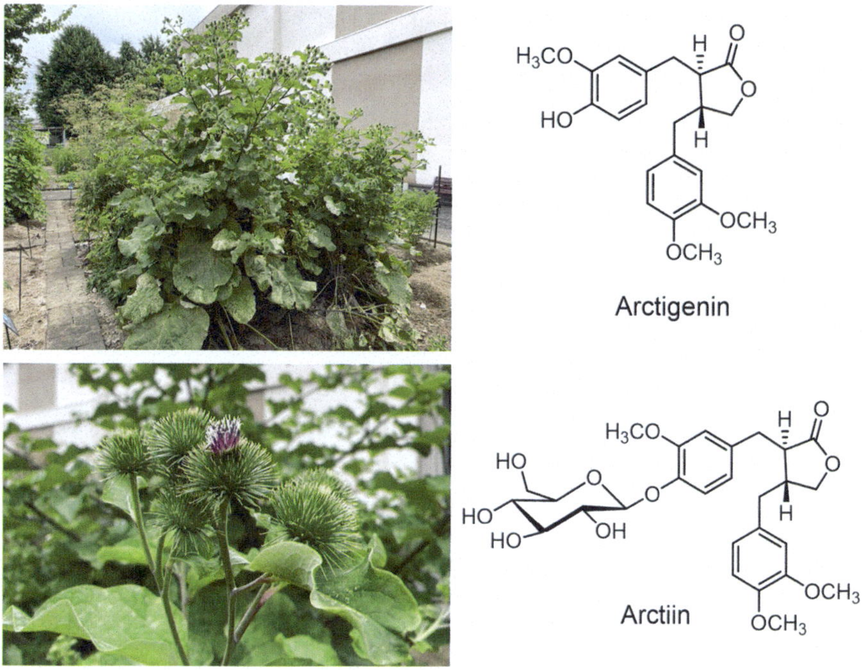

Fig. 1.11 Photographs of *Articum lappa* and chemical structures of arctigenin and arctiin

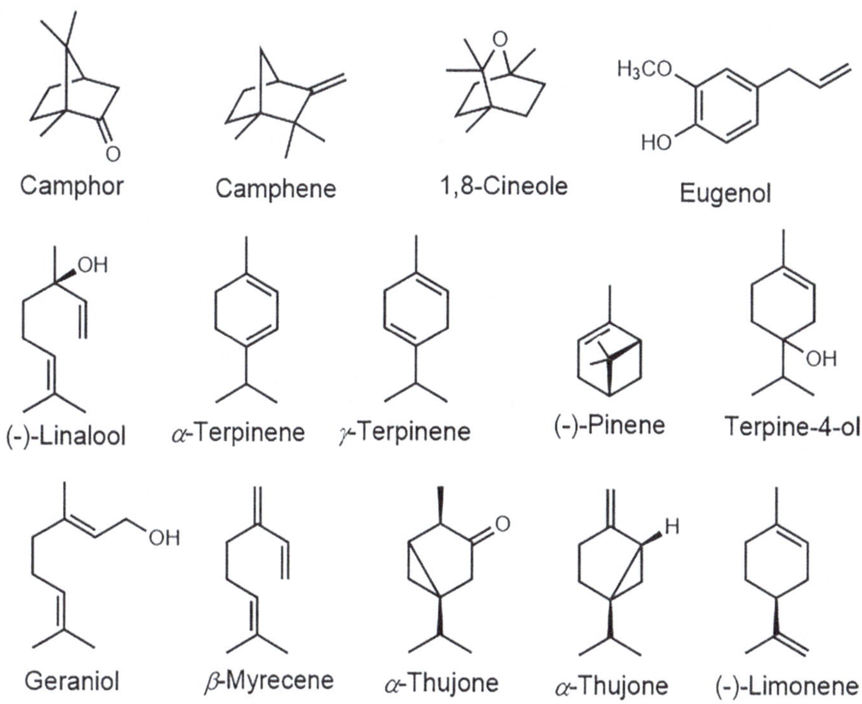

Fig. 1.12 Chemical structures of main components in essential oils of Asteraceae plants

it also contains some allergic and toxic plant species thus great care should be taken when handlining and using these species. Detailed analysis of safety and possible toxicities are necessary in future.

References

Abad MJ, Bedoya LM, Apaza L, Bermejo P (2012) The Artemisia L. genus: a review of bioactive essential oils. Molecules 17:2542–2566. https://doi.org/10.3390/molecules17032542

Abu-Darwish MS, Cabral C, Gonçalves MJ et al (2015) Artemisia herba-alba essential oil from Buseirah (South Jordan): chemical characterization and assessment of safe antifungal and anti-inflammatory doses. J Ethnopharmacol 174:153–160. https://doi.org/10.1016/j.jep.2015.08.005

Ahmad J, Khan I, Blundell R et al (2020) Stevia rebaudiana Bertoni.: an updated review of its health benefits, industrial applications and safety. Trends Food Sci Technol 100:177–189. https://doi.org/10.1016/J.TIFS.2020.04.030

Akkaya MR (2018) Prediction of fatty acid composition of sunflower seeds by near-infrared reflectance spectroscopy. J Food Sci Technol 55:2318. https://doi.org/10.1007/S13197-018-3150-X

Ansari M, Saify Z, Sultana N et al (2013) Malaria and artemisinin derivatives: an updated review. Mini Rev Med Chem 13:1879–1902. https://doi.org/10.2174/13895575113136660097

Atanasov AG, Zotchev SB, Dirsch VM et al (2021) Natural products in drug discovery: advances and opportunities. Nat Rev Drug Discov 20(3):200–216. https://doi.org/10.1038/s41573-020-00114-z

Atawodi SE, Adepoju OA, Nzelibe HC (2017) Antihyperglycaemic and hypolipidemic effect of methanol extracts of ageratum conyzoides L (Asteraceae) in normal and diabetic rats. Trop J Pharm Res 16:989–996. https://doi.org/10.4314/TJPR.V16I5.4

Augustin Y, Staines HM, Krishna S (2020) Artemisinins as a novel anti-cancer therapy: targeting a global cancer pandemic through drug repurposing. Pharmacol Ther 216:107706. https://doi.org/10.1016/J.PHARMTHERA.2020.107706

Awale S, Lu J, Kalauni SK et al (2006) Identification of arctigenin as an antitumor agent having the ability to eliminate the tolerance of cancer cells to nutrient starvation. Cancer Res 66:1751–1757. https://doi.org/10.1158/0008-5472.CAN-05-3143

Ball KR, Kowdley KV (2005) A review of Silybum marianum (milk thistle) as a treatment for alcoholic liver disease. J Clin Gastroenterol 39:520–528. https://doi.org/10.1097/01.MCG.0000165668.79530.A0

Bandeira Reidel RV, Nardoni S, Mancianti F et al (2018) Chemical composition and antifungal activity of essential oils from four Asteraceae plants grown in Egypt. Z Naturforsch C J Biosci 73:313–318. https://doi.org/10.1515/ZNC-2017-0219

Barral-Martinez M, Garcia-Oliveira P, Nuñez-Estevez B et al (2021) Plants of the family Asteraceae: evaluation of biological properties and identification of phenolic compounds. Chem Proc 5:51. https://doi.org/10.3390/CSAC2021-10486

Bohm BA, Stuessy TF (2001) Structural variation of the flavonoids of Asteraceae. In: Flavonoids of the Sunflower Family (Asteraceae). Springer Wien, Vienna, pp 64–122. https://doi.org/10.1007/978-3-7091-6181-4_4

Cazella LN, Glamoclija J, Soković M et al (2019) Antimicrobial activity of essential oil of baccharis dracunculifolia DC (Asteraceae) aerial parts at flowering period. Front Plant Sci 10:27. https://doi.org/10.3389/FPLS.2019.00027/BIBTEX

Ceunen S, Geuns JMC (2013) Steviol glycosides: chemical diversity, metabolism, and function. J Nat Prod 76:1201–1228. https://doi.org/10.1021/NP400203B/ASSET/IMAGES/MEDIUM/NP-2013-00203B_0007.GIF

Chadwick M, Trewin H, Gawthrop F, Wagstaff C (2013) Sesquiterpenoids lactones: benefits to plants and people. Int J Mol Sci 14:12780. https://doi.org/10.3390/IJMS140612780
Chen J, Mangelinckx S, Ma L et al (2014) Caffeoylquinic acid derivatives isolated from the aerial parts of Gynura divaricata and their yeast α-glucosidase and PTP1B inhibitory activity. Fitoterapia 99:1–6. https://doi.org/10.1016/j.fitote.2014.08.015
Cheng W, Li J, You T, Hu C (2005) Anti-inflammatory and immunomodulatory activities of the extracts from the inflorescence of Chrysanthemum indicum Linné. J Ethnopharmacol 101:334–337. https://doi.org/10.1016/J.JEP.2005.04.035
Chicaro P, Pinto E, Colepicolo P et al (2004) Flavonoids from Lychnophora passerina (Asteraceae): potential antioxidants and UV-protectants. Biochem Syst Ecol 32:239–243. https://doi.org/10.1016/J.BSE.2003.08.003
Cifuentes F, Paredes A, Palacios J et al (2016) Hypotensive and antihypertensive effects of a hydroalcoholic extract from Senecio nutans Sch. Bip. (Compositae) in mice: chronotropic and negative inotropic effect, a nifedipine-like action. J Ethnopharmacol 179:367–374. https://doi.org/10.1016/J.JEP.2015.12.048
Corsello MA, Garg NK (2015) Synthetic chemistry fuels interdisciplinary approaches to the production of artemisinin. Nat Prod Rep 32:359–366. https://doi.org/10.1039/c4np00113c
de Souza P, Crestani S, da Silva RDCV et al (2013) Involvement of bradykinin and prostaglandins in the diuretic effects of Achillea millefolium L. (Asteraceae). J Ethnopharmacol 149:157–161. https://doi.org/10.1016/J.JEP.2013.06.015
Devkota HP, Tsushiro K, Watanabe T (2022) Bioactive phenolic compounds from the flowers of Farfugium japonicum (L.) Kitam. var. giganteum (Siebold et Zucc.) Kitam. (Asteraceae). Nat Prod Res 36(15):4036–4039. https://doi.org/10.1080/14786419.2021.1903004
Dib I, Tits M, Angenot L et al (2017) Antihypertensive and vasorelaxant effects of aqueous extract of Artemisia campestris L. from Eastern Morocco. J Ethnopharmacol 206:224–235. https://doi.org/10.1016/J.JEP.2017.05.036
El-karemy ZAR, Mansour RMA, Fayed AA, Saleh NAM (1987) The flavonoids of local members of astereae (compositae). Biochem Syst Ecol 15:53–55. https://doi.org/10.1016/0305-1978(87)90079-2
Emerenciano VP, Militão JSLT, Campos CC et al (2001) Flavonoids as chemotaxonomic markers for Asteraceae. Biochem Syst Ecol 29:947–957. https://doi.org/10.1016/S0305-1978(01)00033-3
Flamm G, Glinsmann W, Kritchevsky D et al (2001) Inulin and oligofructose as dietary fiber: a review of the evidence. Crit Rev Food Sci Nutr 41:353–362. https://doi.org/10.1080/20014091091841
Fradin MS (2013) Insect repellents. In: Comprehensive dermatologic drug therapy. Saunders Elsevier, pp 620–628.e3. https://doi.org/10.1016/B978-1-4377-2003-7.00054-6
He J, Zhang Q, Ma C et al (2021) An effective workflow for differentiating the same genus herbs of Chrysanthemum morifolium flower and chrysanthemum Indicum flower. Front Pharmacol 12:813. https://doi.org/10.3389/FPHAR.2021.575726/BIBTEX
Hristozov D, da Costa FB, Gasteiger J (2007) Sesquiterpene lactones-based classification of the family asteraceae using neural networks and k-nearest neighbors. J Chem Inf Model 47:9–19. https://doi.org/10.1021/CI060046X/ASSET/IMAGES/MEDIUM/CI060046XN00001.GIF
Kashif M, Bano S, Naqvi S et al (2015) Cytotoxic and antioxidant properties of phenolic compounds from Tagetes patula flower. Pharm Biol 53:672–681. https://doi.org/10.3109/13880209.2014.936471
Kazuma K, Takahashi T, Sato K et al (2000) Quinochalcones and flavonoids from fresh florets in different cultivars of Carthamus tinctorius L. Biosci Biotechnol Biochem 64:1588–1599. https://doi.org/10.1271/BBB.64.1588
Khanal A, Devkota HP, Kaundinnyayana S et al (2021) Culinary herbs and spices in Nepal: a review of their traditional uses, chemical constituents, and pharmacological activities. Ethnobot Res Appl 21:1–18

Lakshman HC, Yeasmin T, Gabriel KP (2014) Herbs of asteraceae and their ethano - medicinal uses in dermatological problems. J Biosci 22:127–129. https://doi.org/10.3329/JBS.V22I0.30016
Lanteri CA, Chaorattanakawee S, Lon C et al (2014) Ex vivo activity of endoperoxide antimalarials, including artemisone and arterolane, against multidrug-resistant Plasmodium falciparum isolates from Cambodia. Antimicrob Agents Chemother 58:5831–5840. https://doi.org/10.1128/AAC.02462-14
Ma Z, Woon CYN, Liu CG et al (2021) Repurposing artemisinin and its derivatives as anticancer drugs: a chance or challenge? Front Pharmacol 12:3895. https://doi.org/10.3389/FPHAR.2021.828856/BIBTEX
Mekinic IG, Skroza D, Ljubenkov I et al (2014) Phenolic acids profile, antioxidant and antibacterial activity of chamomile. Common Yarrow and Immortelle (Asteraceae) 9:1745–1748. https://doi.org/10.1177/1934578X1400901222
Menegaes JF, Nunes UR (2020) Safflower: importance, use and economical exploitation. Scientia Agraria Paranaensis 1:1–11. https://doi.org/10.18188/SAP.V19I1.21250
Mensink MA, Frijlink HW, van der Voort MK, Hinrichs WLJ (2015) Inulin, a flexible oligosaccharide I: review of its physicochemical characteristics. Carbohydr Polym 130:405–419. https://doi.org/10.1016/J.CARBPOL.2015.05.026
Michel J, Abd Rani NZ, Husain K (2020) A review on the potential use of medicinal plants from Asteraceae and Lamiaceae plant family in cardiovascular diseases. Front Pharmacol 11:1. https://doi.org/10.3389/FPHAR.2020.00852
Nalewajko-Sieliwoniuk E, Pliszko A, Nazaruk J et al (2019) Comparative analysis of phenolic compounds in four taxa of Erigeron acris s. l. (Asteraceae). Biologia (Bratisl) 74:1569–1577. https://doi.org/10.2478/S11756-019-00332-W/TABLES/3
Newman DJ, Cragg GM (2016) Natural products as sources of new drugs from 1981 to 2014. J Nat Prod 79:629–661. https://doi.org/10.1021/acs.jnatprod.5b01055
Nigam M, Atanassova M, Mishra AP et al (2019) Bioactive compounds and health benefits of artemisia species. Nat Prod Commun 14:1–17. https://doi.org/10.1177/1934578X19850354
Panda SK, Luyten W (2018) Antiparasitic activity in Asteraceae with special attention to ethnobotanical use by the tribes of Odisha. India Parasite 25:10. https://doi.org/10.1051/PARASITE/2018008
Pilorgé E (2020) Sunflower in the global vegetable oil system: situation, specificities and perspectives. OCL 27:34. https://doi.org/10.1051/OCL/2020028
POWO (2022) Plants of the world online. Facilitated by the Royal Botanic Gardens, Kew. http://www.plantsoftheworldonline.org/. Accessed 30 June 2022
Rolnik A, Olas B (2021) The plants of the Asteraceae family as agents in the protection of human health. Int J Mol Sci 22:1–10. https://doi.org/10.3390/IJMS22063009
Sakamoto HT, Laudares EP, Crotti AEM et al (2010) Sesquiterpenes Lactones and Flavonoids from Eremanthus argenteus (Asteraceae). Nat Prod Commun 5:681–684. https://doi.org/10.1177/1934578X1000500504
Salapovic H, Geier J, Reznicek G (2013) Quantification of Sesquiterpene lactones in Asteraceae plant extracts: evaluation of their allergenic potential. Sci Pharm 81:807–818. https://doi.org/10.3797/SCIPHARM.1306-17
Salehi B, Valussi M, Flaviana Bezerra Morais-Braga M et al (2018) Tagetes spp. essential oils and other extracts: chemical characterization and biological activity. Molecules 23:2847. https://doi.org/10.3390/molecules23112847
Shinozaki J, Kenmoku H, Nihei K et al (2016) Cloning and functional analysis of three chalcone synthases from the flowers of safflowers Carthamus tinctorius. Nat Prod Commun 11:787–790. https://doi.org/10.1177/1934578X1601100621
Shoaib M, Shehzad A, Omar M et al (2016) Inulin: properties, health benefits and food applications. Carbohydr Polym 147:444–454. https://doi.org/10.1016/J.CARBPOL.2016.04.020
Sokovic M, Ciric A, Glamoclija J, Skaltsa H (2017) Biological activities of Sesquiterpene lactones isolated from the genus Centaurea L. (Asteraceae). Curr Pharm Des 23:2767–2786. https://doi.org/10.2174/1381612823666170215113927

Spínola V, Castilho PC (2017) Evaluation of Asteraceae herbal extracts in the management of diabetes and obesity. Contribution of caffeoylquinic acids on the inhibition of digestive enzymes activity and formation of advanced glycation end-products (in vitro). Phytochemistry 143:29–35. https://doi.org/10.1016/J.PHYTOCHEM.2017.07.006

Stefani R, Schorr K, Tureta JM et al (2006) Sesquiterpene lactones from Dimerostemma species (Asteraceae) and in vitro potential anti-inflammatory activities. Z Naturforsch C J Biosci 61: 647–652. https://doi.org/10.1515/ZNC-2006-9-1006/MACHINEREADABLECITATION/RIS

Timalsina D, Devkota HP (2021) Eclipta prostrata (L.) L. (Asteraceae): ethnomedicinal uses, chemical constituents, and biological activities. Biomolecules 11:1738. https://doi.org/10.3390/BIOM11111738

Tiwari MK, Chaudhary S (2020) Artemisinin-derived antimalarial endoperoxides from bench-side to bed-side: chronological advancements and future challenges. Med Res Rev 40:1220–1275. https://doi.org/10.1002/med.21657

Valarezo E, Aguilera-Sarmiento R, Meneses MA, Morocho V (2021) Study of essential oils from leaves of Asteraceae family species Ageratina dendroides and Gynoxys verrucosa. J Essential Oil Bearing Plants 24:400–407. https://doi.org/10.1080/0972060X.2021.1948919

Wang T, Zhu Z, Guo Q, Mao P (2013) Variation in major flavonoids glycosides and caffeoylquinic acids during florescence of three Chrysanthemum morifolium Ramat cv. "Hangju" genotypes. Biochem Syst Ecol 47:74–79. https://doi.org/10.1016/j.bse.2012.11.004

Wang D, Bădărau AS, Swamy MK et al (2019) Arctium species secondary metabolites chemodiversity and bioactivities. Front Plant Sci 10:834. https://doi.org/10.3389/FPLS.2019.00834/BIBTEX

Watanabe M, Devkota H, Sugimura K (2021) A guidebook of medicinal Plant Park, 135th Anniversary edn. School of Pharmacy, Kumamoto University, Kumamoto

Willis KJ (ed) (2017) State of the World's plants 2017. Report. Royal Botanic Gardens, Kew

Youssef FS, Eid SY, Alshammari E et al (2020) Chrysanthemum indicum and Chrysanthemum morifolium: chemical composition of their essential oils and their potential use as natural preservatives with antimicrobial and antioxidant activities. Foods 9:1460. https://doi.org/10.3390/FOODS9101460

Chapter 2
Artemisia annua L.: Traditional Uses, Phytochemistry, and Pharmacological Activities

Kaiser Iqbal Wani, Andleeb Zehra, Sadaf Choudhary, M. Naeem, M. Masroor A. Khan, Riyazuddeen Khan, and Tariq Aftab

Abstract *Artemisia annua* L. is an annual, herbaceous, aromatic medicinal plant belonging to the family Asteraceae. It is mentioned in traditional Chinese medicine as a cure for different diseases like fever, hemorrhoid, and malaria. It is native to the mild and temperate climate of Asia but has been naturalized to other countries outside Asia as well. After the discovery of its antimalarial potential by Prof. Tu Youyou in 1972, the World Health Organization has recommended it as an antimalarial. The most common ethnobotanical usage of this plant involves the use of whole plant decoction for the treatment of cold, malaria, and cough. The whole flowering plant is known to be antipyretic, antihelminth, antispasmodic, antiseptic, and antimalarial. The antimalarial activity of this plant is due to artemisinin, a sesquiterpene lactone containing an endoperoxide moiety that acts as a key pharmacore. Artemisinin forms an important part of combinatorial treatment therapy recommended for the treatment of malaria. Artemisinin and its derivatives like artesunate have also been reported to have potent anticancer properties as well. Besides artemisinin, certain other phytochemicals reported in this plant, particularly flavonoids, have been found to have medicinal properties. They have been reported to synergize the activity of artemisinin and its derivatives against malaria. Considering the immense medicinal properties of this plant, immense research is being carried out throughout the world to isolate and characterize the different phytochemicals present in this plant. This chapter comprises the information about the general biology, distribution, and phytochemical composition of *A. annua*, updated information about its medicinal properties and health benefits, and an overview of its safety and toxicity.

Keywords Artemisinin · Antimalarial drug · Glandular trichomes

---

K. I. Wani · A. Zehra · S. Choudhary · M. Naeem · M. M. A. Khan · T. Aftab (✉)
Department of Botany, Aligarh Muslim University, Aligarh, Uttar Pradesh, India

R. Khan
Department of Environmental Sciences, Integral University, Lucknow, Uttar Pradesh, India

H. P. Devkota, T. Aftab (eds.), *Medicinal Plants of the Asteraceae Family*,
https://doi.org/10.1007/978-981-19-6080-2_2

2.1 Introduction

Medicinal plants have been used in the treatment and prevention of numerous diseases since time immemorial. At the beginning of civilization, they were important components of medication whereas nowadays they are instrumental in the manufacturing of drugs. A medicinal plant can be defined as a plant that, in one or more of its organs, contains substances that act as precursors for the synthesis of certain drugs or has therapeutic value. There are certain medicinal plants that have been scientifically proven to have medicinal and therapeutic properties, whereas some plants are regarded as medicinal and used in traditional medicines but have not been subjected to thorough scientific studies to prove their efficacy. Due to the proven effectiveness and safety claims, the consumption of medicinal plants is showing a steep rise (Perez Gutierrez and Baez 2009).

Asteraceae is regarded as one of the largest families of flowering plants which has a cosmopolitan distribution, as its plants are found all over the world, including Antarctica where it is most probably anthropogenically mediated (Smith and Richardson 2011). This family of flowering plants consists of about 13 subfamilies, 1620 genera, and around 23,600 accepted species. It comprises about 10% of total flowering plants and is rivaled only by Orchidaceae in terms of total species. The most distinguishing and diagnostic characteristics of the Asteraceae family are the presence of capitulum inflorescence, inferior ovary, and anthers united in a tube (Fornara 2014).

Artemisia annua L. ("sweet wormwood," "annual wormwood," "Qinghao"; $2n = 36$), belonging to Asteraceae family of flowering plants, is annual, aromatic, herbaceous, and glabrous or sparsely hairy, with an upright brownish colored stem. It naturally grows up to a height of 1 meter, but under cultivated conditions, it may reach up to a height of 2 meters. It usually consists of a single stem with alternate branches and deeply dissected leaves. The inflorescence consists of small capitula arranged in loose panicles with bisexual disc florets at the center and pistillate ray florets at the margins. The pollens are smooth and tricolpate, a typical characteristic of anemophilous species (Stix 1960) whereas the ovary is unilocular and inferior. The glandular secretory trichomes which are the sites of biosynthesis of antimalarial compound artemisinin (Wani et al. 2022), and nonglandular T-shaped trichomes are present on stem, leaves, and inflorescences (Ferreira and Janick 1996) which are easily visible by scanning electron microscopy (Fig. 2.1).

The plant is native to the mild and temperate climate of Asia, most probably China, and has been found as native to Myanmar, Japan, Korea, Northern India, Southern Siberia, and Vietnam; however, it has become naturalized in many countries including some areas of North America (Desrosiers and Weathers 2016). Its cultivation is carried out on a large scale in countries such as China, Tanzania, and Kenya, with small-scale cultivation being done in India and certain countries of South America and South Europe (WHO 2006).

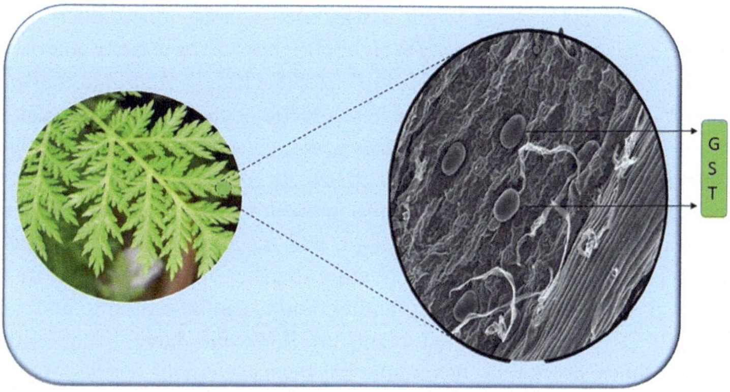

Fig. 2.1 Scanning electron microscopic image of *A. annua* leaf showing Glandular secretory trichomes in high resolution. *GST* glandular secretory trichomes

2.2 Traditional Uses

Artemisia annua has a long history of being used in traditional Chinese medicine as an anti-malarial and antipyretic, dating back to over 2000 years. China has a long history of its cultivation with its remains found in Shengjindian cemetery (about 2400–2000BP based on ^{14}C dating), Xinjiang, China. This archaeological discovery gives an idea about its use in ancient China. As this plant is highly aromatic, the most probable reason for its use in the tombs in cemeteries would have been to eliminate the unpleasant odor of the dead (Liu et al. 2013; Sadiq et al. 2014). The essential oils extracted from this plant have also been used in the perfume industry. A treatise with a description of 224 medicines and their preparation methods was excavated in 1973 and named "Prescriptions for 52 diseases." It is regarded as one of the oldest sources of knowledge about Chinese pharmaceutics (Unschuld 1986). It described *A. annua* (qinghao) as a prescription for the treatment of hemorrhoids (Riddle 2010). Its utility as an herbal remedy against malaria was first described in *A Handbook of Prescriptions for Emergency* by Hong Ge. The described recipes were in the form of infusion, drink, powder, wine, and pills. *A. annua* decoction is also mentioned in the "General Medical Collection of Royal Benevolence" written during the Song Dynasty (Ekiert et al. 2021). Its preparation was recommended for paroxysmal malarial fever in the book *Compendium of Materia Medica* by Shizhen Li. In Chinese literature, its scientific name *Artemisia annua* was given in the twentieth century in the publication *First Chinese Pharmacopoeia* in 1930 (Riddle 2010; Ekiert et al. 2021). However, this plant was not given any attention in western herbal medicine until the start of the twentieth century.

Based on the traditional Chinese medicinal records, *A. annua* has a long-established tradition of being used in the treatment of diseases. In traditional medicine, all the parts of this plant and its extracts are used in the treatment of diseases like fever, jaundice, and dysentery (Ekiert et al. 2021). Its usage in China dates back to over 2000 years when it was used against tuberculosis, and fever caused by malaria and summer heat (WHO 2006; https://www.herbsociety.org/file_download/inline/d52eae8c-be89-497d-94b3-7fc8da4105f1). *A. annua* is also used in the treatment of hemorrhoids, wounds, and infections caused by protozoan species belonging to different genera like *Leishmania*, *Plasmodium*, *Schistosoma*, and *Acanthamoeba* (Alesaeidi and Miraj 2016). It is regarded as an important ethnomedicinal herb, and its most common method of usage is in the form of whole plant decoction, against cold, cough, and malaria. The powder made from its dried leaves has also been used in treating diarrhea, whereas the entire plant is effective as an antiseptic, antipyretic, anthelminthic, antispasmodic, and stimulant (Nigam et al. 2019). The crushed plants of *A. annua* have been used in the preparation of liniment whereas its tincture has been used in the treatment of nervous diseases (Sadiq et al. 2014). In African countries, the tea infusion made from it has been used to treat malaria. Furthermore, *A. annua* has also been associated with increased longevity, hair growth, and brightening of eyes (Hsu 2006). Qinghao has also been described as a food supplement in some of the records (Hsu 2009). Presently, research on this plant is mainly focused on its efficacy against malaria, cancer, and as an antioxidant (Willcox 2009; Ferreira et al. 2010; Wright et al. 2010).

2.3 Phytochemical Composition of *A. annua*

After the discovery of artemisinin, intensive research has been going on with *A. annua* to evaluate its phytochemical composition. It is a type with variable chemical composition, and a vast variety of phytochemicals have been identified (Marinas et al. 2015). The chemical composition of the plant also depends upon the climate in which it's growing (Hwang et al. 2016). Various chemical compounds that have been identified in *A. annua* include essential oil with monoterpenes and sesquiterpenes, coumarins, flavonoids, phenolic acids (Willcox et al. 2004), fatty acids, phytosterols, saponins, tannins, sesquiterpene lactones, and polyalkenes (Ashok and Upadhyaya 2013).

In *A. annua* the essential oils consist of both volatile and nonvolatile components. The main volatile constituents of essential oils which have been identified include camphene, camphene hydrate, β-camphene, 1-camphor, alpha-pinene, β-pinene, isoartemisia ketone, artemisia ketone, artemisia alcohol, cuminal, germacrene D, 1,8-cineole, camphor, betacaryophyllene, and myrcene (WHO 2006; Ferreira and Janick 2009). The major nonvolatile ingredients include flavonoids, coumarins, sesquiterpenoids, proteins like β-glucosidase, β-galactosidase, and steroids such as β-sitosterol, and stigmasterol (WHO 2006; Cafferata et al. 2010; Brown 2010).

The phenolic constituents of *A. annua* consist of phenolic acids, coumarins, flavones, flavonols, and certain miscellaneous compounds. The phenolic acids include coumaric acid, chlorogenic acid, rosmarinic acid, and quinic acid. Coumarins reported in *A. annua* include coumarin, scopolin, aesculetin, scopoletin, and iso-fraxidin. Flavones identified in *A. annua* include luteolin-7-methyl ether,

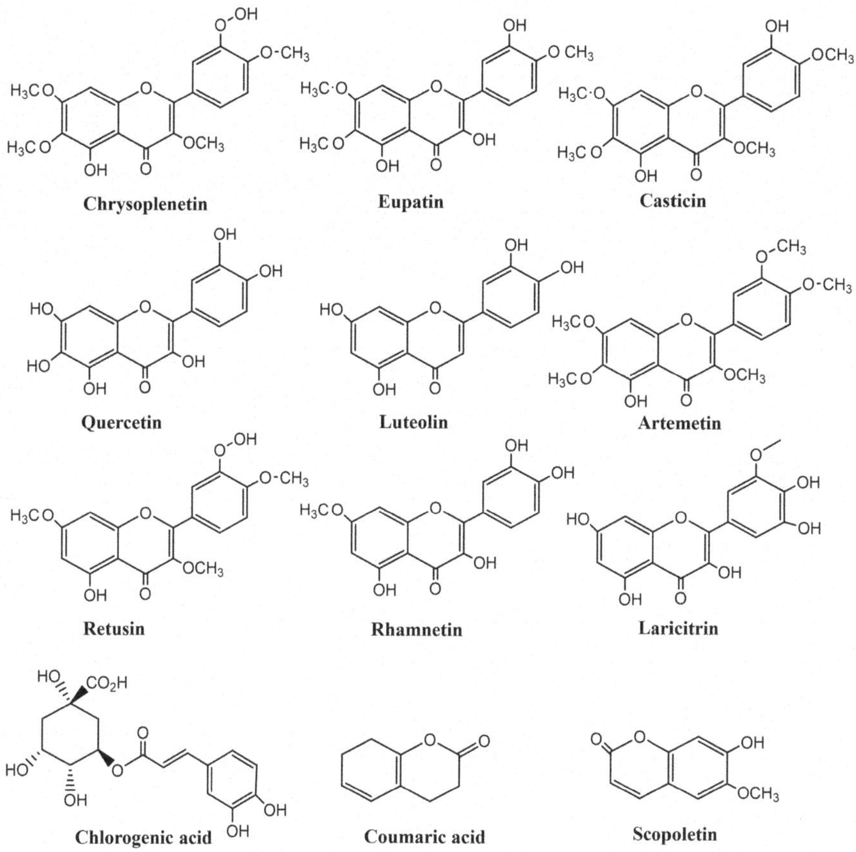

Fig. 2.2 Chemical structure of some phenolic compounds found in *A. annua*

apigenin, acacetin, luteolin, chrysin, cynaroside, chrysoeriol, cirsilineol, cirsimaritin, cirsiliol, and eupatorin. A large number of flavonols have been identified in this plant which includes eupatin, casticin, quercimeritin, artemetin, chrysosplenol C, retusin, mikanin, syringetin, tetra-methoxyflavone, mearnsetin-glucoside, quercetin, kaempferol-6-methoxy glucoside, rhamnetin, chrysosplenol D, astragalin, axillarin, kaempferol, tamarixetin, myricetin, laricitrin, gossypetin-3,-dimethyl ether, mearnsetin, quercetin-3-methyl ether, quercetin-3-glucoside, rutin, isorhamnetin, chrysosplenetin, and isokaempferide (Shatar et al. 2003; Rajeswara Rao et al. 2014; Shahrajabian et al. 2020). The chemical structures of some of the major phenolic compounds found in *A. annua* have been shown in Fig. 2.2.

The majority of the research carried out on *A. annua* revolves around artemisinin and its derivatives like dihydroartemisinin, artemether, arteether, and artesunate due to their antimalarial potential, with less attention given to other chemical constituents like flavonoids. The qualitative profile of the *A. annua* plants as determined by Baraldi et al. (2008) has shown the presence of different flavonoids in leaves and

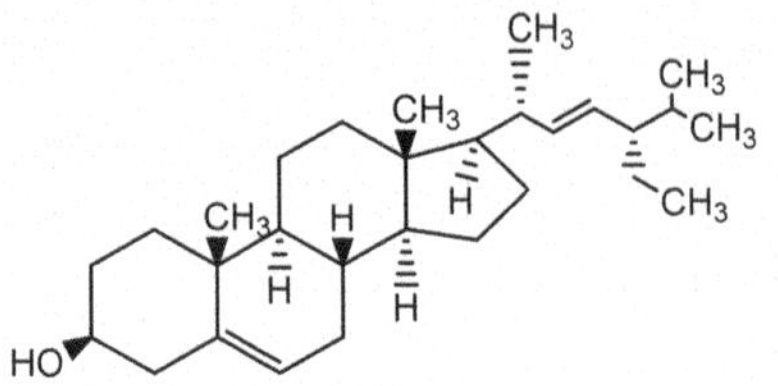

Stigmasterol

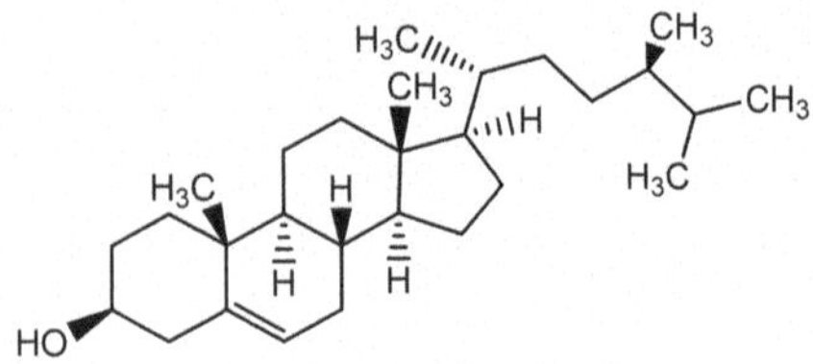

Campasterol

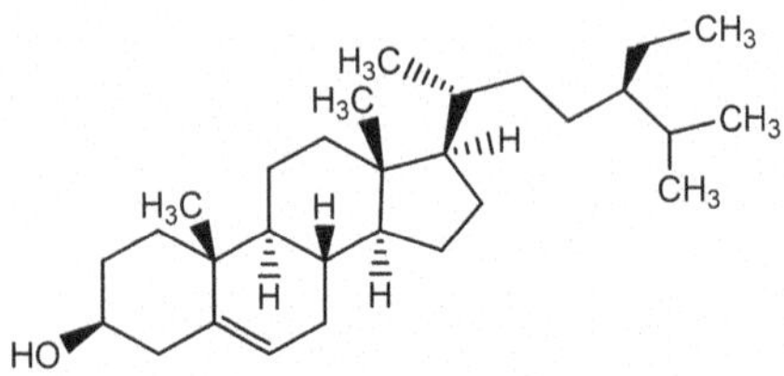

Beta-sitosterol

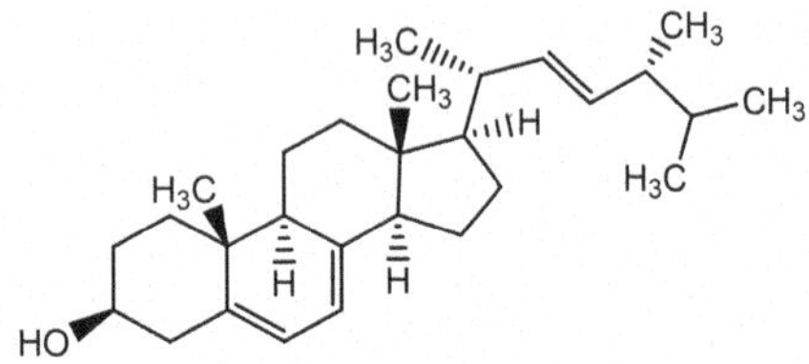

Ergosterol

Fig. 2.3 Chemical structure of phytosterols (stigamsterol, campasterol, and β-sitosterol) found in *A. annua*. Ergosterol is mostly found in fungi

flowers like artemetin, eupatin, chrysoplenetin, and casticin (Mesa et al. 2015). Among these flavonoids, casticin and chrysoplenetin exist in the form of an inseparable mixture, and their content is highest as compared to other flavonoids. Using HPLC and NMR analysis, chrysoplenol-D, camphor, 5-hydroxy-3,6,7-trimethoxy-2-(4′-methoxyphenyl)-4H-chromen-4-one), and 2,4-dihydroxy-6-methoxyacetophenone have also been reported to be present in *A. annua* along with other flavonoids (Kontogianni et al. 2020). Flavonoids have been found to synergize the effect of artemisinin with luteolin showing both antimalarial and antioxidant properties (Ferreira et al. 2005).

Phytosterols are the secondary metabolites present in plants either in free form or can be found esterified with fatty acids or with phenolic acids, with glycosylated sterols also showing their presence in very minute quantities (Ostlund Jr 2007). The most prevalent phytosterols are stigmasterol, campesterol, and β-sitosterol (Fig. 2.3). Ergosterol, which is found in considerable amounts in algae, fungi, and lichens, shows very little or no presence in higher plants. Phytosterols have immunomodulatory and anti-inflammatory properties (Bouic 2002). The β-sitosterol, stigmasterol, campesterol, and ergosterol content per 100-gram dry weight of aerial parts of *A. annua* is about 119.570, 119.538, 17.528, and 0.79 mg, respectively (Ivanescu et al. 2013). The total sterol content in *A. annua* is around 250 mg per 100 g of dry matter.

Among all the phytochemicals present in this plant, sesquiterpene lactones are the most important. The most vital sesquiterpene lactone in this plant is the antimalarial compound artemisinin which is synthesized in the glandular secretory trichomes present in the aerial parts of the plant and characterized by the presence of an endoperoxide bridge as shown in Fig. 2.4 (Aftab et al. 2014; Salehi et al. 2018;

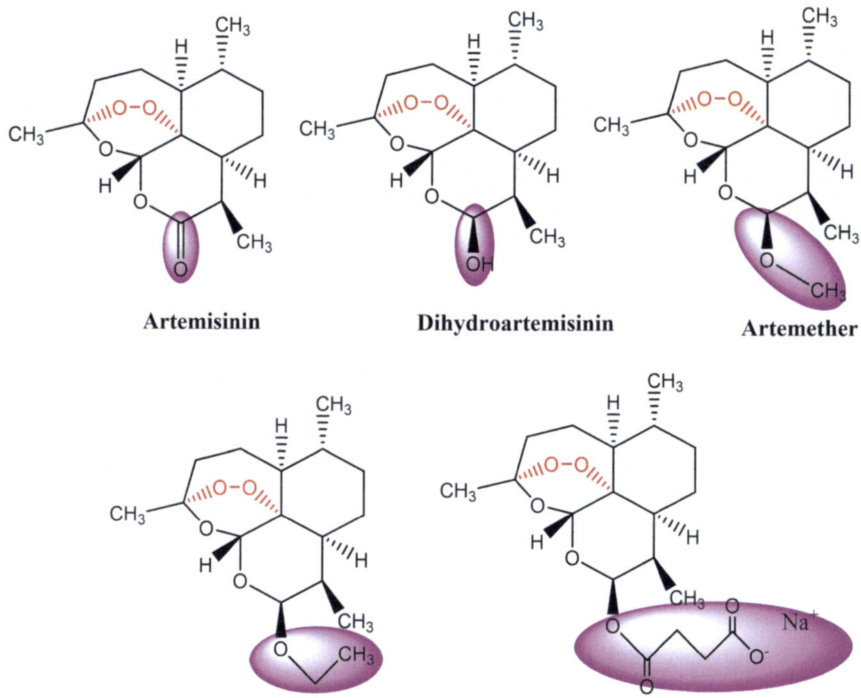

Fig. 2.4 Chemical structure of artemisinin and its derivatives: The endoperoxide bridge (C – O – O – C) shown in red is common to artemisinin and its derivatives and acts as the key pharmacore. The modifications at C_{10} position are highlighted in pink, which are unique to each of these derivatives and determine their solubility and some of their pharmacokinetic properties

Wani et al. 2021). The artemisinin content in the plant is usually low and lies in the range of 0.01–1.50% dry weight (Wani et al. 2021). The credit for the discovery of this lifesaving antimalarial compound goes to Prof. Tu Youyou, a pharmaceutical chemist and a malariologist who discovered it in 1972 by analyzing Chinese herb recipes having antimalarial activities under the project 523 (Su and Miller 2015). She discovered a proper method to isolate the active antimalarial ingredients (ethyl extract) and found that it could inhibit monkey and rodent malaria with 100% efficacy. After that, a clinical trial of the plant extracts was conducted on 21 patients, and a 95–100% success rate was achieved. For this discovery, Prof. Tu Youyou was awarded Nobel Prize in Physiology or Medicine in 2015 (Efferth et al. 2015). Artemisinin-based combination therapies have been endorsed by WHO for the treatment of uncomplicated malaria. These therapies consist of artemisinin or its derivatives in combination with a partner drug like mefloquine (WHO 2015). Various semisynthetic derivatives are also produced from artemisinin which include dihydroartemisinin, artemether, arteether, and artesunate which act as potent antimalarials (Fig. 2.4). Various other sesquiterpenes found in artemisinin include artemisinic acid, epoxyarteannuinic acid, artemisinol, and artemisinin isomers like artemisinin I, II, III, IV, and V (WHO 2006; Das 2012).

2.4 Pharmacological Activities

2.4.1 Antimalarial Activity

The most important constituent of *A. annua* responsible for its antimalarial properties is artemisinin. However, certain other chemical constituents from this plant have been reported to have a synergistic impact on its antimalarial potential. The importance of other chemical constituents of *A. annua* has also increased due to resistance shown by *Plasmodium* against monotherapy which has already emerged more than 25 years ago. Artemisinin is now prescribed in combination with certain other drugs in a practice called Artemisinin-based combination therapies (ACTs). A number of studies have shown the effectiveness of using the whole plant in the form of infusion, powder, or tablets against malaria (El Hadji Omar et al. 2013). This is further confirmed by a study in which a 40 mg dry weight of plant powder (containing nearly 600 μg Artemisinin) equivalent to 24 mg artemisinin/kg live body weight was effective in controlling the infection caused by *Plasmodium chabaudi* in rodents as compared to an equivalent dose of pure artemisinin (Elfawal et al. 2012). The increased efficacy of whole plant extracts is most probably due to the increased artemisinin bioavailability (40-fold) in the mice blood (Elfawal et al. 2012).

The effects of *A. annua* tea on malaria-causing *P. falciparum* has also been studied under in vitro conditions. The herbal tea extract from the plant was tested against chloroquine-sensitive D10 and chloroquine-resistant W2 strains of this parasite (De Donno et al. 2012). From the experimental results, it was found that herbal tea extract had a dose-dependent inhibitory effect on this parasite. Drug-free parasitized and unparasitized erythrocytes were used as blank and positive control. The impact of the herbal tea was three times greater than that of artemisinin alone which implies that there could be a synergistic impact of other constituents like flavonoids on its anti-plasmodial activity (Elford et al. 1987). The presence of other amphiphilic constituents like glucosides, flavonoids, or saponins may also improve its solubility in water. So, the presence of different active ingredients against malaria makes *A. annua* a natural artemisinin combination therapy.

Mouton et al. (2013) conducted an in vitro study to test the claims of synergism enhancing the antimalarial activity of artemisinin using non-polar extracts and tea infusions. Contrary to some earlier studies like De Donno et al. (2012), they reported that the IC_{50} of artemisinin in nonpolar extracts and tea infusions did not show any significant difference as compared to the IC_{50} of pure artemisinin. Pure artemisinin was used as a positive control. They also tested the tea infusions from *A. afra* against *P. falciparum*, which may be regarded as a negative control as *A. afra* contains negligible/does not contain artemisinin but shows the presence of other chemical constituents as reported in *A. annua* (Liu et al. 2010). The infusions of *A. afra* did not show any activity against the *Plasmodium* parasite (Mouton et al. 2013).

The aqueous extracts of *A. annua* and *A. sieberi* have also been found to have an inhibitory impact on β-hematin formation (under in vitro conditions), which is a synthetic analogue of hemozoin formed by *Plasmodium* parasites for its protection against excess ferriprotoporphyrin (IX) accumulation (Akkawi et al. 2014). In this study 2-mercaptopyrimidine (1 mg/mL) and chloroquine (0.1 mg/mL) were dissolved in ultrapure water, and both were used as positive controls, whereas ultrapure water was used as a negative control. They demonstrated that *A. annua* leaf infusions prepared in salt water (0.5 g salt/150 mL H_2O) had greater efficacy in inhibiting β-hematin formation than those prepared in distilled water. However, these extracts had a decreased activity as time progressed so using dried leaf powder may be a better option in vivo.

The antimalarial action of hydroalcoholic and aqueous extracts of *A. annua* was studied in vitro against *P. falciparum* and in vivo in *P. berghei* NK 173-infected mice, with artemisinin (98%) used as a positive control for in vitro analysis, and 140 mg/kg/day for in vivo analysis (Zime-Diawara et al. 2015). Under in vitro conditions, the activity of the abovementioned extracts was similar to that of pure artemisinin at the same dosage. Moreover, the in vivo studies conducted on mice revealed that the aqueous extracts (artemisinin content of 20 mg/kg) have the same efficacy as compared to pure artemisinin (140 mg/kg dosage). Furthermore, the hydroalcoholic extract of *A. annua* (artemisinin content of 20 mg/kg) revealed the best results in comparison to aqueous extracts. The difference in the activity of aqueous and hydroalcoholic extracts is probably due to their composition. The comparison of TLC chromatograms of these two extracts revealed that hydro alcoholic extracts contain more sesquiterpenes and extra flavonoids as compared to aqueous extracts (Zime-Diawara et al. 2015). The increased activity of hydroalcoholic extracts against plasmodium might be potentiated by the presence of these extra flavonoids (Elford et al. 1987) and sesquiterpenes.

More recently a study was conducted using whole leaf extracts from artemisinin-producing homozygous *chi1–1* (flavonoid lacking) and heterozygous *chi1–1* (flavonoid containing) Artemis hybrids of *A. annua* (Czechowski et al. 2019). Moreover, the extracts from RNAi lines impaired in *amorpha-4,11-diene synthase* gene expression and *cyp71av1–1* mutants, both of which are impaired in artemisinin biosynthesis were also tested. Based on their observations, they reported that flavonoids do not add to the anti-plasmodial activity beyond that provided by artemisinin under in vitro conditions.

However, the findings from different studies need to be checked by taking into account the time period for which the plant has been stored, the phase of the life cycle of the parasite tested, and some constituents of the tea infusion may become active after metabolization (Mouton et al. 2013). To detect the components, present in *A. annua* extracts which synergize the anti-plasmodial potential of artemisinin through regulation of its metabolism, a study was carried out by Cai et al. (2017). The extracts from dried plants were eluted using different concentrations of methanol (3%, 50%, and 85%). The pharmacokinetic profiles of artemisinin and monohydroxylated artemisinin (major phase 1 metabolite) were studied in rats after a single oral dose of artemisinin in each *A. annua* extract. Chief components isolated from the methanol extracts were assessed for their enzyme inhibition, and only arteannuin B was found to have a repressive effect on CYP3A4. In order to test the synergism between artemisinin and the other component, mice infected with *P. yoelii* were used, and the pharmacokinetic study was carried out. They reported synergism between artemisinin and arteannuin B which was related to increased artemisinin exposure due to enzyme inhibition (Cai et al. 2017).

2.4.2 Anticancer Activity

Various semisynthetic derivatives of artemisinin like artesunate (Fig. 2.4) have been found to have anticancer properties. Various studies (both in vitro and in vivo) have revealed that artemisinin-type drugs have anticancer activities (Efferth 2017). The anticancer properties of artemisinin and its various derivatives like artesunate, dihydroartemisinin, artemether, and arteether include:

- Reactive oxygen species and nitric oxide-mediated oxidative stress.
- DNA damage and repair.
- Cell death by necrosis, apoptosis, ferroptosis, autophagy, oncosis, etc.
- Inhibition of tumor-related signaling pathways like Wnt/β-catenin pathway; signal transducers like MYC/MAX, NF-κB, mTOR, AP-1, etc.; and angiogenesis.

The cytotoxicity of artemisinin and its different derivatives toward cancer cells has been described in a number of studies carried out in 1990s like Woerdenbag et al. (1993). Various studies have been conducted to test these claims, and vast pieces of evidence are suggesting that artemisinin-type compounds may inhibit tumor cells in vitro. Endoperoxide bridge, which is a characteristic feature of artemisinin has been found to be essential for the anticancer activity of these compounds, as artemisinin-type compounds without it have shown inactivity towards cancer inhibition (Beekman et al. 1998). The antitumor activity of these compounds has been confirmed by in vitro as well as in vivo studies (Efferth 2005; Seo et al. 2015; Subedi et al. 2016), with some studies involving human xenograft tumors transplanted on nude mice (Hou et al. 2008; Chen et al. 2009; Zhang et al. 2012; Tong et al. 2016). However, these athymic (lacking or deteriorated thymus) mice characterized by a lack of body hair have an inhibited immune system with reduced number of T cells. Keeping this disadvantage in mind, some researchers have also used syngeneic models in which rodent tumors are transplanted on rats or mice (Efferth 2017). The activity of artemisinin-type compounds against syngeneic and xenograft tumor models further validated the claims of anticancer activity of these compounds. The anticancer activity of these compounds has been even seen against orthopedically transplanted tumors.

The endoperoxide-bridge in artemisinin (Fig. 2.2) is important for its bioactivity as it produces ROS after its cleavage which results in oxidative stress. The generation of oxidative stress due to ROS by artemisinin-type drugs has been confirmed by a number of studies involving hematopoietic, epithelial, or mesenchymal cell lines from diverse tumor origins (Efferth and Oesch 2004; Du et al. 2010; Zhu et al. 2014; Gerhardt et al. 2015; Jia et al. 2016). The efficacy of ROS-mediated antitumor activity of such drugs was supported by many studies in which certain prooxidants (e.g., vitamin C) amplified artemisinin-mediated cytotoxicity and antioxidants (e.g. vitamin E) decreased the rate of tumor cell death (Efferth and Volm 2005; Noori et al. 2014; Papanikolaou et al. 2014; Beccafico et al. 2015; Lemke et al. 2016).

Artemisinin and its derivatives like artesunate have been reported to have a genotoxic effect as they induce breaks in DNA in a dose-dependent manner as observed by single-cell gel electrophoresis (Li et al. 2008). These findings were further supported by a dose-dependent buildup of γ-H2AX, a histone protein produced upon double-stranded DNA breaks (Li et al. 2008). Polymerase β-deficient cells, cells deficient in nonhomologous end joining (ku80 inactive) and homologous recombination (BRCA2 and XRCC2 inactive), have been found more sensitive to artesunate as compared to wild type cells as these cells lack the required DNA repair pathways (Li et al. 2008). So artesunate induces DNA breaks which makes it a therapeutic compound against cancer cells. A number of other studies have also revealed this anticancer activity of artesunate and other related compounds (Berdelle et al. 2011; Alcântara et al. 2013; Park et al. 2015).

ROS-mediated oxidative stress and DNA damage affect cell division, integrity, and replication which in turn causes cell cycle arrest (usually at G1 or G2 checkpoints) and ultimately cell death. Various artesunate type drugs have also been found to have this impact as observed in a number of studies (Hou et al. 2008; Li et al. 2009; Tin et al. 2012; Zhao et al. 2013; Tran et al. 2014; Lu et al. 2014; Islam and Mohammad 2020)). Other than cell cycle arrest, oxidative stress and DNA damage may also induce apoptosis as has been seen in cancer cells using artesunate, artemisinin, artemether, and arteether (Lai and Singh 1995; Disbrow et al. 2005; Hou et al. 2008; Zhang et al. 2016). Depending upon the cell model, artesunate can induce mitochondrial, intrinsic, or extrinsic FAS-receptor-mediated apoptosis (Efferth et al. 2007; Sieber et al. 2009) with enhanced Fas/CD95 expression, cessation of the mitochondrial membrane potential, release of cytochrome C, poly-(ADP-ribose) polymerase cleavage, and caspase 3/9 activation (Efferth 2017). Cells transfected with *BCL2* gene (encodes a mitochondrial membrane protein that blocks programmed cell death) have been found to be more resistant to artesunate (Efferth et al. 2003). Various artemisinin-type drugs have also been found to induce caspase-independent non-apoptotic cell death-like autophagy (Zhou et al. 2013; Beccafico et al. 2015; Chen et al. 2015). Such drugs have also been found to induce other types of cell death-like necrosis, necroptosis, oncosis, anoikis, and iron-dependent ferroptosis (Du et al. 2010; Dixon et al. 2012; Zhou et al. 2013; Dixon et al. 2014; Zhu et al. 2021). Artemisinin derivatives have also been reported to have anti-angiogenic potential as tested in Zebrafish as reviewed by Wei and Liu (2017).

Artemisinin-related drugs have been shown to affect different signaling pathways in cancer cells as reviewed by Efferth (2017). These drugs were found to inhibit Wnt/β-catenin signaling, BCR/ABL signaling, as well as epidermal growth factor receptors. They also inactivated various transcription factors related to cancer like MYC/MYX, NF-κB, mTOR, CREB, etc. Artesunate has also been found to inhibit metastasis and invasion by targeting certain extracellular proteases (Rasheed et al. 2010). So artesunate may be used as an anti-invasive and anti-metastatic agent and can thus be added to the cancer treatment approaches.

2.4.3 Immunosuppressive Activity

Artemisia annua extracts obtained with hot water as mentioned in traditional Chinese medicine have been extensively used for treating different autoimmune diseases such as systemic lupus –erythematosus and rheumatoid arthritis (Zhao et al. 1998). In order to test the immunosuppressive activity of *A. annua* extracts, Zhang and Sun (2009) used the ethanol extracts of these plants on mice splenocyte proliferation in vitro and specific antibody and cellular immune responses in the ovalbumin-immunized mice. The immunized mice were given ethanol extracts of 0.25, 0.5, and 1.0 mg and cyclosporin A (positive drug) at a single dose of 0.1 mg in 0.2 mL of saline solution at intervals of 7 days. The ethanol extracts of these plants showed a significant inhibitory activity on splenocyte proliferation induced by lipopolysaccharides and concanavalin A under in vitro conditions. The plant extracts also showed inhibitory activity against concanavalin A, lipopolysaccharides, and ovalbumin-induced splenocyte proliferation in ovalbumin immunized mice. These extracts also reduced the ovalbumin-specific antibodies like IgG, IgG1, and IgG2b, with a significant reduction at 1.0 mg (Zhang and Sun 2009). These findings show that *A. annua* has immunosuppressive properties and could be used as immunosuppressants. The immunosuppressive effect of artemisinin has also been found to be effective against IgA nephropathy, an autoimmune kidney disease, with combination treatment of artemisinin and hydroxychloroquine giving better results as compared to their individual application in rats. The combination treatment decreased the deposition of complement 3 and IgA immune complexes (Bai et al. 2019).

2.4.4 Antimicrobial Activity

Artemisia annua is known to produce various secondary metabolites which have antimicrobial properties. The antimicrobial potential of artemisinin isolated from in vitro grown plantlets was tested by Appalasamy et al. (2014) against three gram-positive bacteria *Bacillus thuringiensis*, *Bacillus subtilis*, and *Staphylococcus aureus*; two gram-negative bacteria *Salmonella* sp. and *Escherichia coli*; and *Candida albicans*. They used streptomycin and acetonitrile as positive and negative control, respectively. They reported that artemisinin and a precursor extracted from the in vitro grown plantlets showed inhibitory activity against gram-positive and gram-negative bacteria but not against *C. albicans*. The antimicrobial action was found to be similar to that of streptomycin and the minimum inhibitory concentration was found to be 0.09 mg/mL (Appalasamy et al. 2014). The chloroform, alcohol, and water extracts of *A. annua* have been found to be effective in the treatment of acanthamoebiasis, caused by *Acanthamoeba* sp. (Derda et al. 2016). These extracts were tested against pathogenic *Acanthamoeba castellanii* (309 strain) and *Acanthamoeba* sp. (Ac32 strain). Based on the results, they found that the extracts have inhibitory activity against *Acanthamoeba* both in vitro and in vivo. These extracts also prolonged the survival of infected animals. These results indicate *A. annua* extracts could be used in treating acanthamoebiasis either singly or in combination with other antibiotics.

The extracts from *A. annua* and *A. afra* have been found to have strong bactericidal activity against *Mycobacterium tuberculosis* (Martini et al. 2020). The minimum inhibitory concentration for pure artemisinin was 75 μg/mL. Whereas for *A. annua* it was the extract from 4.81 mg of dried leaves per mL media, which resulted in 39 μg/mL of artemisinin. The bactericidal activity of these extracts is stronger than pure artemisinin which hints at a possible synergism between artemisinin and other compounds in eradicating *M. tuberculosis*. These extracts also show bacteriostatic activity against *M. abscessus* (a nontuberculous mycobacterium that usually affects patients who are immunocompromised) but were not bactericidal (Martini et al. 2020). These results give a strong indication that *A. annua* extracts could be used in combination with other antibiotics for the treatment of tuberculosis and prevention of *M. abscessus* infections.

The antimicrobial activity of methanol, water, ethanol, or acetone extracts of *A. annua* has also been reported against certain periodontopathic microbes like *Prevotella intermedia*, *Fusobacterium nucleatum* subsp. *polymorphum*, *F. nucleatum* subsp. *animalis*, and *Aggregatibacter actinomycetemcomitans* (Kim et al. 2015). Considering the high lipophilicity of *A. annua* essential oils, oil in water type Pickering *A. annua* essential oil nanoemulsions with 20 nm Stober silica nanoparticles as the stabilizing agent were tested on mature *Candida* biofilms (Das et al. 2020). These emulsions showed significantly higher activity as compared to ethanol extracts and Tween 80 stabilized emulsion. These emulsions resulted in peroxide and superoxide related oxidative stress which might be the reason for the antimicrobial activity of *A. annua* essential oil. The nanoliposome-incorporated essential oils from *A. annua* have shown inhibitory activity against *Candida* species with *C. norvegensis* showing the most susceptibility (Risaliti et al. 2020). The *A. annua* silver nanoparticles of various concentrations (10, 20, 30, and 50 μg/mL) were tested against four microbes *Klebsiella pneumonia*, *S. aureus*, *E. coli*, and *B. subtilis*, with levofloxacin (30 μg/disc) used as a positive control (Adoni et al. 2020). These nanoparticles were found to have free radicle scavenging activity and moderate activity against gram-positive and gram-negative bacteria as compared to standard antibiotics.

2.4.5 Anti-parasitic Activity

A number of compounds isolated from *A. annua* have been found to have activity against *Leishmania* species (intracellular protozoans) which are known to cause leishmaniasis. The n-hexane extracts of its seeds and leaves have been found to cause apoptosis in intracellular amastigotes of *L. donovani* (studied by ex vivo macrophage-amastigote model), and it did not show cytotoxic effects on mammalian macrophages (Islamuddin et al. 2012). When these extracts were tested on infected mice, a significant decrease in splenic and hepatic parasitic load, along with reduced spleen weight was observed. Moreover, the essential oils of *A. annua* leaves have shown significant antileishmanial activity against intracellular amastigotes of *L. donovani* under in vitro conditions (Islamuddin et al. 2014). It did not show any cytotoxicity on murine macrophages. The intraperitoneal administration at a dosage of 200 mg/kg body weight to infected BALB/c mice (immunodeficient inbred strain) decreased the parasitic load by around 90% in spleen and liver. As seen in earlier cases, no toxicity was observed, as supported by the normal levels of the serum enzymes. However, a complete cure of visceral leishmaniasis with *A. annua* n-hexane extracts or essential oils has not been observed in these in vivo studies. However, these in vivo studies did not show any complete cure for cutaneous leishmaniasis. Mesa et al. (2017) tested the antileishmanial activity of *A. annua* leaf powder containing gelatin capsules (totum). These capsules showed moderate activity against amastigotes of *L. (Viannia) panamensis*, and no cytotoxicity or genotoxicity was reported in macrophages (U-937) and human lymphocytes in vitro. The infected hamsters given *A. annua* capsules (500 mg/kg/day) for 30 days were cured, with 83.3% success rate. The administration of these capsules (30 g) to two patients with uncomplicated cutaneous leishmaniasis also gave positive results (Mesa et al. 2017).

A study was conducted on *Neospora canum*, a protozoal parasite that infects a variety of mammals and causes abortion in cattle (MR 2009; Das 2012). *N. caninum* tachyzoites were made to infect the cultured host cells (Vero cells or mouse peritoneal macrophages) and then supplemented with artemisinin at 20, 10, 1, 0.1, and 0.01 μg/mL. At 10 or 20 μg/mL artemisinin for 11 days, all microscopic foci of *N. caninum* were eradicated. The same result was obtained when artemisinin was administered at 1 μg/mL for 14 days. Artemether has been reported to be effective against the larval stages of *Schistosoma mansoni* as seen in hamsters and mice. If these organisms are administered with artemether during the first month of infection, then schistosomiasis (bilharzia) does not develop (Das 2012).

2.4.6 Anti-SARS-CoV-2 Activity

The global pandemic of Covid-19 caused by severe acute respiratory syndrome coronavirus 2 (SARS-CoV-2) has caused mayhem in the whole world, and the death toll has crossed the five million mark in November 2021 as per the data provided by John Hopkins University (https://www.bbc.com/news/world-59119731). Despite the intense efforts to distribute various registered vaccines against Covid-19 like Moderna, Pfizer/BioNTech, and J & J vaccines, except Remdesivir no other orally deliverable drug is available currently. The emergence of a new variant B.1.1.529 on 26 November 2021, named omicron by WHO may also pose a serious challenge in controlling its transmissibility (https://www.who.int/news/item/28-11-2021-update-on-omicron). Artemisinin-type compounds have been reported to have antiviral activity (Efferth 2018). The extracts from *A. annua* plants have shown activity against SARS-CoV-1 which is responsible for the SARS outbreak of 2002–2004 (Li et al. 2005). These observations suggest that the extracts of *A. annua* could be effective against SARS-CoV-2. Artemisinins and the extracts from *A. annua* decrease inflammatory cytokines like IL-6 and TNF-α in vivo (Shi et al. 2015; Hunt et al. 2015). These effector molecules can be tricky during the "cytokine storm" suffered by many patients suffering from SARS-CoV-2 (Schett et al. 2020). Fibrosis has also been reported in persons affected by SARS-CoV-2 which can cause damage to different organs, and artemisinin has been found to dampen fibrosis (Larson et al. 2019; Lechowicz et al. 2020). Recently, a study conducted by Cao et al. (2020) found that a number of compounds related to artemisinin act against SARS-CoV-2, with Arteannuin B showing the highest activity against SARS-CoV-2; dihydroartemisinin and artesunate also showed activity against it. These results indicate the potential of *A. annua* extracts, particularly artemisinin in the treatment of Covid-19, which needs further analysis and research for the development of suitable drugs.

2.5 Clinical Studies Involving Artemisinin-Related Compounds in the Treatment of Malaria, Schistosomiasis, Leishmaniasis, and Certain Cancers

Based on the multiple research results of in vitro and in vivo studies, it seems justified to say that the antimalarial activity of *A. annua* is not solely due to artemisinin; it seems its activity is potentiated by the presence of other constituents in the plant as well (Fouda 2010; De Donno et al. 2012). In a clinical trial conducted by Mueller et al. (2000), five patients suffering from malaria were treated with *A. annua* tea, and a rapid reduction of parasitemia was observed in these patients. They conducted another trial with a larger group of 48 malaria patients and 92% of patients showed disappearance of parasitemia within 4 days. These results justify further research for the elucidation of antimalarial properties of *A. annua* preparations.

Newton et al. (2003) conducted a clinical trial in Thailand for 113 clinically severe malaria patients infected by *P. falciparum*. They reported that artesunate has a rate of mortality (12%) as compared to quinine (22%). During the trial, only a few patients (12%) treated with artesunate became hypoglycemic than quinine-administered patients (28%).

An in vivo study was conducted in Katanga province in the Democratic Republic of Congo (Tchandema and Lutgen 2016). Powdered leaf capsules of *A. annua* from two different geographical locations (Luxembourg and from Burundi) and *A. afra* were fed to patients in doses of 15 g, 7.5 g, and 7.5 g, respectively. Despite relatively low doses, all the patients were fever free within 2 days. Around 85%, 76%, and 40% of patients were free from parasites after 7 days which were fed with capsules of *A. annua* from Luxembourg, Burundi, and *A. afra*, respectively (Tchandema and Lutgen 2016).

As compressed leaf tablets of *A. annua* have shown antimalarial potential as seen in rodents, a study was carried out by Daddy et al. (2017) to test the efficacy of dried leaf *A. annua* in patients suffering from severe malaria and non-responsive to artemisinin combination therapy and i.v. artesunate. These patients were administered twice daily with 0.5 g dried leaf *A. annua* per os for 5 days, and the total delivered dose of artemisinin was 55 mg. All these patients who were resistant to artemisinin combination therapy were cured by these compressed leaf tablets.

Schistosomiasis and malaria coinfection has been commonly reported (Moriyasu et al. 2018). In 2015, a clinical trial was conducted to test the tea infusions of *A. annua* and *A. afra* in the treatment of schistosomiasis as compared to praziquantel which is the currently accepted drug for the treatment of Schistosomiasis (Munyangi et al. 2018). The patients who were given *Artemisia* infusions were cleared of *S. mansoni* eggs quickly (14 days) as checked in their fecal smears. The patients who were given praziquantel took more time for the complete eradication of *S. mansoni* eggs and experienced more adverse effects as compared to *Artemisia*-treated patients (Munyangi et al. 2018). These results received a published critique from Argemi et al. (2019) to which a rebuttal was then published by Cornet-Vernet et al. (2019).

As discussed in Sect. 2.4.5 above, Mesa et al. (2017) tested the antileishmanial activity of *A. annua* leaf powder containing gelatin capsules which gave encouraging results in vitro and in vivo (hamster) systems. Looking at these results, they also treated human males with powdered *A. annua* capsules containing about 0.1% artemisinin. Both of these patients were administered a total *A. annua* of 30 g per patient over 20 days. At the end of the trial, the ulcers had shrunk by 20–35%, and complete closure of the ulcers was reported around 45 days after the treatment ended. During the course of this whole trial, no negative effects were reported in these patients, indicating the safety of using *A. annua* leaf powder in the treatment of leishmaniasis (Mesa et al. 2017). But before that large human trials need to be conducted to fully validate these results.

As artemisinin and its derivatives like dihydroartemisinin and artesunate have shown anticancer activities under in vitro and in vivo conditions as discussed above in Sect. 2.4.2. Various clinical trials involving artemisinin and its derivatives have also been conducted to test the efficacy of these compounds in cancer treatment. A clinical trial was conducted involving cisplatin and vinorelbine with or without artesunate injections of 120 mg for a period of 8 days for the treatment of advanced non-small cell lung cancer (Zhang et al. 2008). The control rate of the trial group was 88.2% which was higher as compared to the control rate of 72.7% in the control group. The progression time of the artesunate-treated patients was 24 weeks as compared to 20 weeks for the control group. The longer time to progression of patients treated with artesunate shows the significance of artesunate in the treatment. Moreover, the patients administered with artesunate did not show any increased toxicity (Zhang et al. 2008).

In St. Georg's (University of London, London, UK), Krishna et al. (2015) conducted a randomized, double-blind, placebo-controlled clinical trial of oral artesunate therapy for colorectal cancer. Before the surgery, 12 patients were given 200 mg artesunate orally for 14 days, and 11 patients were given placebo. Out of these patients, 9 artesunate and 11 placebo administered patients completed the trial. The results showed that apoptotic cell fractions of more than 7% were observed in 67% of artesunate-administered patients as compared to only 55% of placebo-treated patients. The incidence of refractory tumors was also more in case placebo administered patients (6 in number) as compared to artesunate administered patients (1 only). The results of this study were promising, and it may be concluded that artesunate shows good anticancer activity and requires further studies so that its anticancer potential may be fully explored.

2.6 Safety and Toxicity

A. annua is botanically accepted as a safe plant. It is listed in the *Handbook of Phytochemical Constituents of GRAS* (Generally Regarded As Safe) herbs and other economic plants by James A. Duke and even has listed dosages of 30 g of dry leaf per day which is much higher than the daily dosage prescribed for malaria treatment (Duke 1992).

The dried leaves of *A. annua* are widely proposed to be used as a medicinal herb malarial treatment. For treatment of malaria and certain other diseases, it has been used in a number of ways like oral administration of capsules of dried leaf powder, compressed dried leaf tablets, and tea infusion. The encapsulation of dried leaves has not shown any significant decrease in the absorption of artemisinin or flavonoids present in it but a few food items such as peanut butter may decrease its absorption (Desrosiers and Weathers 2016). The plant is generally intended for consumption by adults and children that are infected with malaria.

A study was conducted by Abolaji et al. (2013) to evaluate the effect of ethanol extracts of *A. annua* in pregnant Wistar rats which were given 100, 200, and 300 mg/kg body weights of ethanol extracts of *A. annua* leaves. Based on the biochemical and hematological studies, they reported that the extracts did not result in hematotoxicity, hepatotoxicity, and hyperlipidemia. However, at 300 mg/kg dose, 31% malformed fetuses and 21% nonviable fetuses were observed. Embryotoxicity has also been reported in rats due to consumption of certain artemisinin derivatives like dihydroartemisinin, artemether, arteether, and artesunate during the

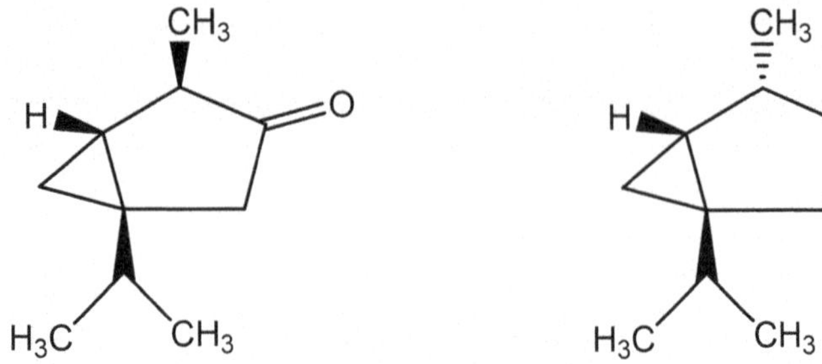

Fig. 2.5 Chemical structure of α-thujone and β-thujone found in *A. absinthium*

organogenesis period, with extended oral dosage (12 days or more) resulting in embryotoxicity in monkeys as well (Clark 2009). These results raise concerns about the usage of these compounds and plant extracts for treatment during the sensitive period of pregnancy in women. These results indicate that although the consumption of leaf extracts may not cause hematotoxicity, hepatotoxicity, and hyperlipidemia, care should be taken during pregnancy due to the possible risk of embryotoxicity due to its consumption beyond the therapeutic dose (Abolaji et al. 2013).

In another study carried out on male Wistar rats, it was found that these extracts may serve as antidiabetic agents, and do not cause hematotoxicity, hepatotoxicity, and testicular toxicity; however, there may be a possible risk of atherosclerosis due to cholesterol buildup as observed in this study (Eteng et al. 2013). So, persons suffering from atherosclerosis should take immense care while using the extracts of *A. annua*.

The genus *Artemisia* has a GRAS status with the United States Food and Drug Administration (FDA), as long as the final product is thujone-free. Thujone is a monoterpene found in certain *Artemisia* species, e.g., *A. absinthium* (Fig. 2.5). At high dosages, thujone acts as a neurotoxin that may cause unconsciousness, convulsions, and death in humans (Cobb 1922). However, *A. annua* has been found to be thujone free and therefore the utilization of *A. annua* is generally considered safe (Tzenkova et al. 2010).

Based on the studies conducted till date, there is not much risk in consuming dried *A. annua* either as a supplement or as part of any other treatment with the exception of the possible presence of heavy metals at concentrations higher than recommended by the FDA. Furthermore, immense care should be taken during its consumption by pregnant women and atherosclerosis patients, and it should not be taken above the therapeutic doses. The issue of excess heavy metal accumulation can be solved by growing these plants in soil and water that does not contain enough of these metals.

2.7 Conclusion

A. annua is an ethnomedicinally important plant, with its medicinal use well established in Chinese pharmacopoeias. It has also obtained a vital place among plant-based advanced therapeutics, particularly against malaria, for which it has undoubtedly become a good hope for treatment. It is a rich source of numerous biologically active constituents, particularly artemisinin. It has the characteristic therapeutic potential against malaria, and besides being antimalarial, it has various other biological activities such as antibacterial, antitumor, and immunosuppressive activities. Nowadays, a lot of research is going on to investigate its anti-cancer and anti-viral activities, particularly against SARS-CoV-2. The elucidation of the mechanism of action of artemisinin, its various derivatives, and different flavonoids against various diseases has become the major area of interest among researchers. The extremely low toxicity associated with *A. annua* is another reason for its wide usage against malaria and certain other diseases. Therefore, *A. annua* is a great option that can be extensively explored for the development of new drugs.

References

Abolaji AO, Eteng MU, Ebong PE, Brisibe EA, Dar A, Kabir N, Choudhary MI (2013) A safety assessment of the antimalarial herb *Artemisia annua* during pregnancy in Wistar rats. Phytother Res 27(5):647–654

Adoni M, Yadam M, Gaddam SA, Rayalacheruvu U, Kotakadi VS (2020) Antimicrobial, antioxidant, and dye degradation properties of biosynthesized silver nanoparticles from *Artemisia annua* L. Lett Appl NanoBioSci 10(1):1981–1992

Aftab T, Ferreira JF, Khan MMA, Naeem M (eds) (2014) Artemisia annua-pharmacology and biotechnology. Springer, Berlin

Akkawi M, Jaber S, Abu-Remeleh Q, Engeu OP, Lutgen P (2014) Investigations of *Artemisia annua* and *Artemisia sieberi* water extracts inhibitory effects on β-hematin formation. Med Aromat Plants 3(150):2167-0412

Alcântara DDFÁ, Ribeiro HF, Cardoso PCDS, Araújo TMT, Burbano RR, Guimarães AC, de Oliveira Bahia M (2013) In vitro evaluation of the cytotoxic and genotoxic effects of artemether, an antimalarial drug, in a gastric cancer cell line (PG100). J Appl Toxicol 33(2):151–156

Alesaeidi S, Miraj S (2016) A systematic review of anti-malarial properties, immunosuppressive properties, anti-inflammatory properties, and anti-cancer properties of *Artemisia annua*. Electron Physician 8(10):3150

Appalasamy S, Lo KY, Ch'ng SJ, Nornadia K, Othman AS, Chan LK (2014) Antimicrobial activity of artemisinin and precursor derived from in vitro plantlets of *Artemisia annua* L. BioMed Res Int

Argemi X, Hansmann Y, Gaudart J, Gillibert A, Caumes E, Jauréguiberry S, Meyer N (2019) Comment on "Effect of *Artemisia annua* and *Artemisia afra* tea infusions on schistosomiasis in a large clinical trial". Phytomedicine 62:152804

Ashok PK, Upadhyaya K (2013) Preliminary phytochemical screening and physico-chemical parameters of *Artemisia absinthium* and *Artemisia annua*. J Pharmacogn Phytochem 1(6):229

Bai L, Li H, Li J, Song J, Zhou Y, Liu B, Zhou J (2019) Immunosuppressive effect of artemisinin and hydroxychloroquine combination therapy on IgA nephropathy via regulating the differentiation of CD4+ T cell subsets in rats. Int Immunopharmacol 70:313–323

Baraldi R, Isacchi B, Predieri S, Marconi G, Vincieri FF, Bilia AR (2008) Distribution of artemisinin and bioactive flavonoids from *Artemisia annua* L. during plant growth. Biochem Syst Ecol 36(5–6):340–348

Beccafico S, Morozzi G, Marchetti MC, Riccardi C, Sidoni A, Donato R, Sorci G (2015) Artesunate induces ROS-and p38 MAPK-mediated apoptosis and counteracts tumor growth in vivo in embryonal rhabdomyosarcoma cells. Carcinogenesis 36(9):1071–1083

Beekman AC, Wierenga PK, Woerdenbag HJ, Van Uden W, Pras N, Konings AW, Wikström HV (1998) Artemisinin-derived sesquiterpene lactones as potential antitumour compounds: cytotoxic action against bone marrow and tumor cells. Planta Med 64(07):615–619

Berdelle N, Nikolova T, Quiros S, Efferth T, Kaina B (2011) Artesunate induces oxidative DNA damage, sustained DNA double-strand breaks, and the ATM/ATR damage response in cancer cells. Mol Cancer Ther 10(12):2224–2233

Bouic PJ (2002) Sterols and sterolins: new drugs for the immune system? Drug Discov Today 7(14):775–778

Brown GD (2010) The biosynthesis of artemisinin (Qinghaosu) and the phytochemistry of *Artemisia annua* L. (Qinghao). Molecules 15(11):7603–7698

Cafferata LF, Gatti WO, Mijailosky S (2010) Secondary gaseous metabolites analyses of wild *Artemisia annua* L. Mol Med Chem 21:48–52

Cai TY, Zhang YR, Ji JB, Xing J (2017) Investigation of the component in *Artemisia annua* L. leading to enhanced antiplasmodial potency of artemisinin via regulation of its metabolism. J Ethnopharmacol 207:86–91

Cao R, Hu H, Li Y, Wang X, Xu M, Liu J, Zhang H, Yan Y, Zhao L, Li W, Zhang T (2020) Anti-SARS-CoV-2 potential of artemisinins in vitro. ACS Infect Dis 6:2524–2531

Chen H, Sun B, Pan S, Jiang H, Sun X (2009) Dihydroartemisinin inhibits growth of pancreatic cancer cells in vitro and in vivo. Anti-Cancer Drugs 20(2):131–140

Chen SS, Hu W, Wang Z, Lou XE, Zhou HJ (2015) p8 attenuates the apoptosis induced by dihydroartemisinin in cancer cells through promoting autophagy. Cancer Biol Ther 16(5): 770–779

Clark RL (2009) Embryotoxicity of the artemisinin antimalarials and potential consequences for use in women in the first trimester. Reprod Toxicol 28(3):285–296

Cobb S (1922) A case of epilepsy with a general discussion of the pathology. Med Clin N Am 5: 1403–1420

Cornet-Vernet L, Munyangi J, Chen L, Towler M, Weathers P (2019) Response to Argemi et al. (2019). Phytomedicine 62:152943

Czechowski T, Rinaldi MA, Famodimu MT, Van Veelen M, Larson TR, Winzer T, Graham IA (2019) Flavonoid versus artemisinin anti-malarial activity in *Artemisia annua* whole-leaf extracts. Front Plant Sci 10:984

Daddy NB, Kalisya LM, Bagire PG, Watt RL, Towler MJ, Weathers PJ (2017) *Artemisia annua* dried leaf tablets treated malaria resistant to ACT and iv artesunate. Phytomedicine 32:37–40

Das S (2012) *Artemisia annua* (Qinghao): a pharmacological review. Int J Pharm Sci Res 3(12): 4573–4577

Das S, Vörös-Horváth B, Bencsik T, Micalizzi G, Mondello L, Horváth G, Széchenyi A (2020) Antimicrobial activity of different *Artemisia* essential oil formulations. Molecules 25(10):2390

De Donno A, Grassi T, Idolo A, Guido M, Papadia P, Caccioppola A, Fanizzi FP (2012) First-time comparison of the in vitro antimalarial activity of *Artemisia annua* herbal tea and artemisinin. Trans R Soc Trop Med Hyg 106(11):696–700

Derda M, Hadaś E, Cholewiński M, Skrzypczak Ł, Grzondziel A, Wojtkowiak-Giera A (2016) *Artemisia annua* L. as a plant with potential use in the treatment of acanthamoebiasis. Parasitol Res 115(4):1635–1639

Desrosiers MR, Weathers PJ (2016) Effect of leaf digestion and artemisinin solubility for use in oral consumption of dried *Artemisia annua* leaves to treat malaria. J Ethnopharmacol 190:313–318. https://doi.org/10.1016/j.jep.2016.06.041

Disbrow GL, Baege AC, Kierpiec KA, Yuan H, Centeno JA, Thibodeaux CA, Schlegel R (2005) Dihydroartemisinin is cytotoxic to papillomavirus-expressing epithelial cells in vitro and in vivo. Cancer Res 65(23):10854–10861

Dixon SJ, Lemberg KM, Lamprecht MR, Skouta R, Zaitsev EM, Gleason CE, Stockwell BR (2012) Ferroptosis: an iron-dependent form of nonapoptotic cell death. Cell 149(5):1060–1072

Dixon SJ, Patel DN, Welsch M, Skouta R, Lee ED, Hayano M, Stockwell BR (2014) Pharmacological inhibition of cystine–glutamate exchange induces endoplasmic reticulum stress and ferroptosis. eLife 3:e02523

Du JH, Zhang HD, Ma ZJ, Ji KM (2010) Artesunate induces oncosis-like cell death in vitro and has antitumor activity against pancreatic cancer xenografts in vivo. Cancer Chemother Pharmacol 65(5):895–902

Duke JA (1992) Handbook of phytochemical constituent grass, herbs and other economic plants. CRC press, London

Efferth T (2005) Mechanistic perspectives for 1, 2, 4-trioxanes in anti-cancer therapy. Drug Resist Updat 8(1–2):85–97

Efferth T (2017) From ancient herb to modern drug: *Artemisia annua* and artemisinin for cancer therapy. In: Seminars in cancer biology, vol 46. Academic Press, London, pp 65–83

Efferth T (2018) Beyond malaria: the inhibition of viruses by artemisinin-type compounds. Biotechnol Adv 36:1730–1737

Efferth T, Oesch F (2004) Oxidative stress response of tumor cells: microarray-based comparison between artemisinins and anthracyclines. Biochem Pharmacol 68(1):3–10

Efferth T, Volm M (2005) Glutathione-related enzymes contribute to resistance of tumor cells and low toxicity in normal organs to artesunate. In Vivo 19(1):225–232

Efferth T, Briehl MM, Tome ME (2003) Role of antioxidant genes for the activity of artesunate against tumor cells. Int J Oncol 23(4):1231–1235

Efferth T, Giaisi M, Merling A, Krammer PH, Li-Weber M (2007) Artesunate induces ROS-mediated apoptosis in doxorubicin-resistant T leukemia cells. PLoS One 2(8):e693

Efferth T, Zacchino S, Georgiev MI, Liu L, Wagner H, Panossian A (2015) Nobel Prize for artemisinin brings phytotherapy into the spotlight. Phytomedicine 22(13):A1–A3

Ekiert H, Świątkowska J, Klin P, Rzepiela A, Szopa A (2021) *Artemisia annua*–importance in traditional medicine and current state of knowledge on the chemistry, biological activity and possible applications. Planta Med 87(8):584–599

El Hadji Omar GP, Mouhamadou D, Bineta DA, Sadikh BA, Mare DD, Ambroise A, Ousmane S (2013) Tea Artemisia annua inhibits Plasmodium falciparum isolates collected in Pikine, Senegal. Afr J Biochem Res 7(7):107–112

Elfawal MA, Towler MJ, Reich NG, Golenbock D, Weathers PJ, Rich SM (2012) Dried whole plant *Artemisia annua* as an antimalarial therapy. PLoS One 7(12):e52746

Elford BC, Roberts MF, Phillipson JD, Wilson RJ (1987) Potentiation of the antimalarial activity of qinghaosu by methoxylated flavones. Trans R Soc Trop Med Hyg 81(3):434–436

Eteng MU, Abolaji AO, Ebong PE, Brisibe EA, Dar A, Kabir N, Iqbal Choudhary M (2013) Biochemical and haematological evaluation of repeated dose exposure of male Wistar rats to an ethanolic extract of *Artemisia annua*. Phytother Res 27(4):602–609

Ferreira JF, Janick J (1996) Distribution of artemisinin in *Artemisia annua*. In: Janick J (ed) Progress in new crops. ASHS Press, Arlington, VA, pp 579–584

Ferreira J, Janick J (2009) Annual wormwood (*Artemisia annua* L.). New crop FactSHEET. Purdue University

Ferreira JF, Laughlin JC, Delabays N, de Magalhães PM (2005) Cultivation and genetics of *Artemisia annua* L. for increased production of the antimalarial artemisinin. Plant Genet Resour 3(2):206–229

Ferreira JFS, Luthria DL, Sasaki T, Heyerick A (2010) Flavonoids from *Artemisia annua* L. as antioxidants and their potential synergism with artemisinin against malaria and cancer. Molecules 15:3135–3170

Fornara F (2014) The molecular genetics of floral transition and flower development. Elsevier, London

Fouda E (2010) Etude clinique sur l'efficacité thérapeutique de l'*Artemisia annua* sur l'accès palustre simple. District de santé de la Cité Verte, Yaoundé, Personal Communication

Gerhardt T, Jones R, Park J, Lu R, Chan HW, Fang Q, Lai H (2015) Effects of antioxidants and pro-oxidants on cytotoxicity of dihydroartemisinin to Molt-4 human leukemia cells. Anticancer Res 35(4):1867–1871

Hou J, Wang D, Zhang R, Wang H (2008) Experimental therapy of hepatoma with artemisinin and its derivatives: in vitro and in vivo activity, chemosensitization, and mechanisms of action. Clin Cancer Res 14(17):5519–5530

Hsu E (2006) The history of qing hao in the Chinese materia medica. Trans R Soc Trop Med Hyg 100:505–508

Hsu E (2009) Diverse biologies and experiential continuities: did the ancient Chinese know that qinghao had anti-malarial properties? Can Bull Med Hist 26:203–213

Hunt S, Yoshida M, Davis CE, Greenhill NS, Davis PF (2015) An extract of the medicinal plant *Artemisia annua* modulates production of inflammatory markers in activated neutrophils. J Inflamm Res 8:9–14

Hwang DI, Won KJ, Kim DY, Yoon SW, Park JH, Kim B, Lee HM (2016) Anti-adipocyte differentiation activity and chemical composition of essential oil from *Artemisia annua*. Nat Prod Commun 11(4):539–542

Islam S, Mohammad F (2020) Triacontanol as a dynamic growth regulator for plants under diverse environmental conditions. Physiol Mol Biol Plants 26(5):871–883

Islamuddin M, Farooque A, Dwarakanath BS, Sahal D, Afrin F (2012) Extracts of *Artemisia annua* leaves and seeds mediate programmed cell death in *Leishmania donovani*. J Med Microbiol 61(12):1709–1718

Islamuddin M, Chouhan G, Want MY, Tyagi M, Abdin MZ, Sahal D, Afrin F (2014) Leishmanicidal activities of *Artemisia annua* leaf essential oil against visceral Leishmaniasis. Front Microbiol 5:626

Ivanescu B, Vlase L, Corciova A (2013, November) Importance of phytosterols and their determination in herbal medicines. In: In 2013 E-health and bioengineering conference (EHB). IEEE, pp 1–4

Jia J, Qin Y, Zhang L, Guo C, Wang Y, Yue X, Qian J (2016) Artemisinin inhibits gallbladder cancer cell lines through triggering cell cycle arrest and apoptosis. Mol Med Rep 13(5): 4461–4468

Kim WS, Choi WJ, Lee S, Kim WJ, Lee DC, Sohn UD et al (2015) Anti-inflammatory, antioxidant and antimicrobial effects of artemisinin extracts from *Artemisia annua* L. Korean J Physiol Pharmacol 19(1):21

Kontogianni VG, Primikyri A, Sakka M, Gerothanassis IP (2020) Simultaneous determination of artemisinin and its analogs and flavonoids in *Artemisia annua* crude extracts with the use of NMR spectroscopy. Magn Reson Chem 58(3):232–244

Krishna S, Ganapathi S, Ster IC, Saeed ME, Cowan M, Finlayson C, Kumar D (2015) A randomised, double blind, placebo-controlled pilot study of oral artesunate therapy for colorectal cancer. EBioMedicine 2(1):82–90

Lai H, Singh NP (1995) Selective cancer cell cytotoxicity from exposure to dihydroartemisinin and holotransferrin. Cancer Lett 91(1):41–46

Larson SA, Dolivo DM, Dominko T (2019) Artesunate inhibits myofibroblast formation via induction of apoptosis and antagonism of pro-fibrotic gene expression in human dermal fibroblasts. Cell Biol Int 43(11):1317–1322

Lechowicz K, Drożdżal S, Machaj F, Rosik J, Szostak B, Zegan-Barańska M, Biernawska J, Dabrowski W, Rotter I, Kotfis K (2020) COVID-19: the potential treatment of pulmonary fibrosis associated with SARS-CoV-2 infection. J Clin Med 9(6):1917

Lemke D, Pledl HW, Zorn M, Jugold M (2016) Slowing down glioblastoma progression in mice by running or the anti-malarial drug dihydroartemisinin? Induction of oxidative stress in murine glioblastoma therapy. Oncotarget 7(35):56713

Li SY, Chen C, Zhang HQ, Guo HY, Wang H, Wang L, Zhang X, Hua SN, Yu J, Xiao PG, Li RS (2005) Identification of natural compounds with antiviral activities against SARS-associated coronavirus. Antivir Res 67(1):18–23

Li PC, Lam E, Roos WP, Zdzienicka MZ, Kaina B, Efferth T (2008) Artesunate derived from traditional Chinese medicine induces DNA damage and repair. Cancer Res 68(11):4347–4351

Li S, Xue F, Cheng Z, Yang X, Wang S, Geng F, Pan L (2009) Effect of artesunate on inhibiting proliferation and inducing apoptosis of SP2/0 myeloma cells through affecting NFκB p65. Int J Hematol 90(4):513–521

Liu NQ, Cao M, Frédérich M, Choi YH, Verpoorte R, van der Kooy F (2010) Metabolomic investigation of the ethnopharmacological use of *Artemisia afra* with NMR spectroscopy and multivariate data analysis. J Ethnopharmacol 128(1):230–235

Liu H, Tian X, Zhang Y, Wang C, Jiang H (2013) The discovery of *Artemisia annua* L. in the Shengjindian cemetery, Xinjiang, China and its implications for early uses of traditional Chinese herbal medicine qinghao. J Ethnopharmacol 146(1):278–286

Lu M, Sun L, Zhou J, Yang J (2014) Dihydroartemisinin induces apoptosis in colorectal cancer cells through the mitochondria-dependent pathway. Tumor Biol 35(6):5307–5314

Marinas IC, Oprea E, Chifiriuc MC, Badea IA, Buleandra M, Lazar V (2015) Chemical composition and antipathogenic activity of *Artemisia annua* essential oil from Romania. Chem Biodivers 12(10):1554–1564

Martini MC, Zhang T, Williams JT, Abramovitch RB, Weathers PJ, Shell SS (2020) *Artemisia annua* and *Artemisia afra* extracts exhibit strong bactericidal activity against Mycobacterium tuberculosis. J Ethnopharmacol 262:113191

Mesa LE, Lutgen P, Velez ID, Segura AM, Robledo SM (2015) *Artemisia annua* L., potential source of molecules with pharmacological activity in human diseases. Am J Phytomed Clin Ther 3(5):436–450

Mesa LE, Vasquez D, Lutgen P, Vélez ID, Restrepo AM, Ortiz I, Robledo SM (2017) In vitro and in vivo antileishmanial activity of *Artemisia annua* L. leaf powder and its potential usefulness in the treatment of uncomplicated cutaneous leishmaniasis in humans. Rev Soc Bras Med Trop 50: 52–60

Moriyasu T, Nakamura R, Deloer S, Senba M, Kubo M, Inoue M, Hamano S (2018) *Schistosoma mansoni* infection suppresses the growth of *Plasmodium yoelii* parasites in the liver and reduces gametocyte infectivity to mosquitoes. PLoS Negl Trop Dis 12(1):e0006197

Mouton J, Jansen O, Frédérich M, van der Kooy F (2013) Is artemisinin the only antiplasmodial compound in the *Artemisia annua* tea infusion? An in vitro study. Planta Med 79(06):468–470

MR VR (2009) Chemical composition and antimicrobial activity of the essential oil of *Artemisia annua* L. from Iran. Pharm Res 1(1)

Mueller MS, Karhagomba IB, Hirt HM, Wemakor E (2000) The potential of *Artemisia annua* L. as a locally produced remedy for malaria in the tropics: agricultural, chemical and clinical aspects. J Ethnopharmacol 73(3):487–493

Munyangi J, Cornet-Vernet L, Idumbo M, Lu C, Lutgen P, Perronne C, Weathers P (2018) Effect of *Artemisia annua* and *Artemisia afra* tea infusions on schistosomiasis in a large clinical trial. Phytomedicine 51:233

Newton PN, Angus BJ, Chierakul W, Dondorp A, Ruangveerayuth R, Silamut K, White NJ (2003) Randomized comparison of artesunate and quinine in the treatment of severe falciparum malaria. Clin Infect Dis 37(1):7–16

Nigam M, Atanassova M, Mishra AP, Pezzani R, Devkota HP, Plygun S, Sharifi-Rad J (2019) Bioactive compounds and health benefits of *Artemisia* species. Nat Prod Commun 14(7). https://doi.org/10.1177/1934578X19850354

Noori S, Hassan ZM, Farsam V (2014) Artemisinin as a Chinese medicine, selectively induces apoptosis in pancreatic tumor cell line. Chin J Integr Med 20(8):618–623

Ostlund RE Jr (2007) Phytosterols, cholesterol absorption and healthy diets. Lipids 42(1):41–45

Papanikolaou X, Johnson S, Garg T, Tian E, Tytarenko R, Zhang Q, Heuck C (2014) Artesunate overcomes drug resistance in multiple myeloma by inducing mitochondrial stress and non-caspase apoptosis. Oncotarget 5(12):4118

Park J, Lai HC, Sasaki T, Singh NP (2015) DNA damage in dihydroartemisinin-resistant Molt-4 cells. Anticancer Res 35(3):1339–1343

Perez Gutierrez RM, Baez EG (2009) Cardioactive agents from plants. Mini Rev Med Chem 9:878–899. https://doi.org/10.2174/138955709788452612

Rajeswara Rao BR, Syamasundar KV, Patel RP (2014) Effect of method of distillation on the yield and chemical composition of *Artemisia annua* essential oil. J Essent Oil Res 26(6):486–491

Rasheed SAK, Efferth T, Asangani IA, Allgayer H (2010) First evidence that the antimalarial drug artesunate inhibits invasion and in vivo metastasis in lung cancer by targeting essential extracellular proteases. Int J Cancer 127(6):1475–1485

Riddle J (2010) Goddesses, elixirs, and witches: plants and sexuality throughout human history. Elsevier Inc., New York

Risaliti L, Pini G, Ascrizzi R, Donato R, Sacco C, Bergonzi MC, Bilia AR (2020) *Artemisia annua* essential oil extraction, characterization, and incorporation in nanoliposomes, smart drug delivery systems against Candida species. J Drug Delivery Sci Technol 59:101849

Sadiq A, Hayat MQ, Ashraf M (2014) Ethnopharmacology of *Artemisia annua* L.: A review. Artemisia Annua Pharmacol Biotechnol 9–25

Salehi M, Karimzadeh G, Naghavi MR, Badi HN, Monfared SR (2018) Expression of key genes affecting artemisinin content in five *Artemisia* species. Sci Rep 8(1):1–11

Schett G, Sticherling M, Neurath MF (2020) COVID-19: risk for cytokine targeting in chronic inflammatory diseases? Nat Rev Immunol 20(5):271–272

Seo EJ, Wiench B, Hamm R, Paulsen M, Zu Y, Fu Y, Efferth T (2015) Cytotoxicity of natural products and derivatives toward MCF-7 cell monolayers and cancer stem-like mammospheres. Phytomedicine 22(4):438–443

Shahrajabian MH, Wenli SUN, Cheng Q (2020) Exploring *Artemisia annua* L., artemisinin and its derivatives, from traditional Chinese wonder medicinal science. Notulae Botanicae Horti Agrobotanici Cluj-Napoca 48(4):1719–1741

Shatar S, Dung NX, Karashawa D (2003) Essential oil composition of some Mongolian Artemisia species. Journal of Essential Oil Bearing Plants 6(3):203–206

Shi C, Li H, Yang Y, Hou L (2015) Anti-inflammatory and immunoregulatory functions of artemisinin and its derivatives. Mediat Inflamm 2015:435713

Sieber S, Gdynia G, Roth W, Bonavida B, Efferth T (2009) Combination treatment of malignant B cells using the anti-CD20 antibody rituximab and the anti-malarial artesunate. Int J Oncol 35(1): 149–158

Smith RIL, Richardson M (2011) Fuegian plants in Antarctica: natural or anthropogenically assisted immigrants? Biol Invasions 13(1):1–5

Stix E (1960) Pollenmorphologische untersuchungen an Compositen. Grana 2(2):41–104

Su XZ, Miller LH (2015) The discovery of artemisinin and the Nobel prize in physiology or medicine. Sci China Life Sci 58:1175–1179

Subedi A, Futamura Y, Nishi M, Ryo A, Watanabe N, Osada H (2016) High-throughput screening identifies artesunate as selective inhibitor of cancer stemness: involvement of mitochondrial metabolism. Biochem Biophys Res Commun 477(4):737–742

Tchandema CK, Lutgen P (2016) In vivo trials on the therapeutic effects of encapsulated *Artemisia annua* and *Artemisia afra*. Glob J Res Anal 5(6):228–234

Tin AS, Sundar SN, Tran KQ, Park AH, Poindexter KM, Firestone GL (2012) Antiproliferative effects of artemisinin on human breast cancer cells requires the downregulated expression of the E2F1 transcription factor and loss of E2F1-target cell cycle genes. Anti-Cancer Drugs 23(4): 370–379

Tong Y, Liu Y, Zheng H, Zheng L, Liu W, Wu J, Lu L (2016) Artemisinin and its derivatives can significantly inhibit lung tumorigenesis and tumor metastasis through Wnt/β-catenin signaling. Oncotarget 7(21):31413

Tran KQ, Tin AS, Firestone GL (2014) Artemisinin triggers a G1 cell cycle arrest of human Ishikawa endometrial cancer cells and inhibits cyclin dependent Kinase-4 promoter activity and expression by disrupting NF-kB transcriptional signaling. Anti-Cancer Drugs 25(3):270

Tzenkova R, Kamenarska Z, Draganov A, Atanassov A (2010) Composition of *Artemisia annua* essential oil obtained from species growing wild in Bulgaria. Biotechnol Biotechnol Equip 24(2):1833–1835

Unschuld PU (1986) Medicine in China: a history of pharmaceutics. University of California Press, Berkeley

Wani KI, Choudhary S, Zehra A, Naeem M, Weathers P, Aftab T (2021) Enhancing artemisinin content in and delivery from *Artemisia annua*: a review of alternative, classical, and transgenic approaches. Planta 254(2):1–15

Wani KI, Zehra A, Choudhary S, Naeem M, Khan M, Khan R, Aftab T (2022) Exogenous strigolactone (GR24) positively regulates growth, photosynthesis, and improves glandular trichome attributes for enhanced artemisinin production in Artemisia annua. J Plant Growth Regul 1–10

Wei T, Liu J (2017) Anti-angiogenic properties of artemisinin derivatives. Int J Mol Med 40(4): 972–978

Willcox M (2009) *Artemisia* species: from traditional medicines to modern antimalarials—and back again. J Altern Complement Med 15:101–109

Willcox M, Bodeker G, Rasoanaivo P, Addae-Kyereme J (eds) (2004) Traditional medicinal plants and malaria, vol 4. CRC Press, Boca Raton

Woerdenbag HJ, Moskal TA, Pras N, Malingré TM, El-Feraly FS, Kampinga HH, Konings AW (1993) Cytotoxicity of artemisinin-related endoperoxides to Ehrlich ascites tumor cells. J Nat Prod 56(6):849–856

World Health Organization (2006) WHO monograph on good agricultural and collection practices (GACP) for *Artemisia annua* L. World Health Organization, Geneva

World Health Organization (2015) Guidelines for the treatment of malaria. World Health Organization

Wright CW, Linley PA, Brun R, Wittlin S, Hsu E (2010) Ancient Chinese methods are remarkably effective for the preparation of artemisinin-rich extracts of Qing Hao with potent antimalarial activity. Molecules 15:804–812

Zhang YX, Sun HX (2009) Immunosuppressive effect of ethanol extract of *Artemisia annua* on specific antibody and cellular responses of mice against ovalbumin. Immunopharmacol Immunotoxicol 31(4):625–630

Zhang ZY, Yu SQ, Miao LY, Huang XY, Zhang XP, Zhu YP, Li DQ (2008) Artesunate combined with vinorelbine plus cisplatin in treatment of advanced non-small cell lung cancer: a randomized controlled trial. Zhong Xi Yi Jie He Xue Bao 6(2):134–138

Zhang CZ, Zhang H, Yun J, Chen GG, San Lai PB (2012) Dihydroartemisinin exhibits antitumor activity toward hepatocellular carcinoma in vitro and in vivo. Biochem Pharmacol 83(9): 1278–1289

Zhang J, Guo L, Zhou X, Dong F, Li L, Cheng Z, Liu J (2016) Dihydroartemisinin induces endothelial cell anoikis through the activation of the JNK signaling pathway. Oncol Lett 12(3):1896–1900

Zhao HF, Zhong JX, Peng SQ, Liu SC (1998) Study of sweet wormwood compound on SLE in mice. Chin J Information Trad Chin Med 5(8):18–19

Zhao F, Wang H, Kunda P, Chen X, Liu QL, Liu T (2013) Artesunate exerts specific cytotoxicity in retinoblastoma cells via CD71. Oncol Rep 30(3):1473–1482

Zhou X, Sun WJ, Wang WM, Chen K, Zheng JH, Lu MD, Zheng ZQ (2013) Artesunate inhibits the growth of gastric cancer cells through the mechanism of promoting oncosis both in vitro and in vivo. Anti-Cancer Drugs 24(9):920–927

Zhu H, Liao SD, Shi JJ, Chang LL, Tong YG, Cao J, Lu JJ (2014) DJ-1 mediates the resistance of cancer cells to dihydroartemisinin through reactive oxygen species removal. Free Radic Biol Med 71:121–132

Zhu S, Yu Q, Huo C, Li Y, He L, Ran B et al (2021) Ferroptosis: a novel mechanism of artemisinin and its derivatives in cancer therapy. Curr Med Chem 28(2):329–345

Zime-Diawara H, Ganfon H, Gbaguidi F, Yemoa A, Bero J, Jansen O, Quetin-Leclercq J (2015) The antimalarial action of aqueous and hydro alcoholic extracts of *Artemisia annua* L. cultivated in Benin: in vitro and in vivo studies. J Chem Pharm Res 7(8):817–823

Chapter 3
Artemisia indica Willd.: Ethnobotany, Phytochemistry, Pharmacological Attributes, and Safety Profile

Rupa Sanyal, Sharmila Pandey, Samapika Nandy, Saikat Dewanjee, Abdel Rahman Al-Tawaha, Ercan Bursal, Protha Biswas, Manoj Kumar, Radha, Abilash Valsala Gopalakrishnan, Md. Habibur Rahman, Mahipal S. Shekhawat, Devendra Kumar Pandey, Tabarak Malik, and Abhijit Dey

Abstract The present book chapter is about the phytochemistry, pharmacological properties, and ethonobotanical, safety, and toxicological aspects of the species, *Artemisia indica*. This genus *Artemisia* belonging to Asteraceae family comprises of more than 400 species, among which *A. indica* (also called mugwort) is a perennial medicinal herb found majorly in cold temperate zones of Asia including Pakistan, China, India, Thailand, Korea, Japan, etc. It is extensively utilized by the traditional people in many countries for treating Malaria, chronic fever, dyspepsia, ringworms,

R. Sanyal · S. Pandey
Department of Botany, Bhairab Ganguly College, Kolkata, West Bengal, India

S. Nandy · P. Biswas · A. Dey (✉)
Department of Life Sciences, Presidency University, Kolkata, West Bengal, India
e-mail: abhijit.dbs@presiuniv.ac.in

S. Dewanjee
Advanced Pharmacognosy Research Laboratory, Department of Pharmaceutical Technology, Jadavpur University, Kolkata, West Bengal, India

A. R. Al-Tawaha
Department of Biological Sciences, Al-Hussein Bin Talal University, Maan, Jordon

E. Bursal
Department of Biochemistry, Mus Alparslan University, Muş, Turkey

M. Kumar
Chemical and Biochemical Processing Division, ICAR - Central Institute for Research on Cotton Technology, Mumbai, Maharashtra, India

Radha
School of Biological and Environmental Sciences, Shoolini University of Biotechnology and Management Sciences, Solan, Himachal Pradesh, India

A. V. Gopalakrishnan
Department of Biomedical Sciences, School of Bio Sciences and Technology, Vellore Institute of Technology (VIT), Vellore, Tamil Nadu, India

H. P. Devkota, T. Aftab (eds.), *Medicinal Plants of the Asteraceae Family*,
https://doi.org/10.1007/978-981-19-6080-2_3

hepatic diseases, and diarrhea-like health problems. Artemisinin is one of the prime phytochemical components accountable for the potential antimalarial activity, found in this species. Besides this, the plant comprises chiefly sesquiterpenes and davanone in its volatile oil. Modern pharmacological studies have revealed that its crude extracts exhibit a wide range of pharmacological activities such as anti-inflammatory, anthelmintic, anticancer, antidiabetic, antimicrobial, antidepressant, and many more as a result of the existence of several bioactive components in it. Few toxicological evaluation have documented on this species but in depth experiments, safety, and toxicological evaluation are needed to explore medicinal properties of this species more.

Keywords *Artemisia indica* · Asteraceae · Traditional uses · Phytochemical components · Artemisinin · Pharmacological properties

3.1 Introduction

Humans have depended on plant kingdom for their daily need which is as old as the humans' existence in the world. The plant kingdom includes all the important medicinal flora, and these are used for ages to treat various diseases. Almost 87% human diseases or problems are cured with naturally occurring compounds, derived from medicinal herbs, and their allied medicines (Hussain 2020). One of such important medicinal plant is the genus *Artemisia,* commonly known as "worm wood." This genus is a member of Asteraceae family comprising more than 400 species (Adewumi et al. 2020). One of the mostly known species of this genus is *Artemisia indica*, commonly called "mugwort" or Indian wormwood. This common perennial herb generally spreads in forest edges, grasslands, abandoned lands, roadside, fallow fields, etc. very quickly. This species is found in India, China, Nepal, Pakistan, Thailand, Japan, Korea, and many more cold-temperate regions of Asia; and in most of the regions, this plant is commonly called as "Titepati" (Jassal et al. 2019; Shimono et al. 2013). It is suitable for medium loamy and light sandy soils with well drainage system. This plant can be sometimes annual or woody subshrub that spreads through rhizomatous root-stock. The stem ranges from

M. H. Rahman
Department of Global Medical Science, Wonju College of Medicine, Yonsei University, Wonju, Gangwon-do, South Korea

M. S. Shekhawat
Plant Biotechnology Unit, KM Government Institute for Postgraduate Studies and Research, Puducherry, India

D. K. Pandey
Department of Biotechnology, Lovely Professional University, Phagwara, Punjab, India

T. Malik
Institute of Health, Jimma University, Jimma, Ethiopia

80–150 cm tall, with many branches. Leaves are either stalkless or short – stalked and ovate, elliptic, or oblong-ovate shaped. Flowers are erect, ovoid or oblong-ovoid, bisexual, and borne in conical. Fruit is achene type and brown obovoid or oblong shaped. These plants are hardier and aromatic. It is generally propagated by seeds, also by stolon (Ken Fern 2022; Flowers of India 2021). This species had been widely utilized traditionally in many countries to treat many diseases and health problems like dyspepsia, hepatic diseases, chronic fever, diarrhea, ringworms, wounds, Malaria, and many more (Adewumi et al. 2020; Nigam et al. 2019; Bhattarai 2020). This plant primarily comprises of sterols, terpenoids, acetylenes, coumarins, flavonoids etc.; among these oxygenated sesquiterpenes and davanone are the main compounds found in its volatile oil (Adewumi et al. 2020; Haider et al. 2014). Because of its various biologically active components, it shows numerous pharmacological activities like antimalaria (main activity because of its component artemisinin), anticancer, anthelmintic, antiseptic, inflammatory, antioxidant, and many more; and these activities can be shown at specific safe doses of its extracts. So, the purpose of this study is to discover the traditional and folk medicinal uses, phytoconstituents, pharmacological activity, clinical studies, and safety and toxicity related to this study of *A. indica* in more details.

3.2 Traditional Uses

One of the most necessary health sources is traditional medicinal system in the entire world (especially the developing countries), derived from beneficial medical plants or herbs. *A. indica* is one of those beneficial perennial herbs, found in the different regions of Western Himalayas with the familiar name "Titepati." Local people of this region used this herb for ages to heal dyspepsia, hepatic diseases, and also for chronic fevers (Adewumi et al. 2020). This plant had majorly utilized by the local peoples of China, Nepal, Pakistan, India, Thailand (especially for Malaria), and several other countries or areas of southeast Asia in medicines to treat various ailments (Nahid et al. 2017). In Nepal, the juice of this plant had been utilized in treatments of abdominal pain, diarrhea, ringworms, wounds, cuts, leech infestation, and dysentery. Different parts of this plant had been eaten here such as cooked young leaves were consumed with Barley, also with rice to add some color and flavors. Here, the dried leaf and flowers were used as insecticides and the juice of the leaf were used to heal the skin ailments (Rashid et al. 2013; Nigam et al. 2019). Local peoples of Eastern Nepal, this species along with some other plant species had been found in a survey to treat many infections or infestations such as for helminths, fever, diphtheria, malaria, scabies, gonorrhea, etc. (Bhattarai 2020). In Okinawa, a famous island of Japan in whole world for the presence of long-lived individuals in abundance, it was utilized as an important food plant with some other plants (Niwano et al. 2009). In Xishuangbanna of China, Dai people uses various plants as mosquito repellent but *A. indica* was found to be the most effective mosquito repellent, especially in the cases of Dengue (Gou et al. 2020). In South Uganda, this plant was used as pesticide or for pest control by the local peoples (Mwine et al. 2011). In northeast areas of Pakistan, leaves of this species had been used medicinally by their indigenous peoples, especially for ear problems (Afzal et al. 2009). In Assam, also it had been found to be used by the local people of Tejpur as antimalarial drug to treat malaria and associated symptoms such as fever, headache, sweating, joint pains, weakness, vomiting, and shivering (Namsa et al. 2011). Fried tender shoots of this plant were also reported to eat as a vegetable food by the indigenous people of Garo in Norkek-Biosphere Reserve of Meghalaya state of India, through a survey (Singh et al. 2012). In another state of India, Darjeeling, it was used to cure asthma, amoebic dysentery, helminthic infections, several skin, and stomach problems (Tiwary et al. 2015). Young leaves had also been consumed as vegetable by the indigenous people of different areas (like Arunachal Pradesh) of India (Joram et al. 2021). The flowering shoots and leaves had been documented to show several pharmacological efficacies such as anthelmintic, antispasmodic, anti-septic, cytotoxic, antimicrobial, and anticancer (Rashid et al. 2013). Usage of *A. indica* traditionally in different places is given below in Table 3.1.

3.3 Chemical Constituents (Fig. 3.1)

A. indica represents a rich source of numerous biologically active components, responsible for various pharmacological activities. Quantitative and qualitative composition of these compounds differ hugely that might be linked with species variation; climatic, geographic, genetic, and environmental conditions; vegetation phase; anatomical part and age of plant; soil; method and season of harvesting; etc. (Nigam et al. 2019). This plant chiefly comprises of sterols, terpenoids, acetylenes, coumarins, flavonoids, etc. Oxygenated terpenes (among all terpenes) and hydrocarbons are present in this plant in abundance out of all the components in its volatile oil (Rather et al. 2017; Adewumi et al. 2020). In its essential oil, total 32 compounds had been reported to be found from Uttarakhand Himalaya, isolated and evaluated by GC and GC/MS method, among which oxygenated sesquiterpenes in 33.83% and davanone in 30.8% were the main compound. Besides this two, monoterpene hydrocarbons in 25.90%, sesquiterpene hydrocarbons in 20.54%, and oxygenated monoterpenes in 15.15% were found. Other major components found were β-elemene, β-pinene, β-myrcene, δ-cadinene, τ-muurolol, germacrene-D, trans-caryophyllene, cymene, limonene, linalool, 1,8-cineol, sabinene, etc. (Haider et al. 2014). Some previous studies in this region had shown that essential oil of *A. indica* were lacking the component davanone; even not present in traces (Shah and Rawat 2008). Besides these components, Artemisia ketone, ascaridole, borneol, transverbenol, alpha thujone, p-cymene, chrysanthenyl acetate, cubebene, caryophyllene oxide, terpineol, eucalyptol, isoascaridole, camphor, etc. were also found to be present in its volatile oil (Satyal et al. 2012; Rashid et al. 2013). Diverse solvent extracts had been isolated from the leaf of this species and assessed by HPLC-MS and HPLC-DAD method which showed flavonoids like casticin,

Table 3.1 Traditional use of *A. indica* in different places

Places	Parts used	Uses	Reference
Ilam, eastern Nepal	Whole plant	Treatment of fever, diphtheria, malaria, scabies, gonorrhea, measles, hyperthermia, sore throat, food poisoning, helminthic and lice infections	Bhattarai (2020)
Norkek biosphere, Meghalaya, India	Tender shoots	Fried and eaten as vegetable	Singh et al. (2012)
Northern Pakistan	Leaves	Used in medicines, especially for ear problems	Afzal et al. (2009)
Xishuangbanna, China	Whole plant	Mosquito repellent, mostly in dengue	Gou et al. (2020)
Tejpur, Assam	Whole plant	For malaria and associated symptoms like joint pain, shivering, weakness, fever, headache, and sweating	Namsa et al. (2011)
Nepal	1. Young leaves 2. Dried leaf and flowers	1. Consumed with rice and barley 2. As insecticides and for treating skin ailments	Rashid et al. 2013; Nigam et al. (2019)
Chuadanga, Bangladesh	Leaves	Treatments of leukorrhea, i.e., white discharge in urine of females	Rahmatullah and Biswas (2012)
Western Himalaya	Whole plant	For treating dyspepsia, hepatic ailments, and chronic fevers	Adewumi et al. (2020)
Swat, Pakistan	Aerial parts	Used as anthelmintic	Ahmad et al. (2013)
Darjeeling, India	Whole plant	Cure asthma, amoebic dysentery, helminthic infections, several skin, and stomach problems	Tiwary et al. (2015)
Khimi, Central Nepal	Especially the young leaves (besides, the whole plant)	Use in medicines	Sigdel et al. (2013)
Okinawa, Japan	Whole plant	As food plant	Niwano et al. (2009)
Gilgit-Baltistan, Pakistan	Different parts of whole plant	As foods, fuels, ornaments and for medicinal use	Hussain (2019)
Arunachal Paradesh, India	Young leaves	Consumed as vegetable	Joram et al. (2021)
Bagmati watershed, Nepal	Leaves	As insecticides to repel moths and other insects in order to prevent infestation of foods, clothes, furniture, etc.	Joshi and Joshi (2004)

cirsilineol, eupatin, and chrysoplenetin in it. These are all polymethoxy flavonoids (Tasdemir et al. 2015). In *A. indica*, exiguaflavanone-A, exiguaflavanone-B, maacklain, 2-(2,4-dihydroxyphenyl)-5,6-methylenedioxybenzofuran, and artemisinin were isolated and discovered to show the antimalarial activity as major

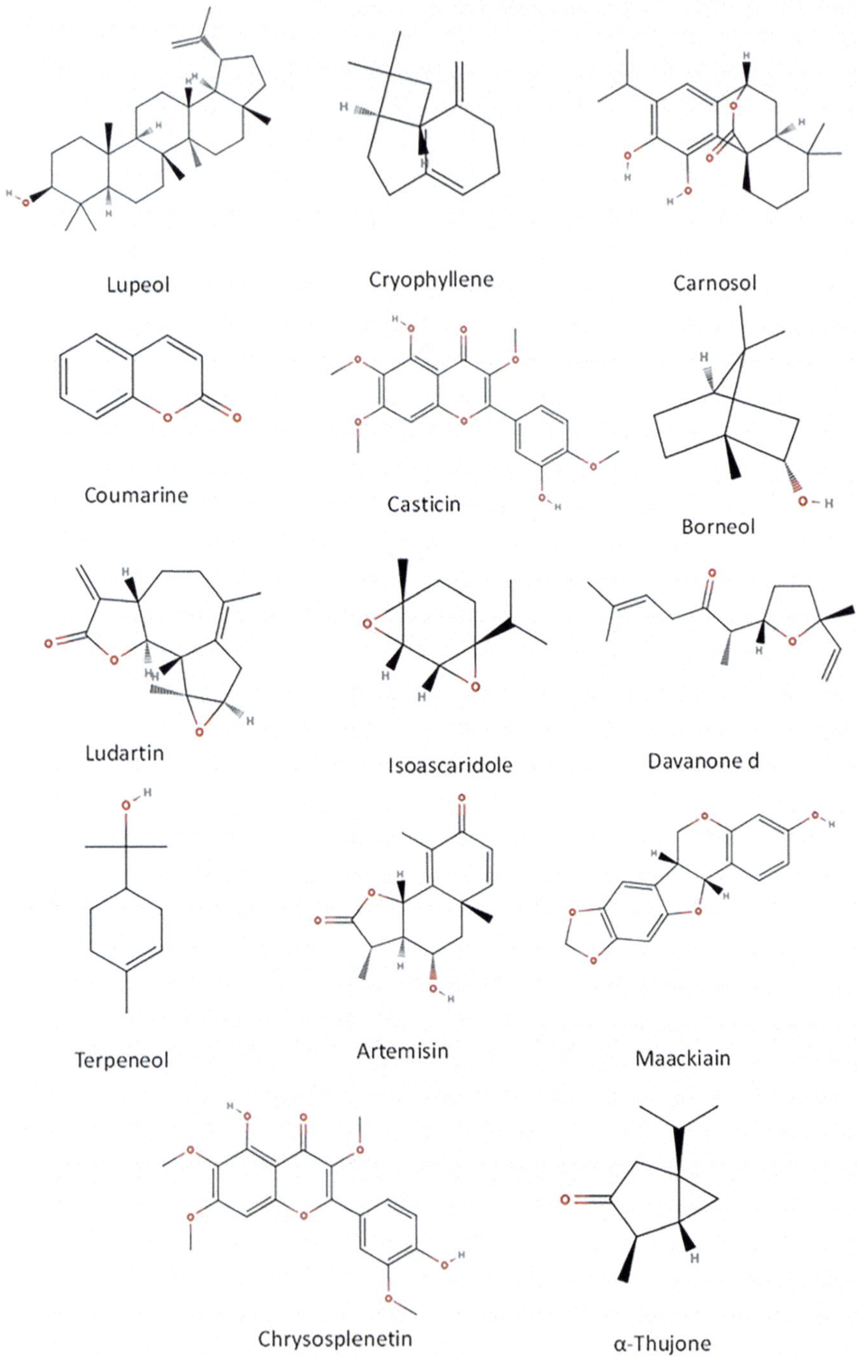

Fig. 3.1 Structures of the Phytochemical Found in *A. indica*

compound responsible for its antimalarial activity (Chanphen et al. 1998; Mannan et al. 2011). From the root and shoot, various compounds had been isolated and identified through bioassay-guided fractionation method such as 5-hydroxy-3,7,4-′-trimethoxyflavone, 6-methoxy-7,8-methylenedioxy coumarin, cis-matricaria ester, lupeol, maackiain, ludartin, cis-matricaria ester, and trans-matricaria ester (Zeng et al. 2015). Various components, isolated from various parts of the plant, are shown below with its pharmacological attributes in Table 3.2.

3.4 Pharmacological Activities

This medicinal plant is industrially very crucial as it exhibits diverse pharmacological properties from its different parts. Some of these activities are mentioned below in brief (Fig. 3.2).

3.4.1 Antimalarial Activity

It is used majorly in malaria, caused by protozoa *Plasmodium sp.*, mostly by *Plasmodium falciparum* (Saxena et al. 2003), as antipyretic to reduce the fevers. An effective antimalarial drug artemisinin, derived from the leaves of its species which is an endoperoxide sesquiterpene lactone (Mannan et al. 2011). This artemisinin had been produced through different cultures such as callus, hairy shoot, and root cultures. One of these cultures had been experimentally proven to show increased amount of artemisinin. In such experiments with hairy root cultures of *A. indica* in liquid medium, it was found that artemisinin content increased (0.042%) in its transformed roots after transformed by T- DNA of *Agrobacterium rhizogenes* strain (Mannan et al. 2008). Another research on *A. indica* showed other two components—exiguaflavanone A and B—to have in vitro antimalarial effect. These components were isolated and purified by the crude MeOH extracts, obtained from the air-dried stem. Then their antimalarial activities were evaluated by their effective concentration (EC_{50}) values (Chanphen et al. 1998).

3.4.2 Anticancer Activity

Cancer is the serious worldwide public health issue, mostly because of its lacking in early detection methods. So anticancer activities generally help to prevent or suppress the carcinogenic development (Chanda and Nagani 2013). Study on 50 μg/mL ethanol extract of leaf of *A. indica* in MTT, trypan blue exclusion assays, and morphological assessment showed that these extracts had inhibited more than 50% of human breast adenocarcinoma (MCF7) growth, human cervix adenocarcinoma

Table 3.2 Various components with its different pharmacological activity, isolated from different parts of *A. indica*

Components	Plant parts	Pharmacological activity	Reference
Lupeol	Shoot	Antiprotozoal, antitumor, apoptotic	Zeng et al. (2015); Gallo and Sarachine (2009)
Caryophyllene	Oil from aerial parts	Hypolipidemic, analgesic, anti-cancer, anti-inflammatory, preventing alcoholic steatohepatitis and osteoclastogenesis	Baldissera et al. (2017); Jassal et al. (2019)
Carnosol	Leaf	Antidiabetic and anti-hyperlipidemic; antioxidant; anti-microbial; antidepressant; anxiolytic	Khan et al. (2017); Nahid et al. (2017); Khan et al. (2016)
Coumarin	Root	Antitumor, anticancer, anticoagulant, antibacterial, anti-inflammatory, apoptotic	Zeng et al. (2015); Xu et al. (2015)
Casticin	Aerial parts	Anti-inflammatory, antitumor, anticancer, immunomodulatory, anti-hyperprolactinemia	Chan et al. (2018); Jassal et al. (2019)
Borneol	Oil from aerial parts	Antibacterial, antifungal, antioxidant, anticancer, antidepressant	Rashid et al. (2013); Hou et al. (2017)
Ludartin	Shoot or aerial parts	Antitumor, apoptotic, anti-inflammatory, recovery in spinal cord injury	Zeng et al. (2015); Xu et al. (2019)
Isoascaridole	Leaf	Cytotoxic, allelopathic, insecticidal, analgesic, antinociceptive activity	Satyal et al. (2012); Jaffal and Abbas (2019)
Davanone	Oil from aerial parts	Insect repellent, fumigant, antispasmodic	Hu et al. (2019); Jassal et al. (2019)
Terpineol	–	Anti-diarrheal, gastro-protective, anti-spasmodic, anti-hyperalgesic activity	Sadraei et al. (2015); Souza et al. (2011); Oliveira et al. (2016); Negreiros et al. (2019)
Artemisin, maccklain, exiguaflavanone	Stem, leaf	Antimalarial	Chanphen et al. (1998); Mannan et al. (2011)
Chrysoplenetin	Leaf	Antimalarial, anti-parasitic, anti-cancer, antioxidant, inhibition of enterovirus 71 replications	Tasdemir et al. (2015); Ferreira et al. (2010); Zhu et al. (2011)
α-Thujone	Aerial parts	Pro-apoptotic, cytotoxic, antiangiogenic, synergistic effects	Torres et al. (2016); Lee et al. (2020)

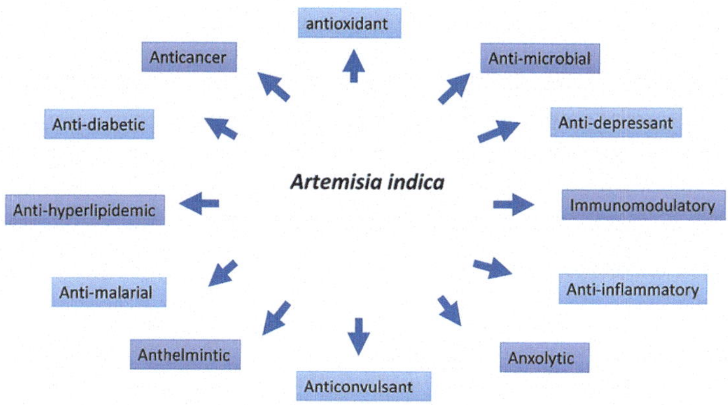

Fig. 3.2 Different pharmacological activities of *A. indica*

(HeLa), and human hepatocarcinoma (HepG2) cell lines (Tiwary et al. 2015). Another study on aerial parts of *A. indica* had been done by sulforhodamine-B assay to know its cytotoxic effects on leukemia (THP1), lung (A-549), liver (HEP-2), and colon (Caco-2) cancer cell lines. This study revealed that essential oil, extracted from these extracts, had significantly inhibited these cell lines' growth in the dilution range of 10–100 μg/mL with 19.5 μg/mL IC_{50} value for Caco-2, 10 μg/mL IC_{50} value for THP-1, 15.5 μg/mL IC_{50} value for HEP-2, and 25 μg/mL IC_{50} value for A-549 (Rashid et al. 2013). Other reports suggest that ethyl acetate extract of this species showed anticancer activity by decreasing the mitochondrial membrane potential and by inducing the DNA damage against lung (A-549), colon (Colo-205), human breast adenocarcinoma (MCF-7), and BHY and Miapaca cancer cell lines. Ludartin and lupeol had been found in this study to be the major component, responsible for this activity (Zeng et al. 2015).

3.4.3 Antidiabetic and Antihyperlipidemic Activity

Diabetes is another major issue (a metabolic disorder) in the world now as insufficient blood sugar regulation creates serious health concern. It is generally distinguished by chronic hyperglycemia, i.e., high blood sugar, because of the disabilities in insulin action and secretion (Salehi et al. 2019). An evaluation study on these activities of *A. indica* in 50 mg/kg diabetic Sprague-Dawley rats induced with streptozotocin had been performed. This study showed that 200 and 400 mg/kg b.w. hydroethanolic crude extracts and 200 mg/kg b.w. chloroform extracts of aerial parts of the plant had reduced the blood glucose levels considerably. It showed antihyperlipidemic (high lipid levels in blood) activities by reducing the cholesterol levels, serum creatinine, lipoproteins, and triglyceride levels (Ahmad et al. 2014). In reports of Jassal et al. (2019), oleanolic acid and carnosol had been found to have antidiabetic activities, whereas caryophyllene had found to treat hyperlipidemia, i.e., high lipid in blood (Jassal et al. 2019). 1–100 mg/kg b.w. carnosol, present in chloroform extracts of *A. indica*, was found in recent studies to lower the blood glucose levels, cholesterol, triglycerides, low-density lipoproteins, and creatinine levels in diabetes-induced rats (Khan et al. 2017).

3.4.4 Anthelmintic Activity

Helminths, the parasitic worms, create the most common infections (helminthic infections) in men. So anthelmintic drugs destroy or expel these parasitic worms which are present in large numbers, majorly in gastrointestinal (GI) tract (Das et al. 2011). Many experiments had been performed to show this activity in *A. indica* plant. One such investigation on crude aqueous, chloroform, and methanolic extracts of this species had been done against the healthy Indian earthworm, *Pheretima posthuma* where 2.5 mg/mL chloroform and methanol extract had shown efficient anthelmintic effect against the earthworm. The former one showed paralysis and death time of 9.67 and 20.67 min respectively, whereas the later one showed 19.67 and 25.33 min for paralysis and death respectively (Sarnim et al. 2013). Another study was done to scientifically prove its activity with 3.75, 6.25, 12.5, 25, and 50 mg/mL concentrations, accompanied by positive control with 10% albendazole and negative control with PBS. This resulted in the highest anthelmintic activity of *A. indica* at 50 mg/mL concentration of ethanolic extracts by different assays such as egg hatch inhibition, larval mortality, and adult worm mortality (Khan et al. 2015).

3.4.5 Antidepressant and Anxiolytic Activity

Antidepressant drugs have been used to decrease the symptoms of psychological disorders, especially the depressive disorders, by rectifying the chemical imbalances of vital neurotransmitters (common link of communication between nerve cells) in the brain that can improve the behavior and mood during these disorders. They generally increase the levels of neurotransmitters (e.g., selective serotonin reuptake inhibitors) around the nerves (Ogbru 2021). An anxiolytic drug has been used to treat anxiety disorders almost in the same way. Some investigations on nonvolatile components of *A. indica* had been done in mouse models to understand its efficacy on the function of central nervous system (CNS). The outcome of these investigations showed that oleanolic acid, carnosol, and ursolic acid, isolated from this species, had induced the anxiolytic (through light or dark box paradigms and elevated plus maze tests) and antidepressant (through forced swim and tail suspension tests) activities in mouse without any signs of toxicity (Khan et al. 2016).

3.4.6 Antimicrobial Activity

Antimicrobial activity is activity of all the active agents that can cease the microbial growth or can inhibit or destroy their colonies or may kill them directly (Elmogahzy 2020). This species has thus shown effectively this antimicrobial activity, especially antibacterial and antifungal activity. Study on antimicrobial activities of *A. indica*, *Tecoma stans*, and *Medicago falcata* in vitro had shown that chloroform, ethyl acetate, and butanol extracts of *A. indica* have 15–20 mm inhibitory effects towards *Pseudomonas aeruginosa*, *Staphylococcus aureus*, and *Escherichia coli* bacteria, whereas it showed 12–14 mm inhibitory activities against bacterium *Salmonella typhi*. Chloroform and n-hexane extracts of *A. indica* had been shown antifungal activities against *Fusarium solani* and *Aspergillus flavus*, respectively, by completely ceasing their growth (Javid et al. 2015). Another investigation to understand the antimicrobial activity of *A. indica* was done on the essential oil, extracted from the aerial parts of the plant. Total 43 phytoconstituents had been found through GC-MS and GC-FID analysis, among which 42.1% artemisia ketone, 8.6% germacrene, 6.1% borneol, and 4.8% cis-chrysanthenyl acetate were the major components to show notable antibacterial activities, mostly against gram-negative bacteria like *Pseudomonas aeruginosa* and *klebsiella pneumonia* and antifungal activities against fungal strains such as *Pseudomonas chrysogenum* and *Aspergillus niger* (Rashid et al. 2013). Leaves of *A. indica* had also been reported to show antileishmanial activity against the protozoan parasite *Leishmania* sp. (which creates leishmaniasis), due to the presence of artemisinin compound in it (Ganguly et al. 2006).

3.4.7 Anti-Inflammatory Activity

Many reports suggested that this species has ability to reduce inflammation. Inflammation generally occurs in response to the cell death, degeneration, cancer, ischemia, and tissue injury or due to the invasion of microbes in the body (Azab et al. 2016). Research on anti-inflammatory action of methanol extract of aerial parts of *A. indica* by carrageenin-induced rat paw edema, a well-established model of assessing chronic inflammation, was done. This research had revealed that methanolic (100, 200, and 400 mg/kg) extracts had notably inhibited the carrageenin-induced paw edema in rats. Flavonoids had been reported to be responsible for this anti-inflammatory activity as it can inhibit the prostaglandin synthesizing enzymes (Mansouri et al. 2015; Sagar et al. 2010). Other reports suggested that saponins, tannins, casticin, and terpenoids (like carnosol, caryophyllene), present in *A. indica* extracts, showed anti-inflammatory activity because of its effect of inhibition on inflammation (Ruwali et al. 2015; Jassal et al. 2019).

3.4.8 Immunomodulatory Activity

Modulation of immune system (defense system of our body) denotes any change in the immune response such as expression, stimulation, inhibition, or amplification of any phase of the immune response. So immunomodulators are used to modify to alter this immune response of the system (Abood et al. 2014). Major information is reported about its immunomodulatory action but in vitro experiments are few. One experiment on methanolic fresh aerial parts of this plant had been performed through B and T lymphocyte proliferation assay in lymphocyte culture of chicken. The results suggested that 200 μg/mL extracts had significantly increased the regulation of B cell proliferation at 11.76% and T cell proliferation at 12.018% in the presence of LPS (B cell mitogen) and Con A, respectively (Ruwali et al. 2015). Casticin had been reported as one of the most responsible phyto-constituent for this activity in *A. indica* (Jassal et al. 2019).

3.4.9 Antioxidant Activity

Prevention of developing reactive oxygen species and their reactions, thereby inhibition or limitation of oxidation of nutrients like proteins, lipids, is called antioxidant activity (Guclu et al. 2021). These reactive oxygen species can also bring about necrosis, apoptosis, and oxidative stress. Thus, antioxidants can prevent all these distress from these harmful substances. Methanolic extracts of leaves of *A. indica* had been evaluated in vitro by DPPH free radical scavenging assay and revealed to consist antioxidant activity because of the existence of phenol and flavonoid in significant quantity (Nahid et al. 2017). Essential oil, collected from its aerial parts, was also reportedly found to have antioxidant activities, after studying through hydroxyl and DPPH radical scavenging assays (Rashid et al. 2013).

3.4.10 Anticonvulsant Activity

Anticonvulsant drugs, also known as anti-epileptic or anti-seizure drugs, have been used to prevent the convulsions or seizures by either decreasing the excitation or enhancing the inhibition. *A. indica* was investigated for its anticonvulsant activity and found as an effective drug for it. Few studies had been done on non-volatile compounds such as ursolic acid, oleanolic acid, and carnosol which had been assessed for this effect on GABA-A receptors in mouse models of convulsions, induced by pentylenetetrazole (PTZ). The result of this assessment showed that 10 mg/kg carnosol, 30 mg/kg ursolic acid, and 100 mg/kg oleanolic acid had effectively increased the span of onset and reduced the span of clonic-tonic seizures and also gave 100%, 83.37%, and 66.7% protection, respectively. In this assessment, 1 mg/kg diazepam had been used as reference drug (Khan et al. 2016).

3.5 Studies Related to Safety and Toxicity

Medicinal plants or natural herbs have been the chief remedy for almost 100 years ago to treat numerous diseases, and 25% of the medicines (those are recently used/ modern medicines) are primarily utilized by these herbs. Generally, these herbs have been regarded safe because of its long history of use in traditional medicines, but still nowadays this safety has been a major concern for both of the general public and the national health authorities. On the other hand, toxicity is the capability of substances to cause toxic or harmful effects in living beings. It can be acute, chronic, subacute, and subchronic type (Mensah et al. 2019). So besides of safety, toxic doses of these herbs must have assessed and are also the major concern. Extracts of *A. indica* have also specific safe and toxic doses of showing particular pharmacological activities. No such toxicity of this plant was found, but in some cases dermatitis or different allergies can occur in people (Foster and Duke 2020). Some reports suggested that crude hydro-methanolic extracts at 200 mg/kg body weight and 400 mg/kg body weight; chloroform extracts at 200 mg/kg body weight of *A. indica* had shown to decrease the blood glucose levels. So, these doses were safe and did not produce any harmful changes or effects in the diabetic rats induced by streptozotocin, but the extract at 2000 mg/kg body weight dose had shown few irritations and escape behavior in rats (Ahmad et al. 2014). Other studies on anthelmintic activity of this species, 6.25 mg/L methanol and chloroform extracts and 2.5 mg/L chloroform extracts were found to show significant positive effect against *Pheretima posthuma*, but the toxic doses were not evaluated that much (Sarnim et al. 2013). In an in vitro study, 0.3–1.6 mg/mL methanolic extracts had also shown 54.72% cytotoxicity in chicken lymphocytes, and 0.2 mg/mL extracts had shown complete cell viability (Ruwali et al. 2015).

3.6 Conclusions and Future Recommendations

Humans rely on medicinal plants for their daily need for years and investigations on these plants have been in attention more for scientists in recent days. *Artemisia* is one of such genera that is very beneficial medicinal plant, but the species, *A. indica*, is not that much explored till now. So, this chapter covers its history of traditional uses in different places for different diseases and its photochemistry, vast pharmacological activities, and clinical studies with safety and toxicity. Various phytochemical components have been isolated from the plant and identified which has enormous health benefits. Several studies on the extracts of *A. indica* have shown that it has numerous pharmacological properties such as antimalarial, antitumor, anticancer, anthelmintic, antihyperglycemic, anti-inflammatory, anticonvulsant, antidepressant, and many more. Many in vitro and in vivo experiments are mentioned regarding to its different pharmacological properties. This plant is generally is considered only as a weed and has been neglected but many more medicinal uses, and experiments should be performed and established on it. In recent times, research interests have been shifted towards the exploration of biologically active components or natural components for the health benefits so various phytochemical components from different plant parts of *A. Indica* should be assessed more to know and understand it's health benefits. To use it more efficiently and before the clinical applications, the toxicity or safety levels or doses should also need to be evaluated. When this plant has already gained attention in the world because of its traditional uses to combat malaria, this information summarized in this chapter thereby is proposed to act as a reference tool to the people interested in this plant and to others for understand the importance of this beneficial plant, also for performing several experiments and research on it to understand it more.

References

Abood WN, Fahmi I, Abdulla MA, Ismail S (2014) Immunomodulatory effect of an isolated fraction from *Tinospora crispa* on intracellular expression of INF-γ, IL-6 and IL-8. BMC Complement Altern Med 14(205). https://doi.org/10.1186/1472-6882-14-205

Adewumi OA, Singh V, Singh G (2020) Chemical composition, traditional uses and biological activities of artemisia species. J Pharmacogn Phytochem 9(5):1124–1140

Afzal S, Afzal N, Awan MR, Khan TS, Gilani A, Khanum R, Tariq S (2009) Ethno-botanical studies from Northern Pakistan. J Ayub Med Coll Abbottabad 21(1):52–57

Ahmad N, Anwar S, Fazal H, Abbasi BH (2013) Medicinal plants used in indigenous therapy by people of Madyan Valley in district Swat, Pakistan. Int J Med Aromat Plants 3(1):47–54

Ahmad W, Khan I, Khan MA, Ahmad M, Subhan F, Karim N (2014) Evaluation of antidiabetic and antihyperlipidemic activity of Artemisia indica Linn (aeriel parts) in Streptozotocin induced diabetic rats. J Ethnopharmacol 151(1):618–623. https://doi.org/10.1016/j.jep.2013.11.012

Azab A, Nassar A, Azab AN (2016) Anti-inflammatory activity of natural products. Molecules 21(10):1321. https://doi.org/10.3390/molecules21101321

Baldissera MD, Souza CF, Grando TH, Doleski PH, Boligon AA, Stefani LM, Monteiro SG (2017) Hypolipidemic effect of β-caryophyllene to treat hyperlipidemic rats. Naunyn Schmiedeberg's Arch Pharmacol 390(2):215–223

Bhattarai KR (2020) Ethnobotanical survey on plants used in Mai municipality of Ilam district, eastern Nepal. Banko Janakari 30(2):11–35. https://doi.org/10.3126/banko.v30i2.33476

Chan EWC, Wong SK, Chan HT (2018) Casticin from Vitex species: a short review on its anticancer and anti-inflammatory properties. J Integr Med 16(3):147–152. https://doi.org/10.1016/j.joim.2018.03.001

Chanda S, Nagani K (2013) Vitro and in vivo methods for anticancer activity evaluation and some Indian medicinal plants possessing anticancer properties: an overview. J Pharmacogn Phytochem 2(2):140–152

Chanphen R, Thebtaranonth Y, Wanauppathamkul S, Yuthavong Y (1998) Antimalarial principles from *Artemisia indica*. J Nat Prod 61(9):1146–1147. https://doi.org/10.1021/np980041x

Das SS, Dey M, Ghosh AK (2011) Determination of anthelmintic activity of the leaf and bark extract of Tamarindus indica Linn. Indian J Pharm Sci 73(1):104–107. https://doi.org/10.4103/0250-474X.89768

Elmogahzy YE (2020) Finished fibrous assemblies. In: Engineering textiles, Integrating the design and manufacture of textile products. The Textile Institute book series, 2nd edn. Woodhead Publishing, Duxford, pp 275–298. https://doi.org/10.1016/B978-0-08-102488-1.00011-3

Ferreira JFS, Luthria DL, Sasaki T, Heyerick A (2010) Flavonoids from *Artemisia annua* L. as antioxidants and their potential synergism with artemisinin against malaria and cancer. Molecules 15(5):3135–3170. https://doi.org/10.3390/molecules15053135

Flowers of India (2021) Artemisia indica – Indian Wormwood. http://www.flowersofindia.net/catalog/slides/Indian%20Wormwood.html. Accessed 17 Nov 2021

Foster S, Duke JA (2020) A field guide to medicinal plants. Eastern and Central N. America. Houghton Mifflin Co. 1990 a concise book dealing with almost 500 species. A line drawing of each plant is included plus colour photographs of about 100 species. Very good as a field guide, it only gives brief details about the plants medicinal properties

Gallo MBC, Sarachine MJ (2009) Biological activities of Lupeol. Int j res pharm biomed sci 3(1): 46–66

Ganguly S, Bandyopadhyay S, Bera A, Chatterjee M (2006) Antipromastigote activity of an ethanolic extract of leaves of *Artemisia indica*. Indian J Pharmacol 38(1):64–65

Gou Y, Li Z, Fan R, Guo C, Wang L, Sun H, Li J, Zhou C, Wang C, Wang Y (2020) Ethnobotanical survey and evaluation of traditional mosquito repellent plants of Dai people in Xishuangbanna, Yunnan Province, China. J Ethnopharmacol 262

Guclu G, Kelebek H, Selli S (2021) Chapter 26 - Antioxidant activity in olive oils. In: Preedy VR, Watson RR (eds) Olives and olive oil in health and disease prevention, 2nd edn. Academic Press, Amsterdam, pp 313–325. https://doi.org/10.1016/B978-0-12-819528-4.00031-6

Haider SZ, Mohan M, Andola HC (2014) Constituents of *Artemisia indica* Willd. from Uttarakhand Himalaya: a source of davanone. Pharm Res 6(3):257–259. https://doi.org/10.4103/0974-8490.132607

Hou T, Li X, Peng C (2017) Borneol enhances the antidepressant effects of asiaticoside by promoting its distribution into the brain. Neurosci Lett 646:56–61. https://doi.org/10.1016/j.neulet.2017.02.068

Hu J, Wang W, Dai J, Zhu L (2019) Chemical composition and biological activity against Tribolium castaneum (Coleoptera: Tenebrionidae) of Artemisia brachyloba essential oil. Ind Crop Prod 128:29–37

Hussain A (2019) Distribution and molecualr phylogeny of Artemisia plants from Gilgit-Baltistan, Pakistan. http://173.208.131.244:9060/xmlui/handle/123456789/4631

Hussain A (2020) The genus Artemisia (Asteraceae): a review on its Ethnomedicinal prominence and taxonomy with emphasis on foliar anatomy, morphology, and molecular phylogeny. Proc Pak Acad Sci B Life Environ Sci 57(1):1–28

Jaffal SM, Abbas MA (2019) Antinociceptive action of *Achillea biebersteinii* methanolic flower extract is mediated by interaction with cholinergic receptor in mouse pain models. Inflammopharmacology 27:961–968. https://doi.org/10.1007/s10787-018-0524-7

Jassal PS, Nahid A, Kumar N, Sharma M (2019) Pharmaceutical applications of Artemisia: an overview. Think India J 22(16):597–610

Javid T, Adnan M, Tariq A, Akhtar B, Ullah R, Salam NMAE (2015) Antimicrobial activity of three medicinal plants (*Artemisia indica, Medicago falcate* and *Tecoma stans*). Afr J Tradit Complement Altern Med 12(3):91–96. https://doi.org/10.4314/ajtcam.v12i3.11

Joram A, Mahanta D, Das AK (2021) Documentation of wild edible plants of ethnobotanical significance associated with selected tribes of Arunachal Pradesh, India. East Himalayan Society for Spermatophyte Taxonomy. Pleione 15(2):127–140. https://doi.org/10.26679/Pleione.15.2.2021.123-140

Joshi AR, Joshi K (2004) Insecticidal plants of the Bagmati watershed, Nepal: ethnobotany and traditional uses. Bionotes 6(2):37–39

Ken Fern (2022) Tropical plants database. https://tropical.theferns.info/viewtropical.php?id=Artemisia+indica. Accessed 15 Oct 2022

Khan S, Afshan K, Mirza B, Miller JE, Manan A, Irum S, Rizvi SS, Qayyum M (2015) Anthelmintic properties of extracts from Artemisia plants against nematodes. Trop Biomed 32(2):257–268

Khan I, Karim N, Ahmad W, Abdelhalim A, Chebib M (2016) GABA-A receptor modulation and anticonvulsant, anxiolytic, and antidepressant activities of constituents from *Artemisia indica* Linn. Evid Based Complement Alternat Med 2016:1215393. https://doi.org/10.1155/2016/1215393

Khan I, Ahmad W, Karim N, Ahmad M, Khan M, Tariq SA, Sultana N, Shah R, Khan A, Abdelhalim A (2017) Antidiabetic activity and histopathological analysis of carnosol isolated from *Artemisia indica Linn* in streptozotocin-induced diabetic rats. Med Chem Res 26:335–343. https://doi.org/10.1007/s00044-016-1750-4

Lee JY, Park H, Lim W, Song G (2020) Therapeutic potential of α,β-thujone through metabolic reprogramming and caspase-dependent apoptosis in ovarian cancer cells. J Cell Physiol 236(2):1545–1558. https://doi.org/10.1002/jcp.30086

Mannan A, Shaheen N, Arshad W, Qureshi RA, Zia M, Mirza B (2008) Hairy roots induction and artemisinin analysis in *Artemisia dubia* and *Artemisia indica*. Afr J Biotechnol 7(18):3288–3292

Mannan A, Ahmed I, Arshad W, Hussain I, Mirza B (2011) Effects of vegetative and flowering stages on the biosynthesis of artemisinin in *Artemisia* species. Arch Pharm Res 34:1657–1661. https://doi.org/10.1007/s12272-011-1010-6

Mansouri MT, Hemmati AA, Naghizadeh B, Mard SA, Rezaie A, Ghorbanzadeh B (2015) A study of the mechanisms underlying the anti-inflammatory effect of ellagic acid in carrageenan-induced paw edema in rats. Indian J Pharmacol 47(3):292–298. https://doi.org/10.4103/0253-7613.157127

Mensah MLK, Komlaga G, Forkuo AD, Firempong C, Anning AK, Dickson RA (2019) Toxicity and safety implications of herbal medicines used in Africa. In: Builders PF (ed) Herbal medicines. Intech Open, London. https://doi.org/10.5772/intechopen.72437

Mwine JP, Damme PV, Gerard K, Charles K (2011) Ethnobotanical survey of pesticidal plants used in South Uganda: case study of Masaka district. J Med Plants Res 5(7):1155–1163

Nahid A, Neelabh C, Navneet K (2017) Antioxidant and antimicrobial potentials of Artemisia Indica collected from the Nepal region. J Pharm Sci Res 9(10):1822–1826

Namsa ND, Mandal M, Tangjang S (2011) Anti-malarial herbal remedies of Northeast India, Assam: an ethnobotanical survey. J Ethnopharmacol 133(2):1–572. https://doi.org/10.1016/j.jep.2010.10.036

Negreiros PDS, Costa DSD, Silva VGD, Lima IBDC, Nunes DB, Sousa FBDM, Araújo TDSL, Medeiros JVR, Santos RFD, Oliveira RDCM (2019) Antidiarrheal activity of α-terpineol in mice. Biomed Pharmacother 110:631–640. https://doi.org/10.1016/j.biopha.2018.11.131

Nigam M, Atanassova M, Mishra AP, Pezzani R, Devkota HP, Plygun S, Salehi B, Setzer WN, Sharifi-Rad J (2019) Bioactive compounds and health benefits of Artemisia species. Nat Prod Commun 14(7). https://doi.org/10.1177/1934578X19850354

Niwano Y, Beppu F, Shimada T, Kyan R, Yasura K, Tamaki M, Nishino M, Midorikawa Y, Hamada H (2009) Extensive screening for plant foodstuffs in Okinawa, Japan with anti-obese activity on adipocytes. Plant Foods Hum Nutr 64:6–10

Ogbru AG (2021) Adult ADHD Medications. 2021-6-9. https://www.rxlist.com/adhd_prescription_drugs/drugscondition.htm

Oliveira GB, Brito RG, Santos PL, Araújo-Filho HG, Quintans JSS, Menezes PP, Serafini MR, Carvalho YMBG, Silva JC, Almeida JRGS, Scotti L, Scotti MT, Shanmugam S, Thangaraj P, Araújo AAS, Quintans-Júnior LS (2016) α-Terpineol, a monoterpene alcohol, complexed with β-cyclodextrin exerts antihyperalgesic effect in animal model for fibromyalgia aided with docking study. Chem Biol Interact 254:54–62. https://doi.org/10.1016/j.cbi.2016.05.029

Rahmatullah M, Biswas KR (2012) Traditional medicinal practices of a Sardar healer of the Sardar (Dhangor) Community of Bangladesh. J Altern Complement Med 18(1):10–19. https://doi.org/10.1089/acm.2011.0395

Rashid S, Rather MA, Shah WA, Bhat BA (2013) Chemical composition, antimicrobial, cytotoxic and antioxidant activities of the essential oil of Artemisia indica Willd. Food Chem 138(1): 693–700. https://doi.org/10.1016/j.foodchem.2012.10.102

Rather MA, Dar BA, Shah WA, Prabhakar A, Bindu K, Banday JA, Qurishi MA (2017) Comprehensive GC–FID, GC–MS and FT-IR spectroscopic analysis of the volatile aroma constituents of Artemisia indica and Artemisia vestita essential oils. Arab J Chem 10(2):S3798–S3803. https://doi.org/10.1016/j.arabjc.2014.05.017

Ruwali P, Ambwani TK, Gautam P, Thapliyal A (2015) Qualitative and quantitative phytochemical analysis of Artemisia indica Willd. J Chem Pharm Res 7(4):942–949

Sadraei H, Asghari G, Kasiri F (2015) Comparison of antispasmodic effects of Dracocephalum kotschyi essential oil, limonene and α-terpineol. Res Pharm Sci 10(2):109–116

Sagar MK, Ashok PK, Chopra H, Upadhyaya K (2010) Phytochemical and pharmacological potential of Artemisia indica in experimental animal models. Pharmacologyonline 2:1–4

Salehi B, Ata A, Kumar NVA, Sharopov F, Ramírez-Alarcón K, Ruiz-Ortega A, Ayatollahi SA, Fokou PVT, Kobarfard F, Zakaria ZA, Iriti M, Taheri Y, Martorell M, Sureda A, Setzer WN, Durazzo A, Lucarini M, Capasso ASR, Ostrander EA, Rahman AU, Choudhary MI, Cho WC, Sharifi-Rad J (2019) Antidiabetic potential of medicinal plants and their active components. Biomol Ther 9(10):551. https://doi.org/10.3390/biom9100551

Sarnim G, Sanjay ST, Roshan A, Vedamurthy AB, Joy HH (2013) Artemisia indica extracts as anthelminthic agent against Pheretima posthuma. Int J Pharm Pharm Sci 5(3):259–262

Satyal P, Paudel P, Kafle A, Pokharel SK, Lamichhane B, Dosoky NS, Moriarity DM, Setzer WN (2012) Bioactivities of volatile components from Nepalese Artemisia species. Nat Prod Commun 7(12):1651–1658

Saxena S, Pant N, Jain DC, Bhakuni RS (2003) Antimalarial agents from plant sources. Curr Sci 85(9):1314–1329. http://www.jstor.org/stable/24108135

Shah GC (2010) Terpenoid diversity in some *Artemisia* species of Uttarakhand Himalaya. Indian Perfumer 54:17–19

Shah GC, Rawat TS (2008) Chemical constituents of *Artemisia indica* Willd. Oil. Indian Perfumer 52:27–29

Shimono Y, Hayakawa H, Kurokawa S, Nishida T, Ikeda H, Futagami N (2013) Phylogeography of Mugwort (*Artemisia indica*), a native Pioneer herb in Japan. J Hered 104(6):830–841. https://doi.org/10.1093/jhered/est054

Sigdel SR, Rokaya MB, Timsina B (2013) Plant inventory and ethnobotanical study of Khimti Hydropower Project, Central Nepal. Sci World J 11(11):105–112

Singh B, Sinha BK, Phukan SJ, Borthakur SK, Singh VN (2012) Wild edible plants used by Garo tribes of Nokrek biosphere reserve in Meghalaya, India. Indian J Tradit Knowl 11:166–171

Souza R, Cardoso M, Menezes C, Silva J, De Sousa D, Batista J (2011) Gastroprotective activity of α-terpineol in two experimental models of gastric ulcer in rats. Daru 19(4):277–281

Tasdemir D, Tierney M, Sen R, Bergonzi MC, Demirci B, Bilia AR, Baser KH, Brun R, Chatterjee M (2015) Antiprotozoal effect of Artemisia indica extracts and essential oil. Planta Med 81(12–13):1029–1037

Tiwary BK, Bihani S, Kumar A, Chakraborty R, Ghosh R (2015) The in vitro cytotoxic activity of ethno-pharmacological important plants of Darjeeling district of West Bengal against different human cancer cell lines. BMC Complement Altern Med 15:22. https://doi.org/10.1186/s12906-015-0543-5

Torres Y, Vargas D, Uribe C, Carrasco C, Torres C, Rocha R, Oyarzún C, Martín RS, Quezada C (2016) Pro-apoptotic and anti-angiogenic properties of the α /β-thujone fraction from Thuja occidentalis on glioblastoma cells. J Neurooncol 128(1):9–19

Xu L, Wu YL, Zhao XY, Zhang W (2015) The study on biological and pharmacological activity of Coumarins. Asia-Pac Power Energy Eng Conf 135–138

Xu W, Miao S, Feng Y (2019) Ludartin exhibits therapeutic effect on spinal cord injury through inhibition of apoptosis and inflammation. Bangladesh J Pharmacol 14(1):54–60. https://doi.org/10.3329/bjp.v14i1.38725

Zeng YT, Jiang JM, Lao HY, Guo JW, Lun YN, Yang M (2015) Antitumor and apoptotic activities of the chemical constituents from the ethyl acetate extract of Artemisia indica. Mol Med Rep 11:2234–2240. https://doi.org/10.3892/mmr.2014.3012

Zhu QC, Wang Y, Liu YP, Zhang RQ, Li X, Su WH, Long F, Luo XD, Peng T (2011) Inhibition of enterovirus 71 replication by chrysosplenetin and penduletin. Eur J Pharm Sci 44(3):392–398

Chapter 4
Arnica montana L.: Traditional Uses, Bioactive Chemical Constituents, and Pharmacological Activities

Nisha Gyawali, Asmita Rayamajhi, Deepa Karki, Tamlal Pokhrel, and Achyut Adhikari

Abstract *Arnica montana*, is a hemicryptophyte plant, belongs to the Asteraceae family. It is a medicinally significant herb that is used in traditional medicine systems in many countries. Flowers, roots, and rhizomes of *Arnica* are traditionally used for the topical treatments of various ailments such as bruises, sprains, backache, rheumatic arthritis, and phlebitis. Sesquiterpene lactones, flavonoids, fatty acids, thymol derivatives, and chlorogenic acid are the main bioactive phytochemicals. Extract and compounds from *A. montana* exhibited several pharmacological activities: anti-inflammatory, anticancer, antioxidant, antimicrobial, antiplatelets, and immunomodulatory activities. Helenalin and dihydrohelinalin are mainly responsible for their anti-inflammatory properties. The clinical trial using gel, cream, oil, ointment, and homeopathic dilutions revealed significant effects in relieving postoperative pains, surgical complications, swelling, edema, and ecchymosis. Different clinical trials using randomized placebo-controlled, randomized double-blinded, and open multicenter trials against different diseases reflect the medicinal importance of this plant. The aim of this chapter is to insight knowledge about the traditional uses, chemical compositions, pharmacological activities, and clinical trials of the plant *Arnica montana*. In vitro in in vitro.

Keywords *Arnica montana* · Homeopathy · Sesquiterpene lactones · Anti-inflammatory · Placebo-controlled

4.1 Introduction

The species *Arnica montana*, mostly distributed in Europe, belongs to the Asteraceae family. It comprises two subspecies: one is *Arnica montana* ssp. *montana* distributed in Central Europe and Scandinavia, and the other is *Arnica montana* ssp. *atlantica* distributed in southern France Portugal and Spain. It is commonly

N. Gyawali · A. Rayamajhi · D. Karki · T. Pokhrel · A. Adhikari (✉)
Central Department of Chemistry, Tribhuvan University, Kathmandu, Nepal

H. P. Devkota, T. Aftab (eds.), *Medicinal Plants of the Asteraceae Family*,
https://doi.org/10.1007/978-981-19-6080-2_4

recognized as leopard's bane, wolf's bane, and mountain tobacco by local people. Most frequently it is also identified as mountain snuff, mountain arnica, and sneezing tree. It is a flowering perennial plant that extent up to 30–60 cm tall. It has an aromatic fragrance. *Arnica* is derived from the Latin word "Ptarnica," which means "sneeze-making" (Maryna et al. 2019). It blooms in July and August with beautiful yellow daisy-like flowers. The length of beam flowers teeth is no longer than 1 mm or lies between 1 and 2 mm. Similarly, the dimension of the second flower differs from 4.9 to 5.7 cm (Kriplani et al. 2017). It grows in damp, grassy highland meadows in Europe, Northern Asia, and Siberia's mountains and hills. It is a reliable bioindicator for nutrient-deficient soils (Hollmann et al. 2020). The active constituents of this plant are mainly flavonoids (quercetin along with its derivatives like quercetin-3-glyco galacturonic and quercetin-3-mono-glucosideo) (Ganzera et al. 2008), sesquiterpene lactones (arnicolide, 11α, 13-dihydro-helenaline and helenaline) (Jürgens et al. 2022), alcohols (arnidiol, isoarnilenediol, arnilenediol), tannins, essential oil, carotenoids, inulin, and fatty acids (Macêdo et al. 2004). Among these, the main active compounds in *A. montana* flower heads are sesquiterpene lactones of the pseudo guaianolide group that is capable of the development of anti-inflammatory drugs (Kriplani et al. 2017). *Arnica* has been utilized in the homeopathic medicine of treatment for millennia. It is used to treat 66 different pathological conditions; however, most favorably treats contusions, wounds, rheumatism, and inflammation (Kriplani et al. 2017). Regrettably, several European countries have considered this species in endangered list including Bosnia and Herzegovina, Sweden, Spain, Hungary, Croatia, Slovenia, Germany, Lithuania, and Luxembourg. Most of the Europen countries has increased cultivation for its protection, but harvesting is prohibited in Italy (Aiello et al. 2012; Kawakami et al. 2011; Stanik et al. 2020). *A. montana* thrives at elevations of 500–2500 m, less fertile grasslands and peat bogs, as well as in soils with lower pH value. It is herbaceous plant with bushy stems, dark green basal, and lower cauline leaves (Kriplani et al. 2017; Aiello et al. 2012). A fruit is seed like cylindrical shaped having shiny whitish color and a plumose pap-pus (Kriplani et al. 2017).

Arnica montana is a medicinally significant herb which is exploited in homeopathy and pharmacy. For medicinal uses, various plant components such as rhizomes, roots, inflorescences, and leaves are harvested. Terpenoids, sesquiterpenes, sesquiterpene lactones, essential oils, phenolic acids, and flavonoids, particularly chlorogenic acids, are all abundant in *Arnica* (Ganzera et al. 2008). Various factors such as habitat, temperature, altitude, different climate conditions, and maturity period affect the chemical composition of this plant (Spitaler et al. 2008). The origin and versatility of ethno-pharmacological categories on medicinal herb *Arnica* offers an interesting field of drug discovery (Obón et al. 2012). *Arnica* leaves, flowers, and roots have been used for therapeutic cures in human beings and veterinarian phytotherapy. Raw materials of this plant are generally used in cosmetics, certain European liquors, and natural flavoring agents (Kouzi and Nuzum 2007). *Arnica*'s tolerance and efficacy also demonstrated that it is a significant treatment modality for relief in pain, posttraumatic edema, and postoperative context, both in formulating gel and homeopathic dilutions for oral administration (Iannitti et al. 2016).

The purpose of this chapter is to sum up evidence of using *Arnica* as an alternative approach for traditional use, chemical composition, ethnopharmacology, and therapeutic drugs that promote future research and development on this valuable medicinal herb. It will provide brief descriptions from clinical trials testing of *Arnica* on relief in pain, surgery, implant placements, and seroma reduction along with a discussion of possible mechanisms of action, safety, toxicity, and it's adverse effects.

4.2 Traditional Uses

Arnica montana is a perennial herbaceous, flowering plant, which has been used as a conventional therapeutic agent for thousands of years. This plant has been used predominantly to medicate various ailments such as backaches, sprains, superficial inflammations, injuries, wounds, and veterinary treatments (Garcia-Oliveira et al. 2021). The whole part of this plant is used as a herbal medication in homeopathy for treating 66 different pathological conditions (Kriplani et al. 2017). However, flower and root extracts have been utilized most frequently in the treatment of various health conditions. Extracts prepared from the root are applied externally on the bruises, rheumatic pains, sprains, and phlebitis to reduce the inflammations and as an immune system activator (Šutovská et al. 2014). Fresh as well as dried flower extracts are extensively used in modified form of tinctures, ointments, creams, gels as an immediate treatment of the sprains, bruises, contusions, and trauma pain in homeopathy medicine (Pieroni and Giusti 2009; Pljevljakušić et al. 2014; Vidic et al. 2016). The topical use of *Arnica* for abrasion and sequelae resulting from accidental injuries like hematoma, sprains, bruises, and dropsy owing to fractures, muscular pain and joint difficulties, insect bite irritation, and phlebitis has also been recognized by the German Commission E (Raza 2021).

A. montana constitutes a higher proportion of sesquiterpene lactones and is traditionally famous for its anti-inflammatory effects in most European countries (Lass et al. 2008). In British Columbia of Canada, veterinarians most commonly make the traditional use of herbal ointments as a protective band for wounds. An herbal ointment is prepared traditionally by mixing mashed flowers and leaves along with bee glue (Lans et al. 2007). People in South Africa use the leaves extract as a traditional healer to get relief from toothache and as a mouth cleanser (Ashu Agbor and Naidoo 2015). In the USA, dried flower is frequently consumed spice and has a traditional medicinal value as a diuretic, to induce sweating and as a stimulant (Sharma et al. 2016). Ongoing through different research articles, it was found that *Arnica* is typically considered safe in foods. This also validated from the Food and Drug Administration. It has been utilized in beverages: alcoholic and nonalcoholic, chilled desserts, confectionery, gelatins, and custards as a flavoring agent, in the form of hair tonic as anti-dandruff, fragrance, and other cosmetics. If significant doses of *Arnica* are consumed, the toxin helenalin present in it causes skin irritation and acute gastroenteritis resulting from internal bleeding of the intestinal tract. As a result, *Arnica* extracts or decoctions are not suggested for its oral consumptions, while some expertise of homeopathy recommends highly reduced concentration. This diluted solution helps to reduce mild fever and treats cold, bronchitis, epilepsy, and sore throat (Denisow-Pietrzyk et al. 2019; Kawakami et al. 2011; Kouzi and Nuzum 2007; Šutovská et al. 2014).

4.3 Chemical Constituents

Arnica montana, an herbaceous perennial plant, mostly grows in the mountain region of Europe (Zucker 2008). Different parts of *A. montana* possess bioactive phytochemicals like flavonoids (FVs), caffeoyl quinic acid derivatives (CQAs), polysterol, phenolic compounds, and sesquiterpene lactones (SQLs) (Oana Teodora et al. 2016; Clauser et al. 2014). Due to these abundant bioactive phytochemicals, *A. montana* has important medicinal values.

The short-chain esters: sesquiterpene lactones, helenalin, and dihydro-helenalin are the main chemicals found in *Arnica* (Fig. 4.1). Helenalin comprises an endocyclic, unsaturated ketone (cyclopentenone), which gives hetero-Michael addition with thiols and act as an NF-kB singling which serves as an anti-inflammatory (Widen et al. 2017). Removing one of the two Michael acceptors of helenalin as compared with parent natural compounds, such as cyclopenten (producing 2,3-dihydrohelenalin) or -methylene-butyrolactone (generating 11,13-dihydrohelenalin; plenolin), dramatically reduces cytotoxicity (Lee et al. 1978). Some important phenolic compounds extracted from flowers are chlorogenic acid, 3,5-dicaffeoylquinic acid and 1-methoxyoxaloyl 3,5-dicaffeoylquinic acid which shows antioxidant activities (Clauser et al. 2014; Ganzera et al. 2008). Kimel et al. analyzed some active compounds from the group of phenolic acids such as derivatives of benzoic acid, p-hydroxybenzoic, protocatechuic, gallic, and vanillic; cinnamic acid derivatives, ferulic and caffeic; and phenolic acids of an ester nature, namely, caffeic acid esters of quinic acid (CQA), chlorogenic acid (5-O-CQA), cynarin (1,3-O-CQA), and isochlorogenic.

Flavonoids show antioxidant and antimicrobial activities. Flavonoid compounds belonging to the group of flavones are luteolin, luteolin 7-O-glucoside, apigenin, and apigenin 7-O-glucoside. Similarly, the group of flavonols are kaempferol, astragalin (kaempferol 3-Oglucoside), hyperoside (quercetin 3-O-galactoside), quercetin, isoquercetin (quercetin-3-O-glucoside), and isorhamnetin.

Cumene, 2,6-diisopropylanisole, decanal, and 1,2,2,3-tetramethylcyclopent-3-enol (Fig. 4.2) are major volatile compounds from the flower of the *Arnica montana* (Sugier et al. 2019). The essential oil of *A. montana* L. Achenes also known for the bioactive compounds consists of 20–23% monoterpene, (10%) sesquiterpene, and 7% aliphatic aldehyde as major constituents (Sugier et al. 2019). There are more than 40 chemical compounds including phenolic acid, organic acids, and fatty acids found in the essential oil of the root of *A. monata* (Petrova et al. 2015). Some major chemical constituents of the essential oil obtained from root and rhizome of

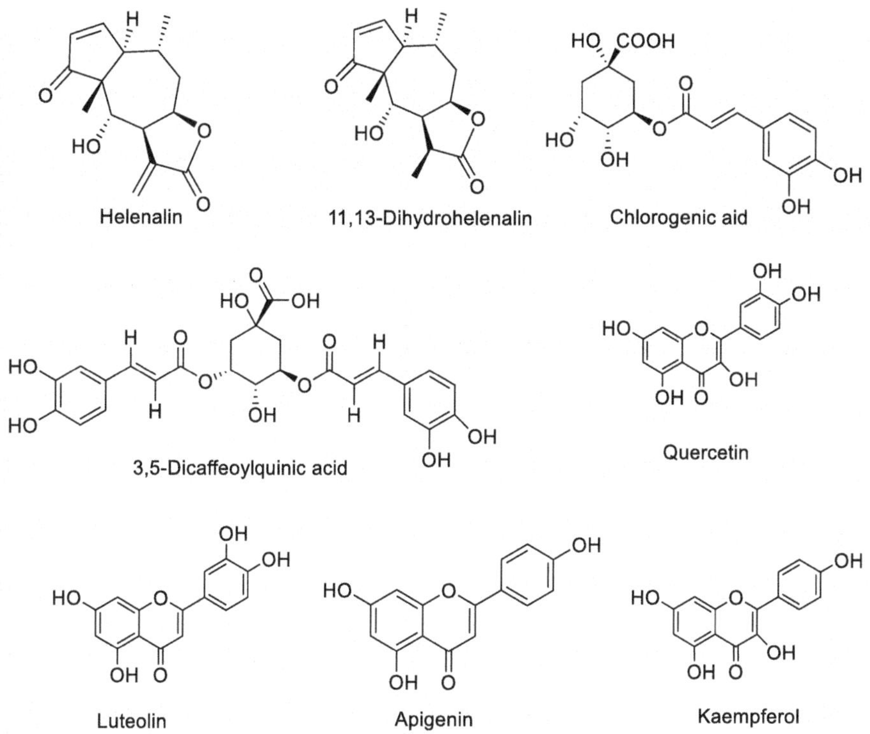

Fig. 4.1 Chemical compositions of lactones, phenols, and flavonoids in *Arnica montana*

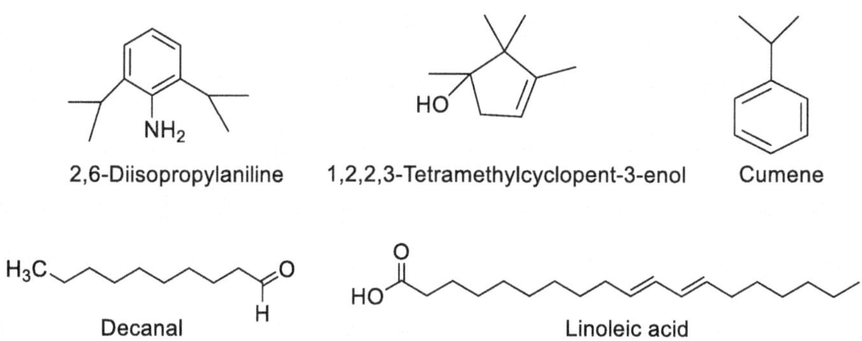

Fig. 4.2 Volatile chemical compounds of *Arnica montana* essential oil

A. monata are 2,5-dimethoxy-p-cymene, thymol methyl ether, p-methoxyheptanophenone, 2,6-diisopropylanisole, etc. (Pljevljakušić et al. 2012).

4.4 Pharmacological Activities

A. montana is a medicinally significant herb frequently used in pharmacy, homeopathy, and cosmetics. Different parts of plants are rich sources of chemical constituents and essential oils such as sesquiterpenes, terpenoids, sesquiterpene lactones, flavonoids, phenolic acid, and most particularly chlorogenic acids (Sugier et al. 2013). The presence of these chemical constituents mediates various pharmacological properties which includes anti-inflammatory, antioxidant, anti-cancerous, anti-hemorrhage, anti-osteoarthrities, antiplatelet, and immunomodulatory activities (Greinwald et al. 2022; Macêdo et al. 2004; Sugier et al. 2013).

4.4.1 Anti-inflammatory Activity

In 2008, an experiment was performed to evaluate the effect of phonophoresis along with *Arnica montana* on the acute inflammation induced in rat muscle. A 90-day-old Wistar male rat model was taken under study, and tibialis anterior muscle was surgically lesions to induce inflammation. Treatment was carried out in four different groups with 10 in each as a control group with no treatment, US group upon treatment with Ultrasound, US+A group treated with both US and *arnica* gel massage, and a group with only *arnica* gel. A 3-minute session treatment per day was initiated 24 h of lesions in each group for 3 days. Quantitative analysis of the isolated inflamed muscles showed a condensed mononuclear cells (macrophages) and light weight polymorphonuclear cells (neutrophils) in all groups than the control group. A histopathological study showed groups with the US acted as a pro-inflammation rather than anti-inflammation which might be due to the stimulatory effect of the US. This study finally revealed *arnica* gel is alone effective in the treatment of inflammations due to acute muscle lesions (Alfredo et al. 2009).

In 2009, phyto-medicine prepared from *A. montana* flowers was subjected to learn the mechanism that inhibits transcription factor AP-1 and NF-kB. This factor regulates the genes: MMP1 and MMP13. Bovine and human chondrocytes cells were taken for the experiment. Four different *Arnica* formulations were applied on MMP1 and MMP13 gene expression arisen from the induction of IL-1β by using real-time qPCR. An electrophoretic mobility shift binding assay was performed to study DNA binding activity of AP-1 and NF-kB. Both human and bovine chondrocytes were treated with 20 ng/mL IL-1β for 4 h to stimulate MMP1 and MMP13 gene expression. MMP1 and MMP13 expression levels in stimulated chondrocytes showed 67 and 213 fold greater, respectively, than in untreated cells. Treatment of different doses of *arnica* before the activation by IL-1β showed significantly decreased levels of MMP1 and MMP13 mRNA. The effect of the gene expression was dose-dependent. Among different formulations, central European *arnica* extract at concentrations of 0.2 and 0.5 μL/mL showed the most effective inhibitory effect. Similarly, AP-1 and NF-kB DNA binding activity was inhibited by *Arnica* extract in dose-dependent manner. Among all, the European *Arnica* tincture showed the highest inhibitory activity at 5 μL/mL. Helenalin isobutyrate was taken as a positive control in both of the mechanisms. This study concluded that the degradation of MMP1 and MMP13 is an attractive mechanism for the treatment of inflammation due to osteoarthritis (Jäger et al. 2009). A similar type of study was carried out to evaluate the anti-inflammatory activity due to LPS stimulations in J774 murine macrophages. This study showed that *arnica* was effective in the reduction of iNOS ($P < 0.05$) and COX-2 ($P < 0.01$) protein level and inhibition in the production of IL-12(2.8 fold, $P < 0.001$) and inhibited nuclear translocation of NF-kB which are the main pro-inflammatory (Verma et al. 2010).

In 2020, another experiment was carried out to study the anti-inflammatory effect of the *A. montana* in skin burn mice induced by UVB radiation. This study was conducted on 25 male Swiss mice of 90 days old. Mice were enclosed in a polypropylene box with the controlled conditions of light/dark cycle of alternating 12 h. The study was carried out with division of mice into five groups: control group (no UVB + no treatment), vehicle group (no UVB + ointment without active principle), third, fourth, and fifth group with UVB radiations along with ointment without active, with active principle *Arnica*, and dexamethasone, respectively. UVB radiation was exposed for 20 h. After 16 h of treatments, mouse ear was further processed for the biochemical assays. Topical treatment of the *arnica* ointment reduced edema in mouse and myeloperoxidase activity induced by UVB radiations and inhibited marked inflammatory response of the NF-kB cytokine transcription factor. This study shows *Arnica* ointment ($p < 0.001$) as effective as positive control dexamethasone ($p < 0.05$) in treatment of the skin inflammation (da Silva Prade et al. 2020).

4.4.2 Anticancer Activity

Sugier et al. in 2019 evaluated the anticancer activity from essential oil of *A. montana* L. achenes. Human anaplastic astrocytoma MOGGCCM and glioblastoma multiforme T98G cell lines were chosen to study anticancer activity. The essential oil at varying concentrations (0, 0.5, 1, 2 μL/mL) was applied to cell lines. 0.5 μL/mL showed effective induction of cell death in MOGGCCM cell line as compared to T98G cell line. IC50 values also reveal the same case: 1.6 and 1.8 value for MOGGCCM cell line and 2.1 and 2.0 for T98G cell line. This study showed extreme proportion of 2, 5-dimethoxy-p-cymene in the essential oil which played an important role of anticancer potential against MOGGCCM and T98G cell line (Sugier et al. 2019).

Similarly, Sugier et al. in 2020 further reported the anticancer activity from essential oil of *A. montana. L.* roots and rhizomes. This study followed a similar procedure to the study conducted lately in 2019 (Sugier et al. 2019). Essential oil of rhizomes and roots at Various concentrations (0, 0.5, 1, 2 μL/mL) were subjected to MOGGCCM and T98G cell lines where 1 μL/mL showed efficient effect on apoptosis of the cell line. At the same time, there is a low level effect of necrosis. Induced apoptosis at a level of 28.5–32.3% was more effective which revealed that chemical components such as 2,5-dimethoxy-p-cymene, 2,6-diisopropylanisole, thymol methyl ether, and p-methoxyheptanophenone were responsible for anticancer activity (Sugier et al. 2020).

4.4.3 Antioxidant Activity

In 2012, Craciunescu et al. evaluated the antioxidant activities of *A. montana* flower from ethanolic extract. Antioxidant activity was examined by using Trolox equivalent antioxidant capacity (TEAC), oxygen radical absorbance capacity (ORAC), and DPPH free radical assays. Obtained IC50 value of DPPH was 0.63 ± 0.07 mg/mL; TEAC and ORAC were 486.06 ± 20.63 μmol Trolox equivalents/g extract and 682.22 ± 17.32 μmol of Trolox equivalents/g extract. *Arnica* extracts were found to be rich in flavonoids and polyphenolic compounds which mediated antioxidant capacity (Craciunescu et al. 2012).

In 2016, Vidic et al. demonstrated the antioxidant capacity from essential oil of *A. montana* flower heads. Antioxidant potential was determined by using ABTS, DPPH, reducing power, and phosphomolybdenum assay while carvacrol, caryophyllene oxide (natural antioxidant), and BHT (synthetic antioxidant) as a standard reference. Its essential oil showed the highest value of DPPH (IC50 = 4.79 mg/mL) than standard carvacrol (IC50 = 14.38 mg/mL). Also, phosphomolybdenum assay showed better antioxidant potential 55.69 mg (AAE)/g equivalent to the DPPH method. The study showed antioxidant property is associated with the presence of fatty acids and phenolic compounds and their derivatives (Vidic et al. 2016).

4.4.4 Antimicrobial Activity

Recently, in 2021 Nieto-trujillo et al. demonstrated in vitro antibacterial potential of the methanolic extract of *A. montana* seed. Kirby-Bauer method was used to test against *S. aureus* (ATCC25923) and *E. coli* (ATCC25922) strains by using vancomycin(1 μg/disk) and chloramphenicol (1 μg/disk) as a positive control, respectively. The diluted fraction as negative control while sterile water was used as growth control. All the fractions (1–8 μg/disk) of extracts showed remarkable growth inhibition of the *E. coli* and *S. aureus*. 8 μg/disk for three different fractions (4 AM, 5 AM, and 6 AM) showed maximum inhibition percentage (14.48%, 16.31%, and 17.57%, respectively) of *E. coli*. The highest percentage of inhibition for *S. aureus* was revealed by 4 AM (16.8%) and 6 AM (20.48%) fractions. Three different fractions consisting of potent SMs: gallic acid, quercetin, verbascoside, parthenolide, and sesquiterpene lactone contents are directly correlated with its antibacterial activity (Nieto-Trujillo et al. 2021).

4.4.5 Antiplatelet Activities

In 2015, Rywaniak et al. evaluated in vitro antiplatelet properties of the polyphenolic isolated from *A. montana* flowers. Healthy human blood was collected from the age group 35.7 ± 10.3 years. Blood incubated with 1% CellFix was used as a positive control and Flow cytometry was used to determine blood platelet viability. 7.5 and 15 mg/mL of *arnica* flower extract showed promising inhibitory action on ADP-stimulated platelet aggregations in blood and PRP and VASP phosphorylation without any toxic effect. This study showed that antiplatelet activity is due to the interaction of the polyphenolic compound of *Arnica* with the platelet surface membrane $P2Y_{12}$ receptor (Rywaniak et al. 2015).

4.4.6 Immunomodulatory and Wound Healing

In 2016, Marzotto et al. examined an in vitro model for evaluating the effect of the *A. montana* on its gene expression relating the wound healing mechanism. Interleukin-4 (IL-4) polarized THP-1 macrophage cell line was used for the analysis. Mother tincture of one centesimal dilution prepared in 30% ethanol and test was carried by incubating 24 h with 2, 3, 9, and 15 centesimal and control, respectively. The control solution used was a 100× diluted solution from the 1c standard *Arnica* solution. 8 week-old, wild-type C57BL/6 J mice were chosen to isolate macrophages from its bone marrow, and wound healing property was tested. The concentration of fibronectin rises up significantly with different centesimal *A. montana* dilution (13.9–39.6% with p value <0.05). This study revealed the healing mechanism of the wound was a bit faster in presence of *A. montana* (99.3 ± 0.1%) than control solvent (98.8 ± 0.7) in IL-4-induced cells. The recovery process was relatively efficient (30%) over control solvent. Also, concluded release of fibronectin is associated with the therapeutic role of wound repairing (Marzotto et al. 2016).

4.5 Clinical Studies

Clinical studies have been performed on *A. montana* aiming to act as a remedy for acute and chronic health problems. *A. montana* has been the source of extensive clinical trials, the researcher had made ample contrasting trial ointment showed repair and regression of postoperative edema and ecchymosis.

The double-blind placebo-controlled study, with 4-arm parallel group phase involved 570 patients. *Arnica* tincture spray (41.5 mg) was also tested in combination with hydroxyethyl salicylate (HES; 12.5 mg) as compared to *Arnica* (41.5 mg), HES (12.5 mg), and placebo for treatment of ankle joint distortion related pain. Combined application of *Arnica* and HES (4–5 times per day) in 50 patients showed immediate recovery. This was assessed by visual analogue scale (VAS). In conclusion, this research revealed that *Arnica* can work in conjunction with other drugs, such as HES, to alleviate pain associated with sprained ankle joint deformity (Kučera et al. 2011). In a 2-week double-blind randomized research, 16 healthy participants were given a 595-nm pulsed-dye laser to create 7-mm standard bruises on their upper inner arms, with contrasting results. When comparing VAS bruise ratings, 20% *Arnica* gel reduced bruising when compared to placebo and a gel containing 1% vitamin K and 0.3% retinol. But, there was no significant difference on comparing 5% vitamin K gel (Leu et al. 2010). *Arnica* cream administered immediately after completing calf raises and 24–48 h later failed to improve leg pain, mobility, or muscle tenderness when compared to placebo in a randomized double-blind trial including 53 participants (Adkison et al. 2010). Arnica 6D tablet, *Arnica* ointment, and placebo were used as research and control medications. Hand surgery (endoscopic carpal tunnel release) was diagnosed in 37 patients for 2 weeks as part of the inclusion criteria. In addition, topical *Arnica* along with homeopathic dilutions greatly minimized postoperative pain. The *Arnica*-treated group experienced a significant reduction in pain in contrast to the placebo group (Jeffrey and Belcher 2002).

Using prospective double-blind, randomized placebo-controlled trial trials conducted over a 20 month period with 55 patients, researchers found that the effect of two homeopathic remedies on postmastectomy seroma production was reduced when used in breast reconstructive surgery. Compared to 6.1–6.4 days, the drain removal time was reduced by 2.4 days in this study (Lotan et al. 2020). Clinical study on ecchymosis resulted after rhinoplastic surgery was carried out among 74 patients for 10 days. A randomized, placebo-controlled, and double-blinded fashion was conducted to design the study. The extent and intensity of postoperative ecchymosis in rhinoplasty surgery were reduced in this study (Chaiet and Marcus 2016). Similarly, the study on ecchymosis and edema found that local *Arnica* and mucopolysaccharide polysulfate cream treatments reduced ecchymosis and edema throughout the postoperative phase in 108 patients for 10 days. In open rhinoplastic surgery, mucopolysaccharide polysulfate ointment showed repair and regression of postoperative edema and ecchymosis (Simsek et al. 2016).

Thus, the prospective randomized double-blinded, randomized placebo method and open multicenter trials were performed in different patients in a controlled manner with homeopathy for effective outcomes.

4.6 Studies Related to Safety and Toxicity

However, dosing is not standardized throughout the products, resulting in differences in concentration levels (Cameron and Chrubasik 2013). Doses and dilutions were usually well-tolerated and employed with few patients experiencing adverse effects in clinical trials of *Arnica* from mild to moderate (Daane 2001). Topical creams should only be used on closed skin (Reddy et al. 2013). Although the use of homeopathic remedies is on the rise, these treatments are frequently seen as safe and risk-free by patients. The increasing and widespread use of alternative substances found in *A. montana* and materials in the treatment of skin diseases has been accompanied by heated debates between orthodox academic medicine, which is often portrayed as aggressive, toxic, harmful, and alternative medicine, which is represented as natural, bland, and free of side effects but is condemned as unscientific and ineffective (Reider et al. 2001). In clinical trials burning, reddening, itching, and urticaria were observed in the group receiving hydroxyethyl salicylate (Kučera et al. 2011). Racing hurt occurring as a cardiovascular event observed in both homeopathy and placebo groups (Cornu et al. 2010). Severe gastroenteritis ("Final Report on the Safety Assessment of Arnica Montana Extract and Arnica Montana," 2001), anxiety, rapid heart rate, muscle weakness, and mortality have been documented after consuming *A. montana* containing drugs ("Final Report on the Safety Assessment of Arnica Montana Extract and Arnica Montana," 2001). Even though homeopathy products and topical applications contain many toxic ingredients in *Arnica*, they are usually safe to use while breastfeeding but they should not be used on broken skin, as it can cause allergic skin reactions and cross-reactions in people who are allergic to these plants (Bethesda 2006).

The greater propensity of blood platelets to clump and induce cell damage owing to cytotoxic effects may both result in enhanced medium transparency (monitored as increased aggregation) in optical (turbidimetric) aggregation as a result of reduced impedance (resistance) detected in impedance (electrical) analysis, as well as whole blood aggregation, could be due to decreased platelet adherence and clumping on the electrodes or antiplatelet cytotoxic effects (Rywaniak et al. 2015).

4.7 Conclusions and Future Recommendations

Arnica montana is an aromatic herbaceous plant. It is used as a traditional remedy for various ailments, especially in European countries. It is used as an alternative treatment for various pathological conditions such as pain, stiffness, sprain, bruises, and rheumatoid arthritis. *A. montana* is rich in active biochemicals such as: sesquiterpene lactones, flavonoids, and fatty acids with therapeutic benefits possessing potent pharmacological properties including anti-inflammatory, anticancer, antioxidant, antimicrobial, antiplatelet, and anti-osteoarthritis drugs. Different clinical investigations are frequently conducted on homeopathic treatment. These studies have shown that gel, pills, cream, ointments, and spray have promising pain-relieving effects. In an open multicenter trial, randomized double-blind, and placebo-controlled studied, the efficacy of *Arnica* on pain and postoperative recovery was assessed.

The chemical constituents present in roots, rhizomes, and leaves were affected by different ecology and climatic conditions. Sesquiterpene lactone is the main compound showing anti-inflammatory property. *A. montana* is used for topical formulation in the form of creams, ointments, and gels with 20–25% of this plant extract. Homeopathic practitioners recommend oral administration of a diluted solution of arnica extracts, but more studies are needed to confirm oral formulations of this plant extract. Helenalin is a toxic chemical compound showing various side effects. Therefore, researchers must focus on the toxicity level present in *A. montana* and the methods for the isolation of toxic compounds. If toxic compounds are isolated, it will be helpful for the preparation of oral herbal medication. *A. montana*, a medicinal plant, could be a source of new pharmaceuticals. Different sections of plants have medicinal effects but still, some portions of this plant are not investigated thoroughly. As the need for pharmaceuticals for various ailments grows, researchers should continue to look for novel and effective drugs through procedures that demonstrate a high level of medical value. Utilizing *A. montana*, which is found in several parts of the country, should be commercially farmed to aid in the industrialization and employment of residents, improving the country's economic status.

References

Adkison JD, Bauer DW, Chang T et al (2010) The effect of topical arnica on muscle pain. Ann Pharmacother 44(10):1579–1584. https://doi.org/10.1345/aph.1P071

Aiello N, Bontempo R, Vender C et al (2012) Morpho-quantitative and qualitative traits of Arnica montana L wild accessions of Trentino, Italy. Ind Crops Prod 40:199–203. https://linkinghub.elsevier.com/retrieve/pii/S0926669012001513

Alfredo PP, Anaruma CA, Pião AC et al (2009) Effects of phonophoresis with Arnica montana onto acute inflammatory process in rat skeletal muscles: an experimental study. Ultrasonics 49(4–5):466–471. https://doi.org/10.1016/j.ultras.2008.12.002

Anon (2001) Final report on the Safety assessment of Arnica montana extract and Arnica montana. Int J Toxicol 20(2):1–11. https://doi.org/10.1080/10915810160233712

Ashu Agbor M, Naidoo S (2015) Ethnomedicinal plants used by traditional healers to treat oral health problems in Cameroon. Evid Based Complement Alternat Med 2015:e649832. https://doi.org/10.1155/2015/649832

Bethesda (2006) Drug and Lactation Database (LactMed). National Library of Medicine (US). https://www.ncbi.nlm.nih.gov/pubmed/30000888

Cameron M, Chrubasik S (2013) Topical herbal therapies for treating osteoarthritis. Cochrane Database Syst Rev 2013:CD010538. https://doi.org/10.1002/14651858.CD010538

Chaiet SR, Marcus BC (2016) Perioperative Arnica montana for reduction of ecchymosis in rhinoplasty surgery. Ann Plast Surg 76(5):477–482. https://doi.org/10.1097/SAP.0000000000000312

Clauser M, Aiello N, Scartezzini F et al (2014) Differences in the chemical composition of Arnica montana flowers from wild populations of North Italy. Nat Prod Commun 9(1):3–6. https://doi.org/10.1177/1934578X1400900102

Cornu C, Joseph P, Gaillard S et al (2010) No effect of a homeopathic combination of *Arnica montana* and *Bryonia alba* on bleeding, inflammation, and ischaemia after aortic valve surgery. Br Clin Pharmacol 69(2):136–142. https://doi.org/10.1111/j.1365-2125.2009.03574.x

Craciunescu O, Constantin D, Gaspar A et al (2012) Evaluation of antioxidant and cytoprotective activities of Arnica montana L. and Artemisia absinthium L. ethanolic extracts. Chem Central J 6(1):97. https://doi.org/10.1186/1752-153X-6-97

da Silva Prade J, Bálsamo EC, Machado FR et al (2020) Anti-inflammatory effect of Arnica montana in a UVB radiation-induced skin-burn model in mice. Cutan Ocul Toxicol 39(2): 126–133. https://doi.org/10.1080/15569527.2020.1743998

Daane SP (2001) Potential for danger with Arnica Montana. Ann Plast Surg 46(3):349–350. https://doi.org/10.1097/00000637-200103000-00033

Denisow-Pietrzyk M, Pietrzyk L, Denisow B (2019) Asteraceae as a potential environmental factors of allergy. Environ Sci Pollut Res Int 26(7):6290–6300. https://doi.org/10.1007/s11356-019-04146-w

Ganzera M, Egger C, Zidorn C et al (2008) Quantitative analysis of flavonoids and phenolic acids in Arnica montana L. by micellar electrokinetic capillary chromatography. Anal Chim Acta 614(2):196–200. https://doi.org/10.1016/j.aca.2008.03.023

Garcia-Oliveira P, Barral M, Carpena M et al (2021) Traditional plants from Asteraceae family as potential candidates for functional food industry. Food Funct 12(7):2850–2873. https://doi.org/10.1039/D0FO03433A

Greinwald A, Hartmann M, Heilmann J et al (2022) Soil and vegetation drive Sesquiterpene lactone content and profile in Arnica montana L. flower heads from Apuseni-Mountains, Romania. Front Plant Sci 13:813939. https://doi.org/10.3389/fpls.2022.813939

Hollmann V, Donath T, Grammel F et al (2020) From nutrients to competition processes: habitat specific threats to Arnica montana L. populations in Hesse, Germany. PLoS One 15(5): e0233709. https://doi.org/10.1371/journal.pone.0233709

Iannitti T, Morales-Medina JC, Bellavite P et al (2016) Effectiveness and safety of Arnica montana in post-surgical setting, pain and inflammation. Am J Ther 23(1):e184. https://doi.org/10.1097/MJT.0000000000000036

Jäger C, Hrenn A, Zwingmann J et al (2009) Phytomedicines prepared from *Arnica* flowers inhibit the transcription factors AP-1 and NF-κB and modulate the activity of MMP1 and MMP13 in human and bovine chondrocytes. Plant Media 75912:1319–1325. https://doi.org/10.1055/s-0029-1185668

Jeffrey SLA, Belcher HJCR (2002) Use of Arnica to relieve pain after carpal-tunnel release surgery. Altern Ther Health Med 8(2):66–68. https://pubmed.ncbi.nlm.nih.gov/11892685/

Jürgens FM, Behrens M, Humpf HU et al (2022) In vitro metabolism of helenalin acetate and 11α,13-dihydrohelenalin acetate: natural Sesquiterpene lactones from arnica. Meta 12(1):88. https://doi.org/10.3390/metabo12010088

Kawakami AP, Sato C, Cardoso TN et al (2011) Inflammatory process modulation by homeopathic *Arnica montana* 6CH: the role of individual variation. Evid Based Contemp Altern Med 2011: 1–12. https://doi.org/10.1155/2011/917541

Kouzi SA, Nuzum DS (2007) Arnica for bruising and swelling. Am J Health Syst Pharm 64(23): 2434–2443. https://doi.org/10.2146/ajhp070155

Kriplani P, Guarve K, Baghael US (2017) *Arnica montana L.* – a plant of healing: Review. J Pharm Pharmacol 64(23):2434–2443. https://doi.org/10.2146/ajhp070155

Kučera M, Kolar P, Barna M, Kučera A, Hladiková M (2011) Arnica/hydroxyethyl salicylate combination spray for ankle distortion: a four-arm randomised double-blind study. Pain Res Treat 2011:365625. https://doi.org/10.1155/2011/365625

Lans C, Turner N, Khan T, Brauer G, Boepple W (2007) Ethnoveterinary medicines used for ruminants in British Columbia, Canada. J Ethnobiol Ethnomed 3(1):11. https://doi.org/10.1186/1746-4269-3-11

Lass C, Vocanson M, Wagner S, Schempp CM et al (2008) Anti-inflammatory and immune-regulatory mechanisms prevent contact hypersensitivity to *Arnica montana* L. Exp Dermatol 17(10):849–857. https://doi.org/10.1111/j.1600-0625.2008.00717.x

Lee KH, Ibuka T, Mar EC, Hall IH (1978) Antitumor agents. 31. Helenalin sym-dimethylethylenediamine reaction products and related derivatives. J Med Chem 21(7): 698–701. https://doi.org/10.1021/jm00205a022

Leu S, Havey J, White LE, Martin N, Yoo SS et al (2010) Accelerated resolution of laser-induced bruising with topical 20% arnica: a rater-blinded randomized controlled trial: topical treatments for bruising. Br J Dermatol 163(3):557–563. https://doi.org/10.1111/j.1365-2133.2010.09813.x

Lotan AM, Gronovich Y, Lysy I, Binenboym R et al (2020) Arnica Montana and Bellis perennis for seroma reduction following mastectomy and immediate breast reconstruction: randomized, double-blind, placebo- controlled trial. Eur J Plast Surg 43(3):285–294. https://doi.org/10.1007/s00238-019-01618-7

Macêdo SB, Ferreira LR, Perazzo FF, Tavares Carvalho JC (2004) Anti- inflammatory activity of *Arnica montana* 6cH: preclinical study in animals. Homeopathy 93(3):84–87. https://doi.org/10.1016/j.homp.2004.02.006

Maryna K, Trush K, Koscova J, Eftimová J (2019) Antimicrobial, antioxidant and some biochemical properties of arnica Montana L. Bull Problems Biol Med 3:268. https://doi.org/10.29254/2077-4214-2019-3-152-268-273

Marzotto M, Bonafini C, Olioso D, Baruzzi A et al (2016) *Arnica montana* stimulates extracellular matrix gene expression in a macrophage cell line differentiated to wound-healing phenotype. PLoS One 11(11):E0166340. https://doi.org/10.1371/journal.pone.0166340

Nieto-Trujillo A, Cruz-Sosa F, Luria-Pérez R, Gutiérrez-Rebolledo GA et al (2021) *Arnica montana* cell culture establishment, and assessment of its cytotoxic, antibacterial, α-amylase inhibitor, and antioxidant in vitro bioactivities. Plan Theory 10(11):2300. https://doi.org/10.3390/plants10112300

Oana Teodora A, Elena T, Lacon E, Cretu R et al (2016) Phytochemical screening and chromatographic fingerprint studies on ethanolic extracts of Arnica montana L. pp 53–60

Obón C, Rivera D, Verde A, Fajardo J et al (2012) Arnica: a multivariate analysis of the botany and ethnopharmacology of a medicinal plant complex in the Iberian Peninsula and the Balearic Islands. J Ethnopharmacol 144(1):44–56. https://doi.org/10.1016/j.jep.2012.08.024

Petrova M, Zayova E, Dincheva I, Badjakov I, Vlahova M (2015) Influence of carbon sources on growth and GC-MS based metabolite profiling of Arnica montana L. hairy roots. Turk J Biol 39(3):469–478. https://doi.org/10.3906/biy-1412-37

Pieroni A, Giusti ME (2009) Alpine ethnobotany in Italy: traditional knowledge of gastronomic and medicinal plants among the Occitan of the upper Varaita Valley, Piedmont. J Ethnobiol Ethnomed 5(1):32. https://doi.org/10.1186/1746-4269-5-32

Pljevljakušić D, Rančić D, Ristić M, Vujisić L, Radanović D, Dajić-Stevanović Z (2012) Rhizome and root yield of the cultivated Arnica montana L., chemical composition and histochemical localization of essential oil. Ind Crop Prod 39:177–189. https://doi.org/10.1016/j.indcrop.2012.02.030

Pljevljakušić D, Janković T, Jelačić S, Novaković M et al (2014) Morphological and chemical characterization of Arnica montana L. under different cultivation models. Ind Crop Prod 52: 233–244. https://doi.org/10.1016/j.indcrop.2013.10.035

Raza A (2021) Clinical efficacy of homeopathic remedy "Arnica montana": a systematic review. Int J Homeopathy Complement Altern Med 13:13

Reddy KK, Grossman L, Rogers GS (2013) Common complementary and alternative therapies with potential use in dermatologic surgery: risks and benefits. J Am Acad Dermatol 68(4):e127–e137. https://doi.org/10.1016/j.jaad.2011.06.030

Reider N, Komericki P, Hausen BM, Fritsch P, Aberer W (2001) The seamy side of natural medicines: contact sensitization to arnica (Arnica montana L.) and Marigold (Calendula officinalis L.): contact allergy to Arnica and Marigold. Contact Dermatitis 45(5):269–272. https://doi.org/10.1034/j.1600-0536.2001.450503.x

Rywaniak J, Luzak B, Podsedek A, Dudzinska D, Rozalski M, Watala C (2015) Comparison of cytotoxic and anti-platelet activities of polyphenolic extracts from Arnica flowers and Juglans regia husks. Platelets 26(2):168–176. https://doi.org/10.3109/09537104.2014.894970

Sharma S, Arif M, Nirala RK, Gupta R, Thakur SC (2016) Cumulative therapeutic effects of phytochemicals in *Arnica montana* flower extract alleviated collagen-induced arthritis: inhibition of both pro-inflammatory mediators and oxidative stress: anti-arthritis effects of *Arnica montana* flower extract. J Sci Food Agric 96(5):1500–1510. https://doi.org/10.1002/jsfa.7252

Simsek G, Sari E, Kilic R, Bayar Muluk N (2016) Topical application of Arnica and mucopolysaccharide polysulfate attenuates periorbital edema and ecchymosis in open rhinoplasty: a randomized controlled clinical study. Plast Reconstr Surg 137(3):530e–535e. https://doi.org/10.1097/01.prs.0000479967.94459.1c

Spitaler R, Winkler A, Lins I, Yanar S, Stuppner H, Zidorn C (2008) Altitudinal variation of phenolic contents in flowering heads of *Arnica montana* cv. ARBO: a 3-year comparison. J Chem Ecol 34(3):369–375. https://doi.org/10.1007/s10886-007-9407-x

Stanik N, Lampei C, Rosenthal G (2020) Summer aridity rather than management shapes fitness-related functional traits of the threatened mountain plant *Arnica montana*. Ecol Evol 10(11): 5069–5078. https://doi.org/10.1002/ece3.6259

Sugier D, Sugier P, Gawalik-Dziki U (2013) Propagation and introduction of *Arnica montana L.* into cultivation: a step to reduce the pressure on endangered and high-valued medicinal plant species. ScientificWorldJournal 2013:e414363. https://doi.org/10.1155/2013/414363

Sugier D, Sugier P, Jakubowicz-Gil J, Winiarczyk K, Kowalski R (2019) Essential oil from *Arnica montana L.* achenes: chemical characteristics and anticancer activity. Molecules 24(22):4158. https://doi.org/10.3390/molecules24224158

Sugier P, Jakubowicz-Gil J, Sugier D, Kowalski R, Gawlik-Dziki U, Kolodziej B, Dziki D (2020) Chemical characteristics and anticancer activity of essential oil from *Arnica montana L.* rhizomes and roots. Molecules 25(6):1284. https://doi.org/10.3390/molecules25061284

Šutovská M, Capek P, Kočmalová M, Pawlaczyk I, Zaczyńska E et al (2014) Characterization and pharmacodynamic properties of *Arnica montana* complex. Int J Biol Macromol 69:214–221. https://doi.org/10.1016/j.ijbiomac.2014.05.051

Verma N, Tripathi SK, Sahu D, Das HP, Das RH (2010) Evaluation of inhibitory activities of plant extracts on production of LPS-stimulated pro-inflammatory mediators in J774 murine macrophages. Mol Cell Biochem 336(1–2):127–135. https://doi.org/10.1007/s11010-009-0263-6

Vidic D, Ćavar Zeljković S, Dizdar M, Maksimović M (2016) Essential oil composition and antioxidant activity of four Asteraceae species from Bosnia. J Essent Oil Res 28(5):445–457. https://doi.org/10.1080/10412905.2016.1150216

Widen JC, Kempema AM, Villalta PW, Harki DA (2017) Targeting NF-κB p65 with a helenalin inspired Bis-electrophile. ACS Chem Biol 12(1):102–113. https://doi.org/10.1021/acschembio.6b00751

Zucker L (2008) Chapter 6—Arnica. In: Deutsch JE, Anderson EZ (eds) Complementary therapies for physical therapy. W.B. Saunders, Philadelphia, PA, pp 75–83. https://doi.org/10.1016/B978-072160111-3.50012-9

Chapter 5
Aster tataricus L.f.: Ethnomedicinal Uses, Phytochemistry, and Pharmacological Activities

Deepak Timalsina and Hari Prasad Devkota

Abstract *Aster tataricus* L.f. (family: Asteraceae) is an important medicinal plant that had been used in traditional Chinese medicine for many centuries. This plant is used in treating asthma, tuberculosis, and bronchitis-related disease to relieve the cough. The main aim of this book chapter is to collect and critically analyze the available information and updates on chemical constituents and medicinal properties of *A. tataricus*. The scientific information was collected from different online database such as Scopus, SciFinder, PubMed, Google Scholar, and books. The active phytochemicals were peptides, flavonoids, phenols, coumarins, etc. These phytochemicals possess different biological activities such as anticancer, anti-inflammatory, anti-asthma, and antidiabetic. Only a few studies have been performed to report the pharmacological activities; however, detail mechanism of action and its clinical significance are yet to be studied. Future study should focus on clinical evaluation of isolated compounds to justify its traditional uses and development of therapeutics.

Keywords *Aster tataricus* · Cyclopeptide · Anti-asthma · Anticancer

5.1 Introduction

Aster tataricus L.f. (tatarian aster) is a plant that belongs to the family Asteraceae, native to the Siberia, Korea, Japan, and eastern Asia (Fig. 5.1). This plant is terrestrial perennial herb of about 1.8 m; its flower is violet-blue showy with yellow center (Shen et al. 2018). The disk flower grows and widens like a funnel. The leaves and stem of this plant may have fuzzy or hairy structures. The plant has underground

D. Timalsina (✉)
Central Department of Chemistry, Tribhuvan University, Kathmandu, Nepal

H. P. Devkota
Headquarters for Admissions and Education, Kumamoto University, Kumamoto, Japan

Graduate School of Pharmaceutical Sciences, Kumamoto University, Kumamoto, Japan
e-mail: devotah@kumamoto-u.ac.jp

H. P. Devkota, T. Aftab (eds.), *Medicinal Plants of the Asteraceae Family*,
https://doi.org/10.1007/978-981-19-6080-2_5

Fig. 5.1 Photographs of *Aster tataricus*

caudex to survive in extreme cold. This plant is native to the northern and southern China. This plant has been reported to be used in traditional Chinese medicine over 2000 years in several complications such as snake bites, tonsils, bronchial infections, pneumonia, and throat infections (Chen et al. 2021). The wider classes of bioactive compounds have been isolated from this plant such as triterpenoids, saponins, flavonoids, lignans caffeoylquinic acids, epifriedelinol, and sterols (Kim et al. 2014; Yu et al. 2015; Zhang et al. 2017). This plant has been widely studied for its anti-inflammatory (Yue et al. 2012), anti-asthma, antidiabetic, and antimicrobial activities (Zhang et al. 2013). The symbiotic relation of *A. tataricus* and fungal endophyte *Cyanodermella asteris* is found to produce astin which possess anticancer and immunosuppressive activities (Schafhauser et al. 2019). A number research articles have been published on their chemical constituents and biological activities; however, detailed understanding is needed by highlighting the research gaps. Hence, the main aim of this book chapter is to collect overall information and update on the progress of research on *A. tataricus* and its chemical constituents and biological activities.

5.2 Traditional Uses

In traditional Chinese medicine, the rhizomes and root of this plant have been used for management of several complications such as asthma, diabetes, and other respiratory diseases (Bown and Herb Society of America 1995; Yu et al. 2015). This plant was widely used to decrease the mucus production in respiratory system and relieving cough and management of asthma (World Health Organization. Regional Office for the Western Pacific 1998; Zhao et al. 2015; Wong et al. 2021). The root of this plant is taken internally with honey to treat chronic bronchitis, tuberculosis and to increase expectorant effect (Yeung 1985).

5.3 Chemical Constituents

The bioactive compounds of different classes such as peptides, terpenoids, flavonoids, and saponins have been isolated from this plant. The bioactive compounds are mainly reported from the roots and rhizomes of this plant. The detail list of the compounds is given in Tables 5.1 and 5.2. The structures of common phenolic compounds are represented in Fig. 5.2.

5.4 Pharmacological Activities

5.4.1 Anticancer Activities

The anticancer activities of *A. tataricus* was evaluated on human oral squamous carcinoma on SCC-9 and NIH/3 T3 cell lines. At a concentration of 640 μg/ml, the ethanolic extract of this plant inhibited the proliferation of SCC9 cell by 50%. Treatment of this extract significantly reduces the cloning ability and affected the cell cycle by increasing the cell at G2/M phase and decreasing the cell at G0/G1 phase indicating its anticancer activities (Wang et al. 2017).

The homogenous polysaccharide ATP-II was isolated from this plant and evaluated its anticancer efficacy in glioma C6 cells. The result showed the decreased proliferation and biological response due to DNA damage. Glioma tumors were reduced by induced apoptosis, increasing the ratio of Bax/Bcl-2 and activating caspase-3, caspase-8, and caspase-9 cascade, and by downregulating Akt (Du et al. 2014). A water-soluble polysaccharide isolated from water extracted *A. tataricus* was studied on human gastric SGC-7901 cells. A study reported the perfect tumor growth inhibitory activities by inducing apoptosis with no cytotoxicity (Zhang et al. 2012).

5.4.2 Hepato-Protective Activities

The protective effect of *A. tataricus* in liver was studied in thioacetamide-induced hepato-fibrosis in Sprague-Dawley rat model. The hepatic stellate cells treated with the extract showed restoration of morphological changes and degradation of fibronectin and collagen. Moreover, it attenuated the increased serum levels of hydroxyproline, aspartate transaminase, and alanine transaminase. It also significantly restored the decreased glutathione levels (Kim et al. 2020, p. 3).

Table 5.1 Bioactive compounds isolated from roots and rhizomes of *A. tataricus*

Class	Compounds	References
Fatty acids	Cetylic acid	Su et al. (2019a)
	Kauran-18-oic acid	Su et al. (2019a)
	α-Linoleic acid	Su et al. (2019a), Tori et al. (2001)
	α-Linolenic acid	Su et al. (2019a)
Amides	Aurantiamide	Ng et al. (2003)
Peptides and cyclopeptides	Astin A–H	Saviano et al. (2004), Rossi et al. (2004), Liu et al. (2012), Xu et al. (2013b), Wang et al. (2014), Zhao et al. (2015), Sun et al. (2018), Li et al. (2018a, b)
	Astin K–P	Xu et al. (2013b), Yu et al. (2015)
	Asterin	Kosemura et al. (1993)
	Tataricins A–B	Xu et al. (2013a)
Ester	Lachnophyllol Lachnophyllol acetate	Su et al. (2019a) Su et al. (2019a)
Flavonoids and other phenolic compounds	Astragaline	Su et al. (2019a)
	Isoquercetin	Su et al. (2019a)
	Jaboticabin acid	Su et al. (2019a)
	Caffeic acid	Su et al. (2019a), Ma et al. (2020)
	3,4-Dicaffeoylquinic acid	Zhao et al. (2015)
	3,5-Dicaffeoylquinic acid	Zhao et al. (2015)
	3-Caffeoylquinic acid	Zhao et al. (2015)
	Methyl chlorogenate	Su et al. (2019a)
	3,5-O-Dicaffeoyl-1-O-methylquinic acid methyl ester	Su et al. (2019a)
	Methyl 4-caffeoylquinate	Su et al. (2019a)
	Helonioside A	Su et al. (2019a)
	Helonioside B	Su et al. (2019a)
	Parispolyside F	Su et al. (2019a)

	Benzyl-O-β-D-glucopyranoside	Su et al. (2019a)
	Isolariciresinol 9-O-β-D-glucopyranoside	Su et al. (2019a)
	Lariciresinol 9-O-β-D-glucopyranoside	Su et al. (2019a)
	Pinoresinol O-β-D-glucopyranoside	Su et al. (2019a)
	Arillanin B	Su et al. (2019a)
Terpenoids and saponins	(3,3-Dimethylbicyclo[2.2.1]hept-2-yl)methyl-O-β-D-Glucopyranoside	Su et al. (2019a)
	Shionoside A_1	Su et al. (2019a)
	Shionoside A_2	Su et al. (2019a)
	Astersaponin A–C	Zhao et al. (2015), Su et al. (2019b)
	Shionoside C	Dongliang and Yu (1993)
	3-O-α-L-Arabinopyranosyl-(1 → 6)-β-D trihydroxyolean-12-en-28-oic acid	Su et al. (2019b)
	Astersaponin A_2	Su et al. (2019b)
	Astersaponin C_2	Su et al. (2019b)
	Astersaponin E–H	Nagao et al. (1988), Dongliang and Yu (1993), Su et al. (2019b)
	Astersaponin G_2	Su et al. (2019b)
	(+)-Spathulenol	Su et al. (2019a)
	Bungeolic acid (ursolic acid)	Su et al. (2019a)
	(+)-Isobauerenol	Su et al. (2019a)
	Shion-22(30)-en-3,21-dione	Zhou et al. (2010)
	Shion-22-methoxy-20(21)-en-3-one	Zhou et al. (2010)
	Shion-22-methoxy-20(21)-en-3β-ol	Zhou et al. (2010)
	Astataricusol A	Zhou et al. (2013)
	Astataricusones A–D	Zhou et al. (2013)
	Astershionones A–F	Zhou et al. (2014)
	Epishionol	Zhou et al. (2013)
	Epifridelinol	Ng et al. (2003), Yin et al. (2016), Su et al. (2019b), Ma et al. (2020)

Table 5.2 Bioactive compounds isolated from whole plant of *A. tataricus*

Phenolic acids	Chlorogenic acid	Ma et al. (2020)
	Protocatechuic acid	Ma et al. (2020)
	Ferulic acid	Zhao et al. (2015), Ma et al. (2020), Wang et al. (2020a)
Coumarins	7-Hydroxycoumarin	Ma et al. (2020)
	Scopoletin	Ng et al. (2003), Ma et al. (2020)
Flavonoid	Luteolin	Ma et al. (2020), Wang et al. (2020a)
	(2R,2"R)-7-O-Methyl-2,3,2″,3″-etrahydrorobustaflavone	Chen et al. (2021)
	2,3,2′,3"-Tetrahydrorobustaflavone	Chen et al. (2021)
	4'-Hydroxyflavone	Chen et al. (2021)
	7,4'-Dihydroxyflavanone	Chen et al. (2021)
	Astragalin	Choi et al. (2009)
	Isoliquiritigenin	Chen et al. (2021)
	Isoquercitrin	Ma et al. (2020)
	Isorhamnetin	Ma et al. (2020)
	Quercetin	Choi et al. (2009), Ma et al. (2020), Wang et al. (2020a)
	Kaempferol	Choi et al. (2009), Ma et al. (2020), Wang et al. (2020a)
	Kaempferol-7-*O*-β-D-glucopyranoside	Ma et al. (2020)
Saccharides	ATP-II	Du et al. (2014)
Terpenoids	Shionoside A	Choi et al. (2009)
	Shionone	Wu et al. (2003), Wang et al. (2012), Yin et al. (2016), Ma et al. (2020), Wang et al. (2020a)

5.4.3 Anti-Inflammatory Activities

The anti-neuroinflammatory activities of root extract of *A. tataricus* was studied by investigating the inhibition activities on inflammatory mediators' production in lipopolysaccharide (LPS)-induced C6 cells. C6 cells treated with *A. tataricus* extract (20 and 40 mg/kg) showed a significant decrease in reactive oxygen species (ROS), malondialdehyde (MDA), nitrate release, translocation of Nf-kB, and glutathione level. The expressions of TNF-α in THP-1 cells induced by LPS were significantly decreased by treating with *A. tataricus* extract indicating its anti-neuroinflammatory activities (Zhang et al. 2017).

The ethanolic extract significantly alleviated the hemorrhage and edema. This extract reduced the anti-inflammatory index with improved histopathology score in bladder of rat. Pyroptosis ratio decreased and cell viability improved. The extract inhibited the expression of NLRP3 and other pyroptosis related protein, both in vitro and in vivo (Wang et al. 2020a). The anti-inflammatory activities of root extract of this plant studied in the ovalbumin induced allergic asthma showed significant

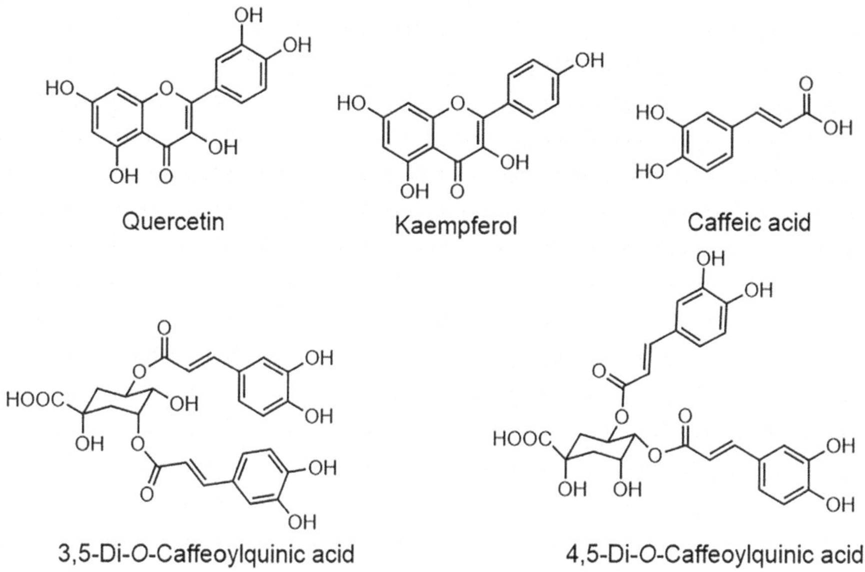

Fig. 5.2 Structures of some common compounds of *Aster tataricus*

decrease in lung injury and expression of other inflammatory cells (Lin et al. 2020; Chen et al. 2020). Similarly, Fengbiasan and Bu-Fei decoction, a formulation that includes *A. tataricus*, inhibited TGF-β1 secretion, lung lesions, neutrophils, and inflammatory factors in a chronic obstructive pulmonary disease (COPD) rats (He et al. 2017; Wang et al. 2020b).

The astersaponin B isolated from the root and rhizomes of *A. tataricus* showed potent inhibition of nitric oxide (NO) formation in murine macrophages and suppressed cyclooxygenase-2 (COX-2) protein level and inducible nitric oxide synthase (iNOS). The anti-inflammatory mechanism involved the attenuation of phosphorylation and NF-κB inhibitor degradation (Su et al. 2019b). The daily oral administration of dry roots and rhizomes to rats was found to regulate the apoptosis and inflammation by decreasing the level of pro-inflammatory cytokines, such as interleukin-6 (IL-6), interleukin-1β (IL-1β), and tumor necrosis factor-α (TNF-α). It also reduced the expression of inducible nitric oxide synthase (iNOS) and cyclooxygenase-2 (COX-2) showing anti-inflammatory activities. In the study of Lee et al., the water extracted shoot of this plant showed similar effect (Lee et al. 2020; Rho et al. 2020). The effect of 4-hydroxyphenyl acetic acid isolated from the roots of *A. tataricus* was studied in seawater aspiration-induced inflammation in rats. The result showed attenuated hypoxia, vascular leak, edema, inflammation, and a decrease in the level of HIF-1α protein. This compound also lowered the inflammatory cytokine level by suppressing hypertonicity (Liu et al. 2014). Similarly, the shinone isolated from the roots of this plant inhibited the production of NLRP3, NF-κB, ASC, caspase-1, and pro-caspase-1, at the mRNA and protein levels in

SV-HUC-1 cell models and in rats, suggesting inhibition of the NLRP3 inflammasome pathway (Wang et al. 2021).

5.4.4 Antidiabetic Activities

The root extract of *A. tataricus* showed a promising effect on lowering concentration of blood glucose level in blood. The retinal leukocytes and vascular permeability were decreased while the activity of superoxide dismutase (SOD), glutathione peroxidase (GSH), and catalase (CAT) was increased. The altered biochemical parameter such as TNF-α, IL10, and NF-kB in the retina were restored after treating with *A. tataricus* extract (Du et al. 2017).

5.4.5 Other Activities

The protective effect on dermal cytotoxicity was studied by Chung et al. This study evaluated the protective effect of *A. tataricus* extract on NaBrO3-induced cytotoxicity in cultured NIH3T3 fibroblast. The extract showed increased cell viability and antioxidant properties comparable to quercetin in NaBrO3-induced cytotoxicity (Chung et al. 2019). The ethanol extract of *A. tataricus* was studied for its activities to alleviate constipation in mice. The result showed the increase in content of fecal water as well as decrease in the fecal remnants. In vitro analysis showed the decrease in contraction, caused by KCl, supported its attenuation of constipation (Wu et al. 2021).

5.4.6 Studies Related to Safety and Toxicity

The cytotoxic activities of robustaflavone, (2R,20'R)-7-O-methyl-2,3,2″,3-″-tetrahydrorobustaflavone, isolated from methanolic extract of *A. tataricus* was studied against PC3, A549, DU145, HepG2, MCF-7, NCI-H1975, and LOVO cell lines for its cytotoxic activities. This compound showed significant inhibition on the proliferation of A549 cells. The apoptotic cell percentage of A549 cells increased when treated with this compound, and the cell scratch assay showed significant decrease in cell migration of A549 cells indicating its cytotoxic activities (Chen et al. 2021).

Astin B, isolated from the *A. tataricus*, provoked the inflammation mediated by oxidative stress. This compound causes increased level of ROS and reduced glutathione, provokes mitochondrial dysfunction, causes induced apoptotic cell death and increased level of phosphorylation c-jun N-terminal kinase. This compound tested in mice in vivo increased liver injury as evidenced from histopathological examination

(Wang et al. 2014). The toxicological study on SD rats demonstrated the mild toxicity of petroleum ether fraction of *A. tataricus* (Peng et al. 2016).

5.4.7 Conclusions and Future Recommendations

Aster tataricus is widely used in traditional Chinese medicine for its activities against respiratory diseases such as asthma, bronchitis, and tuberculosis. Various classes of compounds such as cyclopeptide, flavonoid, phenols, and saponin are of pharmacologic importance. The raw extracts and isolated compounds are studied for their anti-inflammatory, anti-asthma, and antidiabetic activities to justify the traditional importance. However, most of the study are in vitro, and in vivo study is limited within animal model. Detail clinical studies of these bioactive compounds should be carried out to evaluate the safety, toxicity, and efficacy which can direct the development of possible therapeutics from *A. tataricus*.

References

Bown D, Herb Society of America (1995) Encyclopedia of herbs & their uses. Dorling Kindersley; Distributed by Houghton Mifflin, London

Chen Y, Wu H, Li Y et al (2020) Aster tataricus attenuates asthma efficiently by simultaneously inhibiting tracheal ring contraction and inflammation. Biomed Pharmacother 130:110616. https://doi.org/10.1016/j.biopha.2020.110616

Chen T, Yang P, Chen H, Huang B (2021) A new biflavonoids from Aster tataricus induced non-apoptotic cell death in A549 cells. Nat Prod Res 36:1409–1415. https://doi.org/10.1080/14786419.2021.1882456

Choi D-Y, Choi E-J, Jin Q et al (2009) Biological activity of flavonoids isolated from Aster tataricus L. Kor J Pharmacogn 40:123–127

Chung J-H, Lee G-W, Seo Y-M (2019) Protective effect of Aster tataricus L. extract on the dermal cytotoxicity induced by sodium bromate, oxidant of hair dye. Biomed Sci Lett 25:348–356. https://doi.org/10.15616/BSL.2019.25.4.348

Dongliang C, Yu S (1993) Terpenoid glycosides from the roots of Aster tataricus. Phytochemistry 35:173–176. https://doi.org/10.1016/S0031-9422(00)90528-4

Du L, Mei H-F, Yin X, Xing Y-Q (2014) Delayed growth of glioma by a polysaccharide from Aster tataricus involve upregulation of Bax/Bcl-2 ratio, activation of caspase-3/8/9, and downregulation of the Akt. Tumor Biol 35:1819–1825. https://doi.org/10.1007/s13277-013-1243-8

Du H, Zhang M, Yao K, Hu Z (2017) Protective effect of Aster tataricus extract on retinal damage on the virtue of its antioxidant and anti-inflammatory effect in diabetic rat. Biomed Pharmacother 89:617–622. https://doi.org/10.1016/j.biopha.2017.01.179

He X-R, Han S-Y, Li X-H et al (2017) Chinese medicine Bu-Fei decoction attenuates epithelial-mesenchymal transition of non-small cell lung cancer via inhibition of transforming growth factor β1 signaling pathway in vitro and in vivo. J Ethnopharmacol 204:45–57. https://doi.org/10.1016/j.jep.2017.04.008

Kim AR, Jin Q, Jin H-G et al (2014) Phenolic compounds with IL-6 inhibitory activity from Aster yomena. Arch Pharm Res 37:845–851. https://doi.org/10.1007/s12272-013-0236-x

Kim JS, Koppula S, Yum MJ et al (2020) Aster tataricus Linn., suppresses hepatic stellate cell activation and protects against Thioacetamide-induced liver fibrosis in rats. Indian J Pharm Sci 82:110–119. https://doi.org/10.36468/pharmaceutical-sciences.627

Kosemura S, Ogawa T, Totsuka K (1993) Isolation and structure of Asterin, a new halogenated cyclic penta-peptide from Aster tataricus. Tetrahedron Lett 34:1291–1294. https://doi.org/10.1016/S0040-4039(00)91777-5

Lee C-Y, Park H-S, Kong D-H et al (2020) Immunomodulatory effect of the water extract of Aster tataricus through mitogen-activated protein kinase signaling pathway. J Nutr Health 53:452–463

Li F, Guo X-X, Zeng G-Z et al (2018a) Design and synthesis of plant cyclopeptide Astin C analogues and investigation of their immunosuppressive activity. Bioorg Med Chem Lett 28: 2523–2527. https://doi.org/10.1016/j.bmcl.2018.05.050

Li S, Hong Z, Wang Z et al (2018b) The Cyclopeptide Astin C specifically inhibits the innate immune CDN sensor STING. Cell Rep 25:3405–3421.e7. https://doi.org/10.1016/j.celrep.2018.11.097

Lin C-C, Wang Y-Y, Chen S-M et al (2020) Shegan-Mahuang decoction ameliorates asthmatic airway hyperresponsiveness by downregulating Th2/Th17 cells but upregulating CD4+FoxP3+ Tregs. J Ethnopharmacol 253:112656. https://doi.org/10.1016/j.jep.2020.112656

Liu X, Cao P, Zhang C et al (2012) Screening and analyzing potential hepatotoxic compounds in the ethanol extract of Asteris radix by HPLC/DAD/ESI-MSn technique. J Pharm Biomed Anal 67–68:51–62. https://doi.org/10.1016/j.jpba.2012.04.034

Liu Z, Xi R, Zhang Z et al (2014) 4-hydroxyphenylacetic acid attenuated inflammation and edema via suppressing HIF-1α in seawater aspiration-induced lung injury in rats. Int J Mol Sci 15: 12861–12884. https://doi.org/10.3390/ijms150712861

Ma W, Yao G, Huang X et al (2020) Comparison of the active components of Aster tataricus from different regions and related processed products by ultra-high performance liquid chromatography with tandem mass spectrometry. J Sep Sci 43:865–876. https://doi.org/10.1002/jssc.201900814

Nagao T, Okabe H, Yamauchi T (1988) Studies on the constituents of Aster tataricus L. f. I. : structures of Shionosides A and B: monoterpene glycosides isolated from the root. Chem Pharm Bull (Tokyo) 36:571–577. https://doi.org/10.1248/cpb.36.571

Ng TB, Liu F, Lu Y et al (2003) Antioxidant activity of compounds from the medicinal herb Aster tataricus. Comp Biochem Physiol C Toxicol Pharmacol 136:109–115. https://doi.org/10.1016/S1532-0456(03)00170-4

Peng W, Xin R, Luo Y et al (2016) Evaluation of the acute and subchronic toxicity of Aster tataricus L.F. Afr J Tradit Complement Altern Med 13:38–53. https://doi.org/10.4314/ajtcam.v13i6

Rho J, Seo C-S, Park H-S et al (2020) Asteris Radix et Rhizoma suppresses testosterone-induced benign prostatic hyperplasia in rats by regulating apoptosis and inflammation. J Ethnopharmacol 255:112779. https://doi.org/10.1016/j.jep.2020.112779

Rossi F, Zanotti G, Saviano M et al (2004) New antitumour cyclic astin analogues: synthesis, conformation and bioactivity. J Pept Sci 10:92–102. https://doi.org/10.1002/psc.506

Saviano G, Benedetti E, Cozzolino R et al (2004) Influence of conformational flexibility on biological activity in cyclic astin analogues. Biopolymers 76:477–484. https://doi.org/10.1002/bip.20145

Schafhauser T, Jahn L, Kirchner N et al (2019) Antitumor astins originate from the fungal endophyte Cyanodermella asteris living within the medicinal plant Aster tataricus. Proc Natl Acad Sci 116:26909–26917. https://doi.org/10.1073/pnas.1910527116

Shen X, Guo S, Yin Y et al (2018) Complete chloroplast genome sequence and phylogenetic analysis of Aster tataricus. Molecules 23:2426. https://doi.org/10.3390/molecules23102426

Su XD, Jang H-J, Li HX et al (2019a) Identification of potential inflammatory inhibitors from Aster tataricus. Bioorg Chem 92:103208. https://doi.org/10.1016/j.bioorg.2019.103208

Su X-D, Jang H-J, Wang C-Y et al (2019b) Anti-inflammatory potential of Saponins from Aster tataricus via NF-κB/MAPK activation. J Nat Prod 82:1139–1148. https://doi.org/10.1021/acs.jnatprod.8b00856

Sun Y, Li L, Liao M et al (2018) A systematic data acquisition and mining strategy for chemical profiling of Aster tataricus rhizoma (Ziwan) by UHPLC-Q-TOF-MS and the corresponding anti-depressive activity screening. J Pharm Biomed Anal 154:216–226. https://doi.org/10.1016/j.jpba.2018.03.022

Tori M, Murata J, Nakashima K, Sono M (2001) The structure of linoleic acid ester of trans-lachnophyllol isolated from Aster tataricus. Spectroscopy 15:119–123

Wang D, Bai A, Lin X et al (2012) Efficient method for extraction and isolation of shionone from Aster tataricus L. f. by supercritical fluid extraction and high-speed counter-current chromatography. Acta Chromatogr 24:615–625. https://doi.org/10.1556/achrom.24.2012.4.8

Wang L, Li M-D, Cao P-P et al (2014) Astin B, a cyclic pentapeptide from Aster tataricus, induces apoptosis and autophagy in human hepatic L-02 cells. Chem Biol Interact 223:1–9. https://doi.org/10.1016/j.cbi.2014.09.003

Wang R, Xiao S, Niu Z (2017) Anti-cancer activity of Aster tataricus on SCC-9 human oral squamous carcinoma. Afr J Tradit Complement Altern Med 14:142–147. https://doi.org/10.4314/ajtcam.v14i2

Wang X, Fan L, Yin H et al (2020a) Protective effect of Aster tataricus extract on NLRP3-mediated pyroptosis of bladder urothelial cells. J Cell Mol Med 24:13336–13345. https://doi.org/10.1111/jcmm.15952

Wang Y, Su N, Pan S et al (2020b) Fengbaisan suppresses endoplasmic reticulum stress by up-regulating SIRT1 expression to protect rats with chronic obstructive pulmonary diseases. Pharm Biol 58:878–885. https://doi.org/10.1080/13880209.2020.1806335

Wang X, Yin H, Fan L et al (2021) Shionone alleviates NLRP3 inflammasome mediated pyroptosis in interstitial cystitis injury. Int Immunopharmacol 90:107132. https://doi.org/10.1016/j.intimp.2020.107132

Wong LH, Tay L, Goh RMWJ et al (2021) Systematic review: guideline-based approach for the Management of Asthma and Subtypes via Chinese medicine. Evid Based Complement Alternat Med 2021:4319657. https://doi.org/10.1155/2021/4319657

World Health Organization. Regional Office for the Western Pacific (1998) Medicinal plants in the Republic of Korea: information on 150 commonly used medicinal plants. WHO Regional Office for the Western Pacific, Manila

Wu T, Wang G, Chou G et al (2003) Determination of shionone in Radix Asteris by HPLC. Zhongguo Zhong Yao Za Zhi 28:738–740

Wu H, Chen Y, Huang B et al (2021) Aster tataricus alleviates constipation by antagonizing the binding of acetylcholine to muscarinic receptor and inhibiting Ca2+ influx. Biomed Pharmacother 133:111005. https://doi.org/10.1016/j.biopha.2020.111005

Xu H-M, Yi H, Zhou W-B et al (2013a) Tataricins A and B, two novel cyclotetrapeptides from Aster tataricus, and their absolute configuration assignment. Tetrahedron Lett 54:1380–1383. https://doi.org/10.1016/j.tetlet.2012.12.111

Xu H-M, Zeng G-Z, Zhou W-B et al (2013b) Astins K–P, six new chlorinated cyclopentapeptides from Aster tataricus. Tetrahedron 69:7964–7969. https://doi.org/10.1016/j.tet.2013.07.006

Yeung HC (1985) Handbook of Chinese herbal formulas, 2nd edn. Redwing Book Co, Rosemead, CA

Yin D-F, Zhou K, Liu J-T et al (2016) Development and validation of an LC/MS/MS method for simultaneous determination of shionone and epi-friedelinol in rat plasma for pharmacokinetic study after oral administration of Aster tataricus extract. Biomed Chromatogr 30:1112–1117. https://doi.org/10.1002/bmc.3658

Yu P, Cheng S, Xiang J et al (2015) Expectorant, antitussive, anti-inflammatory activities and compositional analysis of Aster tataricus. J Ethnopharmacol 164:328–333. https://doi.org/10.1016/j.jep.2015.02.036

Yue GGL, Chan BCL, Kwok H-F et al (2012) Screening for anti-inflammatory and bronchorelaxant activities of 12 commonly used Chinese herbal medicines. Phytother Res 26:915–925. https://doi.org/10.1002/ptr.3659

Zhang Y, Wang Q, Wang T et al (2012) Inhibition of human gastric carcinoma cell growth in vitro by a polysaccharide from Aster tataricus. Int J Biol Macromol 51:509–513. https://doi.org/10.1016/j.ijbiomac.2012.06.019

Zhang L, Ravipati AS, Koyyalamudi SR et al (2013) Anti-fungal and anti-bacterial activities of ethanol extracts of selected traditional Chinese medicinal herbs. Asian Pac J Trop Med 6:673–681. https://doi.org/10.1016/S1995-7645(13)60117-0

Zhang H, Tian M, He Q et al (2017) Effect of Aster tataricus on production of inflammatory mediators in LPS stimulated rat astrocytoma cell line (C6) and THP-1 cells. Saudi Pharm J 25:370–375. https://doi.org/10.1016/j.jsps.2016.09.001

Zhao D-X, Hu B-Q, Zhang M et al (2015) Simultaneous separation and determination of phenolic acids, pentapeptides, and triterpenoid saponins in the root of Aster tataricus by high-performance liquid chromatography coupled with electrospray ionization quadrupole time-of-flight mass spectrometry. J Sep Sci 38:571–575. https://doi.org/10.1002/jssc.201401008

Zhou WB, Tao JY, Xu HM et al (2010) Three new antiviral triterpenes from Aster tataricus. Z Für Naturforschung B 65:1393–1396. https://doi.org/10.1515/znb-2010-1116

Zhou W-B, Zeng G-Z, Xu H-M et al (2013) Astataricusones A–D and Astataricusol A, five new anti-HBV shionane-type triterpenes from Aster tataricus L. f. Molecules 18:14585–14596. https://doi.org/10.3390/molecules181214585

Zhou W-B, Zeng G-Z, Xu H-M et al (2014) Astershionones A–F, six new anti-HBV shionane-type triterpenes from Aster tataricus. Fitoterapia 93:98–104. https://doi.org/10.1016/j.fitote.2013.12.021

Chapter 6
Atractylodes lancea (Thunb.) DC.: Ethnobotany, Phytochemistry, Pharmacological Attributes, and Safety Profile

Rupa Sanyal, Akash Deep Paul, Tuyelee Das, Saikat Dewanjee, Abdel Rahman Al-Tawaha, Ercan Bursal, Protha Biswas, Manoj Kumar, Radha, Samapika Nandy, Abilash Valsala Gopalakrishnan, Md. Habibur Rahman, Mahipal S. Shekhawat, Devendra Kumar Pandey, Tabarak Malik, and Abhijit Dey

Abstract *Atractylodes lancea* (Thunb.) DC has important implication in traditional medicine and modern implication. Rhizome, being the most important part of the herb, is helpful for numerous diseases we face. Originating in the eastern part of Asia, its rhizome is now used in many important remedies due to its active constituents. As an anti-inflammatory and treatment for spleen disorders, it is an integral

---

R. Sanyal · A. D. Paul
Department of Botany, Bhairab Ganguly College, Kolkata, West Bengal, India

T. Das · P. Biswas · S. Nandy · A. Dey (✉)
Department of Life Sciences, Presidency University, Kolkata, West Bengal, India
e-mail: abhijit.dbs@presiuniv.ac.in

S. Dewanjee
Advanced Pharmacognosy Research Laboratory, Department of Pharmaceutical Technology, Jadavpur University, Kolkata, West Bengal, India

A. R. Al-Tawaha
Department of Biological Sciences, Al-Hussein Bin Talal University, Maan, Jordon

E. Bursal
Department of Biochemistry, Mus Alparslan University, Muş, Turkey

M. Kumar
Chemical and Biochemical Processing Division, ICAR - Central Institute for Research on Cotton Technology, Mumbai, Maharashtra, India

Radha
School of Biological and Environmental Sciences, Shoolini University of Biotechnology and Management Sciences, Solan, Himachal Pradesh, India

A. V. Gopalakrishnan
Department of Biomedical Sciences, School of Bio Sciences and Technology, Vellore Institute of Technology (VIT), Vellore, Tamil Nadu, India

H. P. Devkota, T. Aftab (eds.), *Medicinal Plants of the Asteraceae Family*,
https://doi.org/10.1007/978-981-19-6080-2_6

part of Chinese traditional medicine. Although this plant has been used for centuries, we still don't know much about it. The purpose of this chapter is to discuss the active chemical constituents of *A. lancea*, its pharmaceutical activity, dosage, and toxicity levels and some major studies that have been conducted as well as its potential to become a leading source of medicinal formulations. A thorough yet summarized over view of the plant has been portrayed here with future recommendation for better understanding.

Keywords *Atractylodes lancea Thunb. DC* · Active constituents · Anticancer · Anti-inflammatory · Safety

6.1 Introduction

Plants have always been a primary source of medicine since the evolution of humans. Plants and plant-derived substances are used in different forms as an ailment for diseases. Sometimes as decoction, cooked, or as a part of medicinal formula, plants have showed miracles to cure life-threatening diseases. All animals are naturally adapted for a particular climatic and topographic boundary. This adaption includes usage of plants to its fullest potential. Humans have successfully used plants for treatment of diseases caused by microbes or climatic changes. *Atractylodes lancea* is a perennial herb belonging to the Asteraceae family. It is also common in the eastern region of Asia. It was traditionally used in China, Korea, Japan, and Thailand. The rhizome is the most extensively used part of the plant. In China, it is locally known as "Cangzhu" and grows in Shaanxi Province. It requires cool and dry climate with dry grassland. The soil must be loose sandy soil with good drainage and a semi-shade hillside (Teng et al. 2011). The plant grows up to a meter tall and fruits from August to October. It is dioecious. The roots are either cooked or eaten raw because they are rich in vitamin A and contain 1.5% essential oil. The rhizome which is the most valuable part of this plant has an essential medicinal value. The rhizome can be consumed in several forms, and it has anticancer and anti-inflammatory properties. It is antimicrobial, and researchers are still trying to find out more about its medicinal properties. With no such fetal toxicity found in this plant, it

M. H. Rahman
Department of Global Medical Science, Wonju College of Medicine, Yonsei University, Wonju, Gangwon-do, South Korea

M. S. Shekhawat
Plant Biotechnology Unit, KM Government Institute for Postgraduate Studies and Research, Puducherry, India

D. K. Pandey
Department of Biotechnology, Lovely Professional University, Phagwara, Punjab, India

T. Malik
Institute of Health, Jimma University, Jimma, Ethiopia

is a rich source of essential oils. It contains polysaccharides, polyethylene alkynes, sesquiterpenes, and phytosterols which contribute to the plant's medicinal versatility. This chapter describes the plant's traditional uses along with its chemical constituents. The results of the in vivo and in vitro experiments will certainly help us understand this plant's overall rating.

6.2 Traditional Uses

A. lancea is used comprehensively in traditional medical system of Asia, especially the eastern part of the continent. Countries like Japan and China or south eastern country like Thailand has been using this herb extensively, especially the rhizome. The rhizome is cylindrical and irregularly curved. The dimensions are 3–10 mm in length and 10–25 mm in diameter. The color varies from darker greyish brown to yellow brown. White cotton-like fragrant structures is formed if stored for a long time (Zhang et al. 2020). It has been used in all forms: raw, as a decoction, or even as a component of complicate herbal medicinal formula. This plant cultivated in Hubei and Jiangsu region along with Zhejiang, Anhui, and other places (Qian et al. 2006). It is mostly harvested in spring. According to TCM theory, *A. lancea* is used for digestive disorders, night blindness, influenza rheumatic conditions, and strengthening the spleen (Qian et al. 2006). *Atractylodes lancea* has been generally used in several traditional decoction, for example, Simiao powder, Ermiao powder, Yueju-Wan, and has been a part of several other common decoction (Koonrungsesomboon et al. 2014). The compound found in A lancea is known in China as "Cangzhu" and in Thailand as "Khod-Kha-Mao" and in Japan as "So-jutsu."

A. lancea was used extensively to cure common cold and to eliminate dampness of the skin's surface due to cold wind (Qian et al. 2006). The dried rhizome of *A. lancea* has been used extensively as a treatment for fever and common cold. Traditional Thai medicine uses *A. lancea's* dried rhizome extensively to treat fevers and common colds (Chayamarit 1995). It has always been a component of Thai medicine to cure and mitigate gastrointestinal symptoms which include dyspepsia, flatulence, noninfectious diarrhea, and nausea. The rhizomes of *A. lancea* are used in several Kampo medicines, e.g., Juzen-taiho-to and Saireito (Saiki 2000; Kishida et al. 2007). It seems that plant has been used as a medicinal herb and the part used primarily is the rhizome. Present comprehensive studies have help us reveal the reason why it had such a wide and impactful usage in traditional medicine. It is believed to possess anti-inflammatory, antimicrobial, and even anticancer properties. A list of traditional uses of *A. lancea* rhizome is given in brief in Table 6.1.

Table 6.1 Traditional use of *A. lancea* rhizome (Pharmacognosy 2021)

Formulation	Ailments	Formula
Rhizome of *A. lancea, Ledebouriella* root, and *Asarum* herb	Soreness in limbs, headaches, and fever	Traditional Chinese medicine
Magnolia bark, tangerine peel, and *A. lancea* rhizome	Nausea, vomiting, poor appetite, and dampness of the spleen	Pingwei san
Phellodendron bark, *A. lancea* rhizome, and *Cyathula* root	Swelling and weakness of painful and weak lower limbs	Sanmiao wan
A. lancea rhizome, *Angelica pubescens* root, *Chaenomeles* fruit, mulberry twigs	Painful knee joints due to wind and cold	Traditional Chinese medicine

6.3 Chemical Constituents

There are several bioactive components that are present in *A. lancea*, such as polysaccharides, polyethylene alkynes, sesquiterpenes, and phytosterols. *A. lancea* rhizomes provides good source of volatile oils and several other components. Gas chromatography-mass spectrometry was used to identify them, in which 89.36–95.79% were found to be volatile oils. Sesquiterpene has a higher boiling point than other volatile oils in AL. Chemical profiles of wild *A. lancea* plants showed that the essential oil from *A. lancea* produced significant amounts of sesquiterpenes, with hinesol, β-eudesmol, and elemol being next in terms of quantity (Jia et al. 2004). The unsaturated ketones, esters, furans, benzene, and alcohols with functional groups are all included in alkynes. Some of them are relatively stable which can be easily extracted and separated and is identified with the help of different phytochemical techniques. Polyacetylenes of different types are classified according to their chemical's backbones like diene-diyne types by their alcohol-attached furan ring attached, acetyl-attached, etc. (Zhang et al. 2020). Depending on their chemical properties, compounds can be categorized into several different types. The atractylenolide(s) are sesquiterpene; atractylone, hinesol, and β-eudesmol are sesquiterpenoid; atractylodin is polyethylene alkyne; and sigma sterol and β-sitosterol are phytosterol. Figure 6.1 demonstrates the structure (chemical) of the bioactive components in *A. lancea*. All chemical structures were retrieved from http://www.chemspider.com/:

1. *Atractylenolide I (ATL-I)*: The sesquiterpene atractylenolide I (ATL-I) is soluble in ethanol and methanol. Atractylenolide I (ATL-I) is a sesquiterpene, which is soluble in ethanol as well as methanol. It is also stable without light and moisture. Sesquiterpenes are isolated and identified form the rhizomes of azulene derivatives. The sesquiterpenes that were obtained from *A. lancea* belong to the azulene derivatives, which are aromatic skeleton with 5–7 membered rings. Through cytotoxic experiments, it showed no cytotoxic activity on P388 and A549 (Zhang et al. 2020).

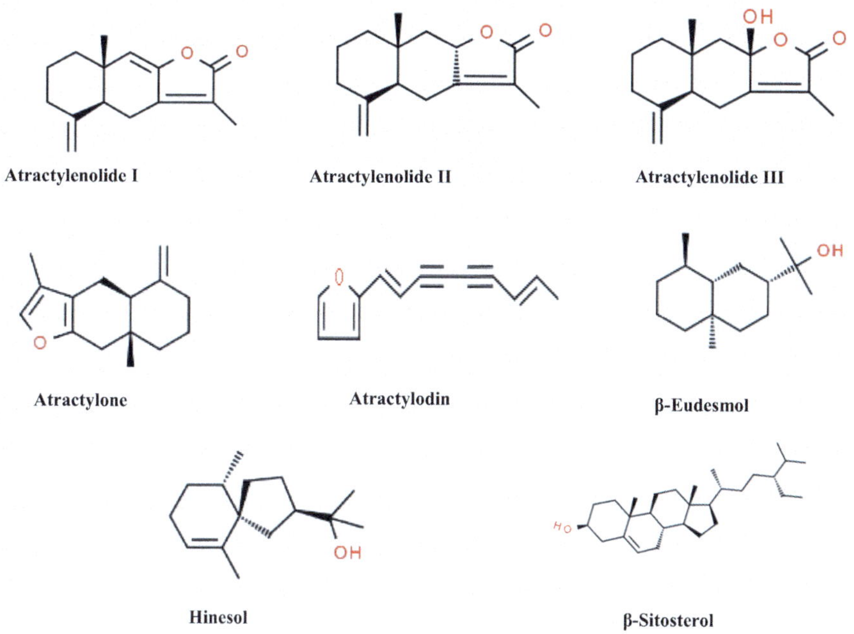

Fig. 6.1 Structures of the main constituents of *A. lancea*

2. *Atractylenolide II (ATL-II),* $C_{15}H_{20}O_2$: Atractylenolide II is also a sesquiterpene extracted from *A. lancea* that has cancer-fighting properties in addition to anti-inflammatory properties. Melting temperature is about 383 °C. It was identified in *A. lancea* in 1998. Signal transduction and transcription activation (STAT3), which becomes overactive due to mutation, suppress the signal and shows anti-melanoma properties (Jun et al. 2018).
3. *Atractylenolide III (ATL-III),* $C_{15}H_{12}O_3$: Atractylenolide III is a sesquiterpene derived from *A. lancea.* The freezing temperature of ATL-III is 392–394 °F. The melting temperature of ATL-III is 200–201 °C. According to Jun et al. (2018), ATL-III was discovered in 2008 in *A. lancea* and can inhibit mast cell proliferation and enhance p53 expression (Jun et al. 2018).
4. *Hinesol,* $C_{15}H_{26}O$: Hinesol, a sesquiterpenoid, comprises about 5–9% of *A. lancea.* Hinesol was first classified in *A. lancea* in 2003. Hinesol inhibits nuclear and DNA fragmentation in human leukemia HL-60 cells (Jun et al. 2018).
5. *β-Eudesmol,* $C_{15}H_{26}O$: β-Eudesmol is a sesquiterpenoid alcohol and also a main constituent of *A. lancea.* The melting temperature of β-eudesmol is about 72–74 ° C. This chemical constituent is a strong inhibitor of CCA and bile cancer (Jun et al. 2018).
6. *Atractylodin (Atr),* $C_{13}H_{10}O$: Atractylodin polyethylene alkyne is derived from *A. lancea* which has anti-inflammatory properties. The melting temperature of

atractylodin is 52 °C. Atr inhibits the inflammatory response induced by LPS (Jun et al. 2018).

7. *Stigma sterol, $C_{29}H_{48}O$*: Stigma sterol, an unsaturated phytosterol, is extracted from *A. lancea*. The melting temperature of it is 160 to 164 °C (Antwi et al. 2017).
8. *Atractylone, $C_{15}H_{2}O$*: Atractylone is one the main constituents of *A. lancea*. The boiling temperature is 285.00–286.00 °C. Atractylone is first identified on *A. lancea* in 2006, which has the total constituent percentage of 9.35% (Guo et al. 2006). With its anti-hepatotoxic effects, it has been used recently to decrease the pulmonary injury due to Influenza A virus. 2018.
9. *β-Sitosterol, $C_{29}H_{50}O$*: The β-sitosterol is a phytosterol which has a melting point of approximately 136–140 °C. It has cytotoxic effect on cancer cells and against several myeloma U266 cells. It also activates AMPK and acetyl-CoA carboxylase (ACC) pathway (Jun et al. 2018). *A. lancea* has several other compounds which include osthol and amino acids and several other water-soluble compounds. It was first isolated by Duan et al. from *A. lancea*. Several other new compounds like phenolic acids, lignans, etc. are also found.

6.4 Pharmacological Activities

With modernization of our medical science studies, the pharmacological activities of *A. lancea* are reviewed by several researchers. In in vitro and in vivo condition, major constituents of the *A. lancea* like have several antiangiogenic, anticancer, anti-inflammatory, and antiallergic activities (Fig. 6.2).

6.4.1 Anticancer Activity

A. lancea has anticancer properties that seem to slow or stop cell proliferation. Both in vivo and in vitro have shown that ATL-1 is highly effective against bladder cancer. By increasing the level of p21 and reducing cyclin B along with CDK1 and Cdc25c. This causes bladder cell proliferation to stop. It inhibits cell cycle in G2/M (Huang et al. 2016). Cholangiocarcinoma (CCA) has the highest mortality and morbidity rate and is the most common type of bile duct cancer particularly with patients in advance stage. This arises within or outside bile duct. The CCA has 11% of the new cases and the treatment works only on patients in early stage. The use of ethanolic extract of *A. lancea* is studied by using positron emission tomography-computed tomography (PET-CT) to monitor CCA male Syrian hamsters. It is observed that the ethanolic extract of A lancea has significant activity against CCA in hamsters when PET-CT is performed using 18F-fluorodeoxyglucose after weeks. (Plengsuriyakarn et al. 2015). The anticancer activity is also shown in several

Fig. 6.2 Pharmacological activities of *A. lancea*

in vivo and in vitro experiments to show the activity against CCA call line. IC_{50}, which stops cell growth by 50%, and, LC_{50} which causes % cell death with more selectivity and potency than 5-fluorouracil. The mice when treated with ethanolic extract of 1000, 3000, and 5000 mg/kg body weight stopped the tumor outgrowth significantly by 10% within 40 days. At highest dose of 5000 mg/kg, the lung metastasis is inhibited by 95%. Ma et al. observed inhibitory effect of human umbilical vein endothelial cells (HUVEC) which has been induced by bFGF and vascular endothelial growth factor (VEGF). Other than that the human liver cancer cell (BEL-7402) is also inhibited by time-to-time dosage of it stood effective against implanted mice tumor cells. The rhizomes of *A. lancea* were used to extract three major chemical constituents which include β-eudesmol (BE), atractylodin (AT), and hinesol (HS), and combinations of them were used to find the cytotoxic activity of human CCA cells. The combination used in experiment were BE:AT, BE:HS, and AT:HS and triple combination of BE:AT:HS. MTT assay was used to evaluate result. The triple combination produced synergistic interaction with a combination index value of approximately 0.519 ± 0.10 and 0.65 ± 0.17 (mean ± SD) which inhibited the cell growth, 50% and 90%, respectively. This makes the chemical constituents of *A. lancea* which is a potential anti-CCA chemotherapeutics with proper dosage with synergistic effect against cholangiocarcinoma cells. Thus, it makes the extracts of *A. lancea* a better alternative for drugs.

6.4.2 Activity in Gastrointestinal System

Gastric cancer has the second highest mortality in cancer death. Combination of plant extract from *A. lancea* can be used to improve the performance of drugs and reducing the side effects. It has been used to increase small intestinal motility and delaying gastric. The action takes place either by inhibition of corticotrophin releasing factor (CRF) and activation of vagal tone or by inhibiting the production of dopamine D and 5-HT3 receptor. When experiment is done on pylorus ligated rats, it showed anti-ulcer action. It may be activated by inhibition of gastric secretion and reduce histamine-induced and stress-induced ulceration (Kubo et al. 1983). β-Eeudesmol also acts as an active constituent to inhibit gastric secretion. This inhibition is stimulated by histamine. The dosage effective against gastric ulceration is 10 mg/kg of body weight. Other than that, hinesol too acts as an anti-ulcer at a suitable dosage of 100 mg/kg of total body weight (Nogami et al. 1986; Koonrungsesomboon et al. 2014). The human gastric cancer cell lines like BGC-823 and SGC-7901 were used with petroleum ether (PE) and ethyl acetate (EtOAc), and water has been used after a cytotoxic and apoptosis assays. The main chemical constituent involved in this reaction was three sesquiterpene lactone compounds that is -b-acetyl-atractylenolide III, atractylenolide I, atractylenolide III, two sesquiterpene compounds that are eudesm-4(15)-ene-7a,11-diol and eudesm-4(15),7-diene-9a11-diol, and one polyacetylene compound that is 4,6,12-tetradecatriene-8,10-diyne1,3,14-triol. The gastric cancer cells growth is inhibited. The MTT assay shows inhibition when the PE, EtOAc, and nBuOH at different fractions as 0, 0.0625, 0.125, 0.25, 0.5, and 1 mg/mL for 24 hours and 48 hours, respectively. The inhibition of BGC-823 cells occurred in a dose-dependent manner, while that of the SGC-7901 was dose and time dependent. PE fraction was the most effective, followed by EtOAc and at last nBuOH fraction. The mitochondrial transmembrane potential decreased in the BGC-823 and SGC-7901 cells after treatment due to decreased fragmentation of apoptotic nuclei (Zhao et al. 2014).

6.4.3 Activities on the Nervous System

The effect of *A. lancea* extract on CNS have not shown any adverse or toxic effect. An experiment with mice has already been administered and the results are positive. The pharmacological activity of rhizome extract has been temporary in terms of side effects, with the highest dosage up to 5000 mg/kg body weight. β-Eudesmol relieves neuromuscular pain as it is a noncompetitive nicotinic acetylcholine (ACh) receptor (nAChR) channel (Acharya et al. 2021). Several tests have already been conducted on animal models to understand the general effects it may cause. Anoxic mice treated with potassium cyanide showed anti-anoxic behavior. When mice were given 1500 mg/kg of *A. lancea* extract, 90% of the mice survived. The control group mice, however, did not survive. The anti-anoxic action of *A. lancea* rhizome extract is due β-eudesmol (Koonrungsesomboon et al. 2014).

6.4.4 Antimicrobial Activity

A. lancea is known to have antibacterial properties as the extracts of the rhizomes are used in traditional medicines. Stigma sterol have an antibiotic effect which successfully suppressed a colony count yielding 98.7% as an adjuvant against both Gram-positive bacteria and Gram-negative bacteria like *Staphylococcus epidermidis*, *Bacillus subtilis*, *Staphylococcus aureus*, and *Micrococcus luteus* and *Escherichia coli*, respectively, during treatment with ampicillin. Mice infected with *Candida* showed prolonged life span when they were administered with oral dosage of 140 mg/kg/d (Zhang et al. 2020). *Trypanosoma congolense* cause a disease called nagana in different animals. Stigma sterol acts as an anti-trypanosoma effectively by reducing sialidase. Against mosquito larvae, it is neurotoxic and acts as larvicidal (Jun et al. 2018). *A. lancea* shows anti-microbial activity against *E. coli*, *S. aureus*, *Candida albicans*, and *Saccharomyces cerevisiae* (Koonrungsesomboon et al. 2014). Essential oil extracted form *showed* a wide range of antibiogram against bacteria. The composition used in the experiment is β-eudesmol, hinesol, elemol, and atractylone at 36.5%, 29.4%, 4.21%, and 4.10%, respectively. Microbes are one of the most common causal organisms for infection, food spoilage, diseases, etc. like *S. aureus* and *E. coli*. by damaging the cell membrane permeability; it has a huge potential to play a crucial role in pharmaceutical and food industry (He et al. 2020).

6.4.5 Anti-Inflammatory Activity

It is a substance that helps to reduce swelling or redness due to infection or wounds possesses anti-inflammatory properties. Atractylodin shows the best results against inflammation like LPS-induced inflammatory responses. Myeloperoxidase (MPO) is a peroxidase and pro-inflammatory biomarker; its activity is suppressed by Atr. Other than that, it checks protein leakage along with infiltration of inflammatory cells. Cytokines like IL-6, TNF-α, and IL-1β are responsible to produce acute phase protein during inflammatory state which is reduced by *A. lancea* (Jun et al. 2018). β-Eudesmol too acts as an anti-inflammatory. Mast cells help to activate several other inflammatory mediators for allergic and hypersensitive reaction. With a 0.2–20 μM dosage, the human mast cells inhibited the production of IL-6 and caspase 1 activity which activates IL-1β inflammatory cytokine. With 20 μM dosage, it showed anti-allergic effects by inhibiting the inhibition of cytokine dans migration of mast cells. The release of tryptase and histamine is also inhibited by β-eudesmol, as these two acts as markers and degranulates the mast cells which leads to allergic reactions. A major transcription factor NF-κB is inhibited in normal human dermal fibroblasts with dosage of 5–20 μM. Atractylenolide III inhibits the production of NO and PGE2 in a dose-dependent manner. Treatment with ATL-III (100 μM) at high concentration inhibits cytokine cyclooxygenase-2 (COX-2) by activating NF-κB and ERK1/2 (Acharya et al. 2021). Studies were conducted to find the effect of *A. lancea* on inflammatory tissue prostaglandin (PGE2) of mice with its paw swelled by using formaldehyde. The volatile oil extracted from *A. lancea* was used on them. The group which was given high dosage showed significant anti-inflammatory results.

6.4.6 Antiangiogenic Activity

Angiogenesis refers to the growth, migration, and differentiation of new endothelial cells that lead to the formation of new blood vessels. This is an essential system in our body and controlled by several chemical signals. If unregulated, it forms cancer cells. Tumors give off a signal that stimulates angiogenesis for blood supply. Diabetic retinopathy and rheumatoid arthritis are also caused due to unregulated angiogenesis. In vitro conditions, β-eudesmol significantly inhibited the formation and proliferation of human umbilical vein endothelial cells (HUVEC), porcine brain microvascular endothelial cells (PBMEC), and human dermal microvascular endothelial cells (HDMEC). Basic fibroblast growth factor (bFGF)-stimulated HUVEC cells were also inhibited from migrating. In in vivo condition, when a group of mice were treated with 0.45 and 0.9 μmol/kg of β-eudesmol, it inhibited granuloma by 18.9% and 45.4%, respectively (Acharya et al. 2021). When zebra fish embryos were treated with sesquiterpenoid extracted from *A. lancea,* β-eudesmol, and atractylodin, it showed inhibition of blood vessel development. With increased concentration, the inhibition becomes much more expressive. *Vegfaa* gene expression were downregulated by β-eudesmol at all concentration. For zebra fish embryos, β-eudesmol and atractylodin were lethal, showing the antiangiogenic property of *A. lancea* extracts (Tshering et al. 2021).

6.4.7 Miscellaneous Activities

Obesity is a major problem today. Lipase inhibition with ethanol extract of *A. lancea* showed an IC50 value of 9.06 μg/mL. The dosage has been evaluated in mice which are introduced to high fat diet and showed a moderate result of 500 mg/kg. In humans, with the help of high-throughput screening, it is observed that atractylodin showed highest lipase inhibition with IC50 of 39.12 μM in pancreatic lipase assay (Jiao et al. 2014). Red flour beetle (*Tribolium castaneum)* is an insect which infests flour. The ethanol extract of *A. lancea* shows repellence and contact activities against it. *Atractylodin* showed the best result, among other chemicals, even at a very low concentration of 0.63 μg/cm^2 with LD_{50} value of 1.83 (Chen et al. 2015) *A. lancea* is a plant with versatile pharmacological application, and with further studies, we can utilize its potential to the fullest.

6.5 Safety and Toxicity

The extracts from *A. lancea* if used need to be safe without any possible side effects. *A. lancea,* which is previously used in traditional medical system, has huge potential in modern medical system, with several acute and subacute dosage testing conducted on rats and mice, to test the range safety profile for *A. lancea*. In Wister rats, it is found to have no severe effect with the highest dosage ranging to 5000 mg/kg of body weight. Few drawbacks were stomach irritation and signs of central nervous system depression which include reduced response to balance and touch and diminished locomotion and alertness. This occurred 1 h after dosing them. All the symptoms were reversible and recovered within 24 h. No adverse effects were observed with maximum tolerated dose (MTD) for standardized *A. lancea* extract is set as 5000 mg/kg body weight (Bangchang et al. 2017). Some tests have been conducted on patients with different diseases with different *A. lancea* formulations, and the results were positive. After centuries of usage in traditional medicines, it has shown no significant adverse effects on humans. An Ayurveda Siriraj herbal recipe like Chantaleela contains *A. lancea* extract (30 mg *A. lancea*/tablet). This herbal remedy has been administered to both males and females, and no adverse effects were observed. "Fufang Cangzhu Tang" which is a Chinese herbal formula with about 15 g of *A. lancea* in 300 mL decoction was given twice for 8 weeks on 32 patients. No adverse effects were observed on volunteered senile patients who already had impaired glucose tolerance and overweight (Nut et al. 2014). *β*-Eudesmol has a MTD and NOAEL of 100 mg/kg of body weight, and the lethal dose is 200 mg/kg of body weight (Wein et al. 2017).

6.6 Conclusions and Future Recommendations

The use of traditional plant extracts has now become a safer alternative to drugs with side effects. With increased number of experiments to test the safety and effectiveness of *A. lancea*, it has shown positive results in terms of reports. *A. lancea* being used as traditional medicine for centuries gives us strong evidence about its safety profile. The trials performed about its effectiveness against CCA cells in both crude and as an oral pharmaceutical product are successful. The chemical constituent β-eudesmol has a huge potential to be an essential therapeutic candidate for chronic inflammatory, cancer, and angiogenic disease. After the in vivo experiments, it is clear that *A. lancea* enhanced the performance of several therapeutic drugs like 5-FU and DOX. It is also a potential remedy for nervous disorders.

Further investigations about the effect of *A. lancea* extracts on normal cells are yet to be studied. In vivo experiments on animals are not sufficient to understand the effects on humans it may cause. Studies should be conducted to find and explore further pharmacological active compounds and find new uses. With rhizome being the most useful part of plant, we have to focus on the already discovered extracts and confirm its mode of action and application. If we can truly cultivate the true potential of *A. lancea*, we can revolutionize the treatment of chronic diseases like cancer with least side effects.

References

Acharya B, Chaijaroenkul W, Na-Bangchang K (2021) Therapeutic potential and pharmacological activities of β-eudesmol. Chem Biol Drug Des 97(4):984–996. https://doi.org/10.1111/cbdd.13823

Antwi AO, Obiri DD, Osafo N (2017) Stigmasterol modulates allergic airway inflammation in guinea pig model of ovalbumin-induced asthma. Mediators Infamm 2017:2953930. https://doi.org/10.1155/2017/2953930

Bangchang KN, Plengsuriyakarn T, Karbwang J (2017) Research and Development of *Atractylodes lancea* (Thunb) DC. as a promising candidate for cholangiocarcinoma chemotherapeutics. Evid Based Complement Alternat Med 2017:5929234. https://doi.org/10.1155/2017/5929234

Chayamarit K (1995) Thai medicinal plants. Department of Forestry, Bangkok

Chen HP, Zheng LS, Yang K, Lei N, Geng ZF, Ma P, Cai Q, Du SS, Deng ZW (2015) Insecticidal and repellant activities of polyacetylenes and lactones derived from *Atractylodes lancea* rhizomes. Chem Biodivers 12(4):593–598. https://doi.org/10.1002/cbdv.201400161

Guo FQ, Huang LF, Zhou SY, Zhang TM, Liang YZ (2006) Comparison of the volatile compounds of Atractylodes, medicinal plants by headspace solid-phase microextraction-gas chromatography- mass spectrometry. Anal Chim Acta 570(1):73–78. https://doi.org/10.1016/j.aca.2006.04.006

He F, Wang W, Wu M, Fang Y, Wang S, Yang Y, Ye C, Xiang F (2020) Antioxidant and antibacterial activities of essential oil from Atractylodes lancea rhizomes. Ind Crops Prod 153:112552. https://doi.org/10.1016/j.bmcl.2016.03.021

Huang HL, Lin TW, Huang YL, Huang RL (2016) Induction of apoptosis and differentiation by atractylenolide-1 isolated from *Atractylodes macrocephala* in human leukemia cells. Bioorg Med Chem Lett 26(8):1905–1909. https://doi.org/10.1016/j.bmcl.2016.03.021

Jia C, Mao D, Zhang W, Sun X (2004) Studies on chemical constituents in essential oil from wild *Atractylodes lancea* in dabie mountains. Zhong Yao Cai 27(8):571–574

Jiao P, Crank JT, Corneliusen B, Yimam M, Hodges M, Hong M, Maurseth C, Oh M, Kim H, Chu M, Jia Q (2014) Lipase inhibition and antiobesity effect of Atractylodes lancea. Planta Med 80(7):577–582. https://doi.org/10.1055/s-0034-1368354

Jun X, Fu P, Lei Y, Cheng P (2018) Pharmacological effects of medicinal components of Atractylodes lancea (Thunb.) DC. Chin Med 13(1):1–10

Kishida Y, Miki H, Nishii T, Inoue T, Nishida S, Yoshikawa H, Sugano N (2007) Therapeutic effects of Saireito (TJ-114), a traditional Japanese herbal medicine, on postoperative edema and inflammation after total hip arthroplasty. Phytomedicine 14(9):581–586. https://doi.org/10.1016/j.phymed.2006.12.024

Koonrungsesomboon N, Nabangchang K, Karbwang J (2014) Therapeutic potential and pharmacological activities of *Atractylodes lancea* (Thunb.) DC. Asian Pac J Trop Med 7(6):421–428. https://doi.org/10.1016/S1995-7645(14)60069-9

Kubo M, Nogami M, Nishimura M, Moriura T, Arichi S (1983) Origins, processing, and qualities of crude drugs (1). Preventive effects of a Chinese crude drug, Zhu, on experimental stomach ulcer and its pharmacological evaluation. 1. Yakugaku Zasshi 103(4):442–448

Nogami M, Moriura T, Kubo M, Tani T (1986) Studies on the origin, processing and quality of crude drugs. II. Pharmacological evaluation of the Chinese crude drug"zhu" in experimental stomach ulcer. (2). Inhibitory effect of extract of *Atractylodes lancea* on gastric secretion. Chem Pharm Bull (Tokyo) 34(9):3854–3860. https://doi.org/10.1248/cpb.34.3854

Pharmacognosy (2021) *Atractylodes rhizome* (Cangzhu)-*Atractylodes lancea* (Thunb.) DC. http://www.epharmacognosy.com/2012/03/atractylodes-rhizome-cangzhu.html

Plengsuriyakarn T, Matsuda N, Karbwang J, Viyanant V, Hirayama K, Na-Bangchang K (2015) Anticancer activity of *Atractylodes lancea* (Thunb.) DC in a hamster model and application of PET-CT for early detection and monitoring progression of cholangiocarcinoma. Asian Pac J Cancer Prev 16(15):6279–6284. https://doi.org/10.7314/APJCP.2015.16.15.6279

Qian S, Wang L, Duan J, Feng H (2006) The research progress in chemical constituents and biological activities of *Atractylodes lancea* DC. Chin Tradit Herbal Drugs 18:8–11

Saiki I (2000) A Kampo medicine "Juzen-taiho-to"-prevention of malignant progression and metastasis of tumor cells and the mechanism of action. Biol Pharm Bull 23(6):677–688. https://doi.org/10.1248/bpb.23.677

Teng Y, Guo H, Liang Z, Shu Z, Li Z, Wu W (2011) Ethnobotanical survey of medicinal plants and their utilization in Shaanxi Province, China. J Med Plant Res 5(9):1762–1778

Tshering G, Pimtong W, Plengsuriyakarn T, Na-Bangchang K (2021) Anti-angiogenic effects of beta-eudesmol and atractylodin in developing zebrafish embryos. Comp Biochem Physiol C Toxicol Pharmacol 243:108980. https://doi.org/10.1016/j.cbpc.2021.108980

Wein L, Savas P, Luen SJ, Virassamy B, Salgado R, Loi S (2017) Clinical validity and utility of tumor-infiltrating lymphocytes in routine clinical practice for breast cancer patients: current and future directions. Front Oncol 7:156. https://doi.org/10.3389/fonc.2017.00156

Zhang W, Zhao Z, Chang L, Cao Y, Wang S, Kang C, Wang H, Zhou L, Huang L, Guo L (2020) Atractylodis Rhizoma: a review of its traditional uses, phytochemistry, pharmacology, toxicology and quality control. J Ethnopharmacol 266:113415. https://doi.org/10.1016/j.jep.2020.113415

Zhao M, Wang Q, Ouyang Z, Han B, Wang W, Wei Y, Wu Y, Yang B (2014) Selective fraction of Atractylodes lancea(Thunb.) DC. and its growth inhibitory effect on human gastric cancer cells. Cytotechnology 66(2):201–208. https://doi.org/10.1007/s10616-013-9559-1

Chapter 7
Carthamus tinctorius L.: Traditional Uses, Phytochemistry, and Pharmacological Activities

Gopal Lamichhane, Hari Prasad Devkota, Kusum Sai, and Prakash Poudel

Abstract *Carthamus tinctorius* L., also known as safflower, is a highly exploited medicinal plant from Asteraceae family. It has been used in agriculture for its oil, food, and fodder and in textile industries as natural dyes. It has also been used in different traditional medicinal systems for the treatment of various diseases in countries around the world. Scientific researches explored its anticoagulant, antihypertensive, cardioprotective, antioxidant, neuroprotective, anti-melanogenic, immune-stimulatory, antitumor, anti-arthritic, anti-inflammatory, anti-obesity, and antidiabetic activities. The presence of various phytoconstituents such as flavonoids, phenyltethanoids glycosides, coumarins, fatty acids, steroids, alkaloids, alkane diols, riboflavin, and safflower polysaccharides are believed to be responsible for its diverse medicinal significances. Hence, this plant not only serves as a source of lead molecule for drug development but it can also be used to develop nutraceuticals and functional foods as well. Furthermore, as a good source of natural yellow and red pigments, it can provide suitable natural alternative to the synthetic coloring agents.

Keywords *Carthamus tinctorius* L. · Safflower · Traditional medicine · Hydroxysafflower A · Safflower yellow

---

G. Lamichhane (✉)
Department of Oriental Pharmacy and Wonkwang-Oriental Medicines Research Institute, Wonkwang University, Iksan, South Korea
e-mail: gopal95@wku.ac.kr

H. P. Devkota
Headquarters for Admissions and Education, Kumamoto University, Kumamoto, Japan

Graduate School of Pharmaceutical Sciences, Kumamoto University, Kumamoto, Japan

Pharmacy Program, Gandaki University, Pokhara, Nepal
e-mail: devkotah@kumamoto-u.ac.jp

K. Sai
Research School of Chemistry, Australian National University, Canberra, ACT, Australia

P. Poudel
Department of Pharmacy, Gandaki University, Pokhara, Nepal
e-mail: prakash.poudel@gandakiuniversity.edu.np

H. P. Devkota, T. Aftab (eds.), *Medicinal Plants of the Asteraceae Family*,
https://doi.org/10.1007/978-981-19-6080-2_7

7.1 Introduction

Carthamus tinctorius L. (Asteraceae), commonly known as safflower (Fig. 7.1), is cultivated around the world, mainly in India, France, the USA, Iran, Egypt, and China. It is a resistant species and can be grown in less-fertile land with adverse environmental conditions like wind, drought, and salinity. This plant is bushy with a height of 100–130 cm. Their leaves are lanceolate with serrated margins while flowers are radial and tubular forming large inflorescence (Adamska and Biernacka 2021).

This plant is a popular annual flowering plant in agricultural industry due to its widespread uses ranging from culinary, textile, coloring, ornamental, edible oil, and as fodder for animal (Menegaes and Nunes 2020). Extensive usefulness of this plant played a role in the economic importance to this plant, mainly for oil production

Fig. 7.1 Photographs of whole plant, flower, and dried flowers of *Carthamus tinctorius*

Fig. 7.2 Photographs of seeds of *Carthamus tinctorius*

from seeds (Fig. 7.2) in Asian countries and for its extract as coloring agent in Brazil and Portugal (Menegaes and Nunes 2020). Flavonoids present in petals of safflower mainly, quinochalcone are used as naturally derived coloring agents. Carthamin in petals is responsible for its red coloration while hydroxysafflower yellow A, safflower yellow A and B, safflomin A and C, and tinctormine yield yellow coloration (Kazuma et al. 2000). They are extracted and used as food and textile dye (Menegaes and Nunes 2020). It has different common names by location and languages. Safflower is commonly known as "Kusum" or "Kusumba" (India, Bangladesh, Pakistan) (Dajue and Mündel 1996), "Honghua" (China), "Benibana" (Japan), "Gurgum" (Bhutan) (Menegaes and Nunes 2020), "Golrang" (Iran) (Asgarpanah and Kazemivash 2013), "Honghwain" (Korea) (Zhang et al. 2016), and "Kafesheh" (Persian) (Delshad et al. 2018).

Safflowers provided a cheap replacement of saffron, which is used as food additive, as a natural dye, and as medicine for promoting circulation of blood in Chinese system of traditional medicine (Fan et al. 2009). The Chinese Pharmacopoeia contains more than 50 patent drugs containing *C. tinctorius*. Some representative example with high popularity in oriental medicinal system are Zhenghonghua oil, Dieda pill, and Qili powder (Fan et al. 2009).

This plant is also exploited in beauty industries to manufacture rough and lipstick in France and Brazil (Menegaes and Nunes 2020). In India, bindi, shampoo, soaps, ointments, lotions, body oil, etc. are produced by using safflower oil and/or color (Menegaes and Nunes 2020). The inflorescence of this plant is highly nutritious as it served as rich source of minerals, high calories, and almost all essential amino acids except tryptophan. It contains vitamin B, C, D, E, riboflavin, and carotene (Menegaes and Nunes 2020).

7.2 Traditional Uses

Safflower is a popular plant in traditional Chinese medicine and known to possess various medicinal properties. The florets of *C. tinctorius* are used for treating stroke, coronary heart disease, and gynecologic diseases and as analgesics. In Korea, it has been used as a nutraceutical, due to its effect in promoting blood circulation and maintaining homeostasis. The medicinal values and health benefits of *C. tinctorius* flowers were found to be documented in ancient scriptures such as the *Compendium of Materia Medica* as being able to "invigorate the blood circulation." In some modern Chinese clinics, Honghua injection (prepared from the aqueous extract of dried florets of *C. tinctorius*) and Danhong injection (extracted from *Salviae miltiorrhizae* Radix et Rhizoma and *Carthami Flos* herb pair) are commonly used for the treatment of coronary heart diseases, angina pectoris, myocardial infarction, ischemic encephalopathy, and cerebral thrombosis (Fan et al. 2009; Bai et al. 2020). Besides these, Chinese folklore system of medicine was reported to use this valuable medicinal plant for treatment of amenorrhea, gastric tumors, as well as both internal and external wounds (Delshad et al. 2018).

Safflower is typically used in Indian traditional medicine to manage scabies, arthritis, and mastalgia. In the Iranian system of traditional medicine, kin patches, baldness, phlegm, and colic are treated by using safflower, while the Persians utilized it for managing diabetes, phlegmatic fever, melancholia, and dropsy. Moreover, water extract of safflower is applied to give sedation in painful menstruation, as a laxative in constipation and also as an anti-inflammatory remedy. Adding further it has been consumed in rheumatism, paralysis, vitiligo, black spots, psoriasis, mouth ulcer, phlegm, scorpion poisoning, and numbness of limb for its ameliorating effect. Its consumption was also believed to improve semen quality (Delshad et al. 2018).

In Egypt, floral stems were used in religious ceremony, and dyes extracted from petals were used in mummification as it was believed to contain preservative property. Carbonized shoots were also used in preparing Kajal, a protective coloring around eyes, in relation to belief of having medicinal properties in it and mystical belief of protecting from bad spirit (Menegaes and Nunes 2020). In Thailand, aqueous extract of safflower flowers has been largely exploited as a hair color promoter (Delshad et al. 2018).

Carthamin, obtained from this plant, is widely exploited since ages as an orange-red dye, commercially important in textile and carpet-weaving industries throughout Eastern Europe, the Middle East, and the Indian subcontinent (Delshad et al. 2018). Safflower as a multipurpose crop was also widely utilized for the production of cooking oil and biodiesel (Menegaes and Nunes 2020). Moreover, tender leaves and shoot were sold as salad and vegetable in India and around. It also served as a decorative plant, fodder for animals, and coloring and flavoring agents in varieties of Italian, French, and British cuisines. In Thailand, water extract of *C. tinctorius* flowers was used as hair color promoter (Delshad et al. 2018).

7.3 Chemical Constituents

C. tinctorius is reported contain more than 200 different compounds in it including flavonoids, phenylethanoid glycosides, coumarins, fatty acids, steroids, alkaloids, alkane diols, riboflavin, and safflower polysaccharides (Fan et al. 2009; Li et al. 2017). Chalcone flavonoids derivatives are the main compounds in the aqueous extract of *C. tinctorius* such as hydroxysafflor yellow A, which is responsible for the main therapeutic effects of safflower. Compounds isolated from safflower till date are listed below, and the structures of some of them are presented in Fig. 7.3.

7.3.1 Flavonoids

Flowers: Carthamin, safflor yellow A, safflor yellow B, saffloflavonesides A, saffloflavonesides B, safflomin A, safflomin B, hydroxysafflor yellow A, tinctormine, safflomin C, isosafflomin C, methylsafflomin C, methylisosafflomin C, precarthamin, saffloquinoside A, saffloquinoside B, saffloquinoside C, saffloquinoside D, saffloquinoside E, cartormin, quercetin, anhydrosafflor yellow B, quercetin-3-O-β-D-glucoside, quercetin-3-O-α-L-rhamnoside-7-O-β-D-glucuronide, quercetin-7-O-β-D-glucoside, quercetin-3,7-di-

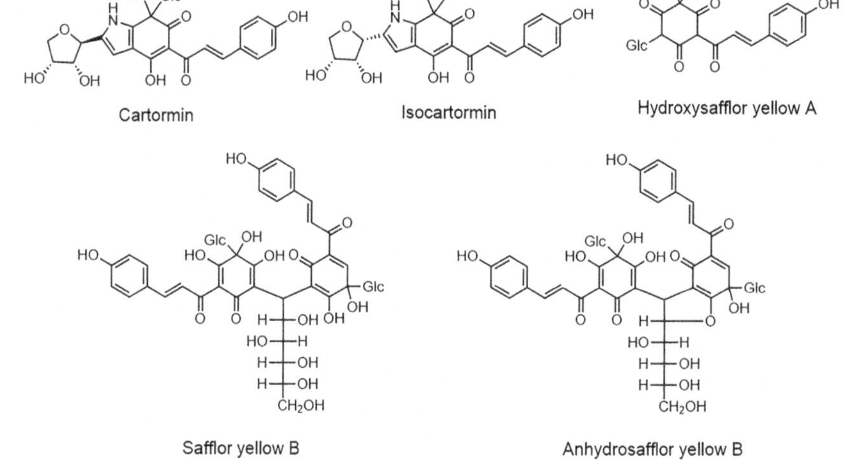

Fig. 7.3 Chemical structures of some of the major compounds of *Carthamus tinctorius*

O-β-D-glucoside, kaempferol-3-O-β-D-glucoside, kaempferol-3-O-β-rutinoside, kaempferol-3-O-β-sophorose, kaempferol-3-O-β-D-glucoside-7-O-β-D-glucuronide, apigenin, scutellarein, rutin, kaempferol, 6-hydroxykaempferol, 6-hydroxykaempferol-3-O-β-D-glucoside, 6-hydroxykaempferol-7-O-β-D-glucoside, 6-hydroxykaempferol-6,7-di-O-β-D-glucoside, (2S)-40,5-dihydroxyl-6,7-di-O-β-D-glucopyranosyl flavanone, 6-hydroxyapigenin-6,7-di-glucoside, 6-hydroxykaempferol-3,6-di-O-β-D-glucoside, 6-Hydroxykaempferol-3,6,7-tri-O-β-D-glucoside, 6-hydroxykaempferol-3,6-di-O-β-D-glucoside-7-O-β-D-glucuronide, 6-hydroxykaempferol-3-O-β-rutinoside-6-O-β-D-glucoside, (2R)-40,5-dihydroxyl-6,7-di-O-β-D-glucopyranosyl flavanone, 6-hydroxykaempferol-3,6-di-O-β-D-glucopyranoside, acacetin (Zhang et al. 2016).

Seeds: Kaempferol 7-O-β-D-glucopyranoside, acacetin 7-O-α-L-rhamnopyranoside, acacetin-7-O-β-D-apiofuranosyl-(1–6)-O-β-D-glucoside (Zhang et al. 2016).

Leaves: Luteolin, luteolin 7-O-β-D-glucopyranoside, luteolin-7-0-(600-O-acetyl)-β-D-glucopyranoside, acacetin-7-O-β-D-glucuronide, apigenin-6,8-di-C-β-D-glucopyranoside, quercetin-7-O-(600-O-acetyl)-β-D-glucopyranoside, quercetin 7-O-β-D-glucopyranoside, isorhamnetin, umbelliferone, daphnoretin (Zhang et al. 2016).

7.3.2 *Alkaloids*

Flowers: 7,8-Dimethyl pyrazino[2,3-g]quinazolin-2,4-(1H,3H) dione, guanosine, safflospermidine A, safflospermidine B, N1,N5,N10-(Z)-tri-p-coumaroylspermidine, N1,N5,N10-(E)-tri-p-coumaroylspermidine, N1,N5-(Z)-N10-(E)-tri-p-coumaroylspermidine (Zhang et al. 2016), safflowerine A (Huang et al. 2017).

Seeds: N-Feruloylserotonin, N-(p-coumaroyl)serotonin, N-(p-coumaroyl)serotonin-O-β-D-glucopyranoside, N-[2-(5-hydroxy-1H-indol-3-yl)ethyl]ferulamide, N-[2-(5-hydroxy-1H-indol-3-yl)ethyl]-p-coumaramide, N,N0-[2,20-(5,50-dihydroxy-4,40-bi-1H-indol-3,30-yl)diethyl]-di-p-coumaramide, N-[2-[30-[2-(p-coumaramido)ethyl]-5,50-dihydroxy-4,40-bi-1H-indol-3-yl]ethyl] ferulamide, N,N0-[2,20-(5,50-dihydroxy-4,40-bi-1H-indol-3,30-yl)diethyl]-diferulamide, N-[2-[5-(β-D-glucosyloxy)-1H-indol-3-yl)ethyl]-pcoumaramide, N-[2-[5-(β-D-glucosyloxy)-1H-indol-3-yl)ethyl]ferulamide, serotobenine, N-feruloyltryptamine, N-(p-coumaroyl)tryptamine (Zhang et al. 2016).

7.3.3 *Polyacetylenes*

Flowers: 4,6-Acetonide-8Z-decaene-4,6-diyne-1-O-β-D-glucopyranoside, 4,6--decadiyne-1-O-β-D-glucopyranoside, (8Z)-decaene-4,6-diyne-1-O-β-D-glucopyranoside, (8Z)-decaene-4,6-diyne-1-ol-1-O-β-D-glucuronyl-(1″-2″-β-D-glucopyranoside, (2E,8Z)-decadiene-4,6-diyne-1-ol-1-O-β-D-glucopyranoside, (2E,8E,10E)-tridecatriene-4,6-diyne-1,12,13-triol-1-O-β-D-glucopyranoside, (2E)-tetradecaene-4,6-diyne-1,10,14-triol-1-O-β-D-glucopyranoside, (2E,8E)-tetradecadiene-4,6-diyne-1,12,14-triol-1-O-β-D-glucopyranoside, (2Z,8Z)-tetradecadiene-4,6-diyne-1,12,14-triol-1-O-β-D-glucopyranoside, (2Z,8E)-tetradecadiene-4,6-diyne-1,12,14-triol-1-O-β-D-glucopyranoside, (2E,8Z)-tetradecadiene-4,6-diyne-1,12,14-triol-1-O-β-D-glucopyranoside, (2E,8E)-tetradecadiene-4,6-diyne-1,11,14-triol, (8E)-decaene-4,6-diyne-1-O-β-D-glucopyranoside (Zhang et al. 2016), (8Z)-decaene-4,6-diyne-1,10-diol-1-O-β-d-glucopyranoside, (8S)-deca-4,6-diyne-1,8-diol-1-O-β-d-glucopyranoside (Baek et al. 2020b) bidenoside C (Baek et al. 2020b) (5R)-5-acetoxy-8,10,12-tetradecatriyne-1-O-β-D-glucopyranoside, (2Z)-decaene-4,6,8-triyne-1-O-β-D-glucopyranoside, (8Z)-1-[(3-O-β-D-glucosyl)-isovaleroyloxy]-8-decaene-4,6-diyne, (8Z)-decaene-1-isovaleroyloxy-4,6-diyne-10-O-β-D-glucopyranoside, and (2E,8E)-decadiene-4,6-diyne-1-O-β-D-glucopyranoside (Li et al. 2021).

Seeds: 1-Tridecene-3,5,7,9,11-pentayne, 11Z-trideca-1,11-diene-3,5,7,9-tetrayne, 11E-trideca-1,11-diene-3,5,7,9-tetrayne, 3E-trideca-1,3-diene-5,7,9,11-tetrayne, 3Z,11Z-Trideca-1,3,11-triene- 5,7,9-triyne, 3Z,11E-trideca-1,3,11-triene-5,7,9-triyne, 3E,11E-trideca-1,3,11-triene- 5,7,9-triyne, 3E,5Z,1lE-trideca-1,3,5,11-tetraene-7,9-diyne, 3Z,5E,1lE-trideca-1,3,5,11-tetraene-7,9-diyne, 3E,5E,1lE-trideca-1,3,5,11-tetraene-7,9-diyne (Zhang et al. 2016).

7.3.4 Organic Acids

Flowers: p-Coumaric acid, p-hydroxybenzoic acid, succinic acid, 4-O-β-D-glucopyranosyloxy-benzoic acid, 4-O-β-D-glucosyl-trans-p-coumaric acid, 4-O-β-D-glucosyl-cis-p-coumaric acid (Zhang et al. 2016), (2E,4E)-dihydrophaseic acid methyl ester-3-O-D-glucopyranoside, (2Z,4E)-dihydrophaseic acid methyl ester-3-O-D-glucopyranoside (Baek et al. 2020a), ethyl 3-O-caffeoylquinic acid (Huang et al. 2017).

Seeds: Oleic acid, linoleic acid, palmitic acid, stearic acid, myristic acid, palmitoleic acid, linolenic acid, caffeic acid, ferulic acid, sinapic acid (Zhang et al. 2016).

7.3.5 Others Compounds

Flowers: Uridine, adenosine, adenine, thymine, uracil, dihydrophaseic acid 3-O-β-D-glucopyranoside, roseoside, sitosterol, syringin, methyl-3-(4-O-β-D-glucopyranosylphenyl) propionate, (−)-4-hydroxybenzoic acid-4-O-[60-O-(20 0-methylbutyryl)-β-D-glucopyranoside], methyl-3-(4-O-β-D-glucopyranosyl-3-

methoxyphenyl) propionate, 2,3-dimethoxy-5-methylphenyl-1-O-β-D-glucopyranoside, 2,6-dimethoxy-4-methylphenyl-1-O-β-D-glucopyranoside, ethyl-3-(4-O-β-D-glucopyranosyl-3-methoxyphenyl), propionate, ethylsyringin, methylsyringin (Zhang et al. 2016), safflower glucopyranoside A, 5-(hydroxymethyl)-2-furancarboxaldehyde, benzyl-O-β-D-glucopyranoside (Huang et al. 2017).

Seeds: Coniferyl alcohol, sinapyl alcohol, secoisolariciresinol, matairesinol, arctigenin, trachelogenin, (15α,20R)-dihydroxypregn-4-en-3-one 60-O-acetyl-20-β-cellobioside, matairesinol 40-O-β-D-apiofuranosyl(1–2)-β-D-glucopyranoside (Zhang et al. 2016) stigmasterol (Chaudhary et al. 2019).

7.4 Pharmacological Activities

7.4.1 Anticoagulant Activity

Safflower seeds extract inhibited ADP-induced human platelet aggregations and further formation of blood clot, providing scientific evidence of use of safflower in removal of blood stasis in traditional Chinese medicine (Lu et al. 2021). Similar finding was observed by Wang et al, showing safflower to affect intrinsic coagulation system to inhibit blood coagulation (Wang et al. 2018). It was also reported that 6-hydroxykaempferol glycoside was obtained in safflower showing antiplatelet aggregative effects (Fan et al. 2009).

7.4.2 Antihypertensive and Cardioprotective Activity

Study had found that hydroxysafflor yellow A, a component of safflower on intravenous administration, dose dependently reduced the mean arterial pressure and heart rate in both normotensive and spontaneously hypertensive rats. The effect seemed to be due to the stimulation of calcium-activated potassium channels and ATP-sensitive potassium channels (Nie et al. 2012). Vascular relaxant effect of hydroxyl safflower A was observed by Bai et al in pulmonary artery of Wister rats by activation of voltage gated potassium channel in vascular smooth muscle cells. This highlights the potential use of hydroxysafflor A in pulmonary arterial hypertension (Bai et al. 2012). Further this ingredient also reduced pulmonary arterial hypertension in monocrotaline-induced hypertensive rats mainly by suppressing inflammation and reduction of oxidative stress (Han et al. 2016). Previous finding showed that safflower yellow, by influencing plasma renin and angiotensin II level, showed hypotensive activity in spontaneously hypertensive rats (Liu et al. 1992). Clinical trial of safflower seeds extract on healthy human volunteers had shown that extract decreased arterial stiffness and lowered blood pressure, vascular cell adhesion protein-1, and low-density lipoprotein level (Koyama et al. 2008). This study was also supported by similar finding by Di et al in mice, showing regulation of vascular tone by safflower injection (Di and Chang 2007). Moreover, N-(p-coumaroyl) serotonin and N-feruloylserotonin, major polyphenols in safflower seeds, had shown to relax femoral arteries. This outlines possible anti-artherogenic mechanism of those polyphenols (Takimoto et al. 2011). The improvement of arterial stiffness and vascular aging as seen in the clinical trial conducted in safflower seed extract further supports the above findings (Suzuki et al. 2010). Some studies also found that safflower can help to protect ischemia-reperfusion injury of the myocardium, cerebrum, and renal tissue by altering molecular regulator like bcl-2, caspase3, and Bax. It was also found to alleviate pulmonary hypertension by induction of endoplasmic reticulum stress pathway in rats (Fan et al. 2012). Safflower injection was also found to ameliorate pressure in pulmonary and carotid artery as well as right ventricular hypertrophy index and helped in remodeling of pulmonary arteries (Chen et al. 2021). Similar reduction in pulmonary hypertension caused during hypoxic state was shown by safflower injection in previous study as well (Zeng et al. 2009). Combination of safflower injection together with alprostadil and sildenafil also seems to show a significant benefit in patients with pulmonary hypertension (Yang et al. 2017). However in one study, safflower oil seemed to speed up renal injury in stroke-prone spontaneously hypertensive rats in comparison to soybean oil (Miyazaki et al. 2000).

7.4.3 *Antioxidant Activity*

Oil of safflower seed contains greater quantity of linoleic acid containing tocopherol, giving it strong antioxidant potential. Oil from safflower seed is considered nutritionally similar to olive oil with antioxidant capacity, which is believed to help in reduction of cholesterol in blood, hence preventing cardiovascular diseases (Menegaes and Nunes 2020). Safflower seed extract showed a remarkable free radical scavenging potential in ferric reducing antioxidant power assay and oxygen radical scavenging assay in a dose-dependent manner (Yu et al. 2013). Moreover, isolated component such as hydroxysafflor yellow A, B, and C, safflor A, and carthamine demonstrated to have antioxidant potential in them (Adamska and Biernacka 2021). Similar strong antioxidant activity was also observed in petals extract of safflowers (Hiramatsu et al. 2009). Honey from safflower flower had shown a strong antioxidant activity on in vitro evaluation by DPPH and ABTS methods and in vivo in Raw 264.7 cells. Honey increased the expression of gene responsible for endogenous antioxidant potential like gene like TXNRD, HO-1, and NQO-1 significantly (Sun et al. 2020a). Serotonin derivatives isolated from safflower cake methanol extract was also found to had strong antioxidant activity (Zhang et al. 1997).

7.4.4 Neurological Disorder

Petals extract and carthamin, a major active constituent of safflower, were found to have neuroprotective effects on cells and animal model of mice and rats (Hiramatsu et al. 2009). Similar neuroprotection was shown by flavonoid rich extract in Parkinson's disease model of cells, mouse, and rats, mainly attributed to its rich anti-inflammatory potential (Ablat et al. 2016; Ren et al. 2016; Lei et al. 2020). Zhang et al found that leaf of safflower ameliorated cognitive function in Alzheimer diseased mice model mainly attributed to its antioxidant and anti-inflammatory potential and excessive astrocyte activation (Zhang et al. 2021). Hydroxysafflower A, one of the most exploited compounds from safflower, had also shown strong neuroprotective effect in cerebral reperfusion-injured mice with its anti-inflammatory potential, inhibiting TLR4 signaling (Lv et al. 2015). This neuroprotection in mice might also be due to its potential to reduce protein oxidation, nitration, controlling 12–15-lipoxinase, and protecting blood brain barrier from interruption (Sun et al. 2012). Cytoprotection was also observed in methylglyoxal induced injury in human microvascular endothelial cell line mainly by its antiglycation effect (Li et al. 2013). It also showed protection in focal cerebral ischemia mainly by controlling cross signaling between JAK2/STAT3 and SOCS3 pathways (Yu et al. 2020). Hydroxysafflower A also reduced effects of brain injuries mainly by favoring superoxide dismutase, ATPase, tissue plasminogen activator while inhibiting production of plasminogen-1 activator inhibitor and malondialdehyde in adjacent tissues (Adamska and Biernacka 2021). This compound was also useful in cerebrovascular injuries due to heat stress as it stimulated proliferation of nerve stem cells while inhibiting cells death due to apoptosis and autophagy (Adamska and Biernacka 2021). This compounds was found to retard synthesis of dopamine and resulting intracellular degradation of brain cells in Parkinson's disease by regulation of α-synuclein (Adamska and Biernacka 2021).

A qualitative electronic survey among 1024 participant from Saudi Arabia had shown that more than three fourth of population use safflower to treat psychological problems (Albaiz 2022). Its uses also seemed to be effective in controlling opiate withdrawal syndrome, and the result was comparable to fluoxetine in mice (Abbasi Maleki 2016).

7.4.5 Anti-melanogenic Activity

Screening of compounds isolated from *C. tinctorius* showed strong inhibition of tyrosinase and diminished production of melanin (Nagatsu et al. 2000). Hydroxysafflower yellow A was also found to be effective in management of hyper/hypo pigmentation. It was found to alter tyrosinase activity by forming complex retarding production of melanin (Adamska and Biernacka 2021).

7.4.6 Effect on Bone Disorders

Safflower yellow pigment on in vitro experiment found to be beneficial in treatment of bone fractures. The benefit was due to augmented angiogenesis and differentiation of bone cell, by upregulation of HIF-1α, VEGF, Ang-2, ALP, Runx2, and OPN-1, through alteration of pVHL/HIF-1α/VEGF signaling (Adamska and Biernacka 2021). Polysaccharide isolated from safflower was also found to be beneficial in treatment of thigh bone head osteonecrosis resulted from prolong steroid use. This activity was due to reduced apoptosis guided by increased expression of Bcl-2 and reduced Bax and caspase-3 proteins (Adamska and Biernacka 2021).

7.4.7 Effects on the Respiratory System

Hydroxy safflower yellow A was found to reduce platelets aggregation in lungs of rats exposed to car exhaust fume by decreasing vessel permeability and platelets count. This compound also alleviated acute lung injury, asthma, chronic obstetric pulmonary disease, and asthma caused by ovalbumin in guinea pigs. Overall this compound had shown protective effect in respiratory system (Adamska and Biernacka 2021).

7.4.8 Effects on the Digestive System

Leaves extract of safflower was found to protect liver from tuberculosis drugs induced injuries by reducing AST, ALT, and total bilirubin. Injection of safflower dried flowers together with *Salvia miltiorrhiza* roots (Danhong injection) was found to alleviate salicylic acid induced gastric mucosal injuries. Injection was also found to reduce proteolytic enzyme pepsin and reactive oxygen species in gastric mucosa. Hydroxysafflower yellow A obtained from safflower was found to hinder fibrosis of liver and protect liver from pathogenic changes (Adamska and Biernacka 2021).

7.4.9 Immunostimulatory Activity

Methanol extract of safflower leaf had shown stimulation of spleen lymphocyte proliferation and nitric oxide production while reducing tumor cell viability demonstrating immune-stimulation (Lee et al. 2008). Polysaccharides from safflower, SF1 and SF2, induced proliferation of B cell and IgM production comparable to those induced by LPS. Both of them also stimulated production of NO in macrophages while SF1 also increased IL-1, IL6, and TNF production comparable to LPS stimulation. These finding demonstrated immunomodulation activity of those constituents (Wakabayashi et al. 1997). Another study by Lee et al showed that safflower leaf, when feed together with food, increased in immunity of chickens (Lee et al. 2007). Major bioactive constituent of safflower, hydroxysafflower A, had prevented anaphylaxis in mice mainly by inhibition of mast cell degranulation by impeding Ca^{2+} transport, and secretion of chemokines and cytokines (Liu et al. 2018).

7.4.10 Antitumor Activity

Safflower polysaccharide control proliferation CTL cell and NK cell line of mice demonstrating its antitumor potential (Shi et al. 2010). These polysaccharides also seemed to be effective in inhibiting metastasis of MCF-7 breast cancer (Luo et al. 2015). Safflower polysaccharide further seemed to induce apoptosis and reduce proliferation in hepatocellular carcinoma by P38MAK pathway (Sun et al. 2020b). Inhibition of growth of tongue squamous cell carcinoma by regulating Bcl2, COX2, Bax, and cleave caspase3 was also shown by safflower polysaccharides (Zhou et al. 2018). Combination of safflower yellow B together with doxorubicin seems to enhance antitumor property in MCF-7 cells (Lin et al. 2021). Safflower yellow also seems to inhibit metastasis of breast cancer by inhibition of invadopodia formation (Fu et al. 2016). Safflower A seems to affect tumor capillary angiogenesis of BGC-823 tumor showing potential anticancer effects (Xi et al. 2012). Serotonin derivatives isolated from this plant had shown potent inhibition of phorbol-12myristate-13-acetate induced Epstein-Barr virus early antigen activation, demonstrating antitumor potential (Nagatsu et al. 2000). Hydroxysafflower yellow A inhibited liver cancer in mice by blocking ERK/MAPK and NF/KB HSYA signaling and p38MAPK phosphorylation. Besides, it also improved immunity thereby reduced viability, proliferation, and migration of HepG2 tumor cells. Moreover, in malignant esophageal cancer cell proliferation, migration was blocked increasing apoptosis. Chemotherapy resistance in mice ovarian cancer was reverted by treatment with hydroxysafflower yellow by opening MAPK signaling in drug-resistant cells. Hydroxysafflower yellow B also inhibited proliferation and survival of human breast cancer cells by arresting cell growth in S phage. This activity was observed due to downregulation of cyclin D1, cyclin E, CDK2, p-PI3K, PI3K, AKT, and Bcl-2 proteins (Adamska and Biernacka 2021).

7.4.11 Anti-arthritis Activity

Safflower seed extract and isolated compounds were found to mitigate cartilage destruction by mediating NF-KB signaling in osteoarthritis (Han et al. 2021). Safflower yellow was found to prevent cartilage degeneration in osteoarthritis by inhibition of inflammation and protection of chondrite mainly by regulation of NF-κB/SIRT1/AMPK and ER stress signaling pathways in rat (Wang et al. 2020). Hydroxysafflower A was found to inhibit inflammation and cholesterol metabolism in cell to control osteoarthritis (Ju et al. 2020).

7.4.12 Anti-inflammatory Activity

Petal aqueous extract and safflower A from *C. tinctorius* demonstrated anti-inflammatory potential by showing inhibition of NO and PGE2 production in RAW264.7 cells through downregulation of iNOS and COX2 expressions (Wang et al. 2010). Hydroalcohol extract of safflower and kaempferol derivative obtained from it had also shown anti-inflammatory potential in animal model of carrageenan and xylene induced ear edema model (Wang et al. 2014). Strong anti-inflammatory and antioxidant action was seen in safflower honey, collected from *Apis mellifera* L. colonies, in Raw 264.7 cells, represented by downregulation of inflammatory gene and increased expression of gene responsible for indigenous antioxidant potential (Sun et al. 2020a). Safflower oil contains high amount of vitamin E (77 times the alpha tocopherol and 100 times the gamma tocopherol than in coconut oil), and a study had found that vitamin E can downregulate expression of intracellular adhesion molecule (ICAM-1) and vascular cell adhesion molecule (VCAM-1), responsible for progression of inflammatory diseases dose dependently. This outlined the possibility that the anti-inflammatory activity of safflower oil might be related to its vitamin E content (Masterjohn 2007). Safflower yellow was found to have anti-inflammatory activity in LPS-stimulated BV2 microglia by converting inflammatory M1 phenotype to anti-inflammatory M2 (Yang et al. 2016). Moreover, aqueous extract of safflower petals and serotonin derivatives in safflowers such as acacetin, cosmosiin, N-feruloyl serotonin, and N-(p-coumaroyl) serotonin had shown strong anti-inflammatory activity in LPS stimulated Raw 264.7 cells. The activity was due to inhibition of production of nitrous oxide and pro-inflammatory cytokines (Wang et al. 2011; Kim et al. 2015). Ointment was also developed by Rakhimov et al, to exploit anti-inflammatory potential of its flowers (Rakhimov et al. 2018). In addition, anti-inflammatory potential of safflower injection, popular traditional remedies in China, was demonstrated in Raw 264.7 cells by Liao et al. (2019). A serotonin derivative isolated from safflower, moschamine, reduced the expression of COX-2, mPEGES-1, iNOS, IL-6, and IL-1β in Raw 264.7 cell showing its strong anti-inflammatory potential (Jo et al. 2017).

7.4.13 Effects on Reproductive Health

Study conducted by Nasiri et al had shown that testosterone level and sperm parameter can be improved by safflower oil supplementation in type 2 diabetic mice model. This was resulted from improvement in leptin, zinc, insulin resistance, inflammation, oxidative stress within body, and increased expression of gene responsible for testosterone synthesis (Nasiri et al. 2021). A similar affirmative finding was obtained by Bahmanpour et al previously on aqueous extract of safflower showing significant improvement in sperm morphology, motility, and count in extract treated male rats (Bahmanpour et al. 2012). However, contradictory results showing that aqueous extract of safflower to have reverse effect causing damage to testicular tissues, evidenced by formulation of multinucleated giant cells in germinal epithelium, decreased seminiferous tubule diameter, and maturation arrest, was observed in NMRI mice model in another study (Mirhoseini et al. 2012).

Hydroxysafflor A injection had shown reduced cyst in female mice with polycystic ovarian syndrome, while it also regulated ovarian cycle. The result was accompanied by reduction of testosterone, follicle stimulating hormone, and increased level of progesterone, luteinizing hormone, anti-Müllerian hormone and estradiol in blood (Adamska and Biernacka 2021).

7.4.14 Anti-obesity Activity

Study had shown that abdominal obesity was reduced on oral administration of safflower oil to male Wister rat model, undergoing aerobic exercise, significantly compared to control (da Silva Pérez et al. 2022). This finding was in accord with previous research by Zhang et al. (2010). They found that safflower feeding ameliorated high-fat diet induced obesity by altering adipogenic gene such as Orexin, Ghrelin, and PPARα (Zhang et al. 2010). Improvement of diet-induced obesity and blood lipid profile was also observed by Crescenzo et al. (2015) on rat while feeding safflower seed oil (Crescenzo et al. 2015). This finding was further supported by finding of Guo et al. (2020), showing reduction of total cholesterol and triglyceride in serum and liver of rat on oil feeding (Guo et al. 2020). Increased expression of LDL receptor and 7α-hydroxylase, responsible for hepatic uptake and biliary excretion, was observed in rat by Sato et al. (2000), indicating potential of safflower oil in alleviating exogenous hypocholesteremia (Sato et al. 2000). Several studies on evaluation of safflower yellow showed anti-obesity effect on high-fat diet-fed mice by improving liver functions, glucose metabolism, insulin sensitivity, peripheral leptin resistance, and increased PPARγ activity (Yan et al. 2020a, b; Lyu et al. 2022). Defatted seed extract of safflower also inhibited adipogenesis in 3 T3-L1 preadipocyte and improved lipid profile in high fat diet fed C57BL/6 J mice model (Hwang et al. 2016). This finding was further supported by study that safflower seed extract reduced lipid deposition in 3 T3-L1 adipocyte significantly compared to control (Yu et al. 2013). A dihydrophasic acid glycoside with strong antiadipogenic activity was isolated from florets of this plant, showing potential of this plant for novel antiadipogenic agent (Baek et al. 2020a). Studies on hydroxysafflower A had proved its effectiveness as anti-obesity agents in mice and rats. This activity was shown mainly as a result of improved function of digestive tract owing to change of intestinal microflora, alleviation of insulin resistance, reduced inflammation, increased expression of antioxidant enzyme in the liver, and synthesis of hormone sensitive lipase responsible for adipocyte proliferation (Adamska and Biernacka 2021).

However, some contradictory finding was also observed by Takeuchi et al showing increased fat accumulation and Santana et al. showing increased cholesterol and LDL-cholesterol in male Wister rat with metabolic syndrome (Takeuchi et al. 1995; Santana et al. 2017).

7.4.15 Antidiabetic Activity

Hydroxysafflower A lowered apoptosis and protects any damage in organ and tissues caused by diabetic complications. It also protected pancreatic β cells from damage caused due to oxidative stress resulting from high blood glucose level. Administration of this compound had shown reduced fasting blood glucose, insulin resistance, LDL, and balanced lipid metabolism. Kidney fibrosis induced due to diabetic complication was also reduced by hydroxysafflower A. Methanol extract of safflower flowers, when given to diabetic rat, reduced symptoms of pancreas dysfunction. The presence of diverse antioxidant compounds in safflower helped to protect pancreatic cells in those diabetic rats (Adamska and Biernacka 2021). Safflower oil helped to increase weight loss and/or glycemic control in women with type 2 diabetes (Norris et al. 2009). Furthermore extract of safflower significantly reduced blood glucose level while increased insulin level in alloxan-induced diabetic rats (Qazi et al. 2014).

7.5 Clinical Studies

Safflower oil on double-blinded, randomized, placebo-controlled trail in 67 patient aged 30–63 year with metabolic syndrome for a 12-week period had shown that the tested group reduced waist circumference, blood pressure, fasting blood sugar, and insulin resistance significantly. Moreover, adiponectin level got increased. This justified beneficial effects of oil on abdominal obesity, blood pressure, and insulin resistance (Ruyvaran et al. 2022). Pilot clinical trial conducted by Koyama et al on 20 healthy male volunteers followed by a large-scale study in 27 males revealed that safflower seed extract supplementation to decreased cardiovascular risk by protecting arterial stiffening, decreasing oxidative stress and pro-inflammatory reaction without any side effect (Koyama et al. 2008; Suzuki et al. 2010). Another trial of "safflower yellow" injections on 448 patients to find out safety and effectiveness in coronary heart diseases like angina pectoris and Xin-blood stagnation syndrome revealed significant benefit of safflower yellow compared to control groups (Qiong et al. 2005). Randomized controlled trial in 127 patients with acute exacerbation of chronic obstructive pulmonary diseases revealed that safflower yellow relieved dyspnea, duration of mechanical ventilation, and period of hospital stay in test subject significantly (Li et al. 2019). A double-blinded study of 40 postmenopausal women in South Korea had showed that safflower tea obtained by using seeds helped to maintain bone health and also reduced incidence of degenerative diseases. Strong antioxidant activity of polyphenol, including serotonin derivatives present in safflower tea was believed to shown those benefits (Cho et al. 2011). An evaluation of safflower peony ointment in 46 healthy volunteers to identify tolerability on cosmeceutical application had shown that the product to be safe for application with incidence of minor side effect in very low population (2.17%) (Lu et al. 2013). A trial on 140 patents in catabolic states obtaining total parenteral nutrition showed that safflower oil emulsion was safe and effective as source of energy when given as emulsion (Wong and Deitel 1981). Safety of emulsion in total parenteral nutrition was further confirmed by Tabrett et al in 23 patients undergoing total parenteral nutrition (Tabrett and Phillips 1982). Evaluation of safflower yellow on acute cerebral infraction among 108 subjects showed significant benefit by decreasing National Institute of Health Stroke Scale, TNF-α, IL-1β, IL-6, and improving hemorheological index (Li et al. 2015).

7.6 Studies Related to Safety and Toxicity

Biochemical evaluation of this plant grown in contaminated area had shown that heavy metals such as mercury and selenium got concentrated in seed of this plant, posing threat for human consumption (Menegaes and Nunes 2020). Carthamus red obtained from safflower up to 200 mg/kg was found to be safe in rat model (Wu et al. 2013). However, long-term administration of safflower flower extract at higher dose (25 and 50 mg/kg) for 35 days showed negative impact on sperm parameters of partially sterile male rats (Bahmanpour et al. 2012). Similar adverse impact on spermatogenesis was observed by Mirhoseini et al. on the treatment of rat with 200 mg/kg of *C. tinctorius* extract. They noticed that extract reduced epithelium vacuolization, sloughing of germs, and detachment. It formed multinucleated giant cells in germinal epithelium and also decreased seminiferous tubule diameter and epithelium height and caused maturation arrest (Mirhoseini et al. 2012). Some also reported pharyngitis and nosebleeds as adverse event to this extract (Adamska and Biernacka 2021). Type 1 allergic reaction was also reported on injection of safflower (Zhang et al. 2012) while intraperitoneal injection (at dose of 180 mg/kg) caused changed in kidney and liver functions (Adamska and Biernacka 2021).

7.7 Conclusions and Future Recommendations

In conclusion, safflower has diverse range of uses owing to its health benefit, food value, and as a coloring agent. Several evidence of health benefits as explained above give sufficient ground to develop safflower as nutraceutical and functional food. It can also serve as source of lead molecule for treatment of different health condition. Further, in the context of increased concern about the use of synthetic coloring agent in food industries and further restriction posed by the European Union and World Health Organization, safflower natural dye can serve as suitable alternatives from nature.

References

Abbasi Maleki S (2016) Effect of ethanolic extract of safflower on naloxone-induced morphine withdrawal signs in mice. Future Nat Prod 2(2):9–15

Ablat N, Lv D et al (2016) Neuroprotective effects of a standardized flavonoid extract from safflower against a rotenone-induced rat model of Parkinson's disease. Molecules 21(9):1107

Adamska I, Biernacka P (2021) Bioactive substances in safflower flowers and their applicability in medicine and health-promoting foods. Int J Food Sci 2021:6657639

Albaiz AS (2022) The use of safflower (Carthamus tinctorius) in treating depression and anxiety. Cureus 14(2):e22278

Asgarpanah J, Kazemivash N (2013) Phytochemistry, pharmacology and medicinal properties of Carthamus tinctorius L. Chin J Integr Med 19(2):153–159

Baek SC, Lee BS et al (2020a) Discovery of dihydrophaseic acid glucosides from the florets of Carthamus tinctorius. Plan Theory 9(7):858

Baek SC, Yi SA et al (2020b) Anti-adipogenic polyacetylene glycosides from the florets of safflower (Carthamus tinctorius). Biomedicine 9(1):91

Bahmanpour S, Vojdani Z et al (2012) Effects of Carthamus tinctorius on semen quality and gonadal hormone levels in partially sterile male rats. Korean J Urol 53(10):705–710

Bai Y, Lu P et al (2012) Hydroxysafflor yellow a (HSYA) from flowers of Carthamus tinctorius L. and its vasodilatation effects on pulmonary artery. Molecules 17(12):14918–14927

Bai X, Wang WX et al (2020) Therapeutic potential of hydroxysafflor yellow A on cardio-cerebrovascular diseases. Frontiers in Pharmacology. Sep 29;11:01265. https://doi.org/10.3389/fphar.2020.01265.

Chaudhary M, Verma V et al (2019) In vitro antiacne and antidandruff activity of extracted stigmasterol from seed waste of safflower (Carthamus tinctorius L.). Plant Sci Today 6(sp1): 568–574

Chen A, Ding S et al (2021) Safflower injection inhibits pulmonary arterial remodeling in a monocrotaline-induced pulmonary arterial hypertension rat model. Z Naturforsch C J Biosci 76(1–2):27–34

Cho S-H, Jang J-H et al (2011) Effects of a safflower tea supplement on antioxidative status and bone markers in postmenopausal women. Nutr Res Pract 5(1):20–27

Crescenzo R, Bianco F et al (2015) Fat quality influences the obesogenic effect of high fat diets. Nutrients 7(11):9475–9491

da Silva Pérez EM, de Alencar NMN et al (2022) Effect of safflower oil (Carthamus tinctorius L.) supplementation in the abdominal adipose tissues and body weight of male Wistar rats undergoing exercise training. Food Chem (Oxf) 4:100083

Dajue L, Mündel H-H (1996) Safflower, Carthamus tinctorius L, Bioversity International

Delshad E, Yousefi M et al (2018) Medical uses of Carthamus tinctorius L.(safflower): a comprehensive review from traditional medicine to modern medicine. Electron Physician 10(4):6672

Di K-p, Chang L-g (2007) Experimental study of inhibition of safflower injection on mesenteric microvascular motion in rabbits DI. Zhongguo Zhong Xi Yi Jie He Za Zhi 27(4):339–342

Fan L, Zhao H-Y et al (2009) Qualitative evaluation and quantitative determination of 10 major active components in Carthamus tinctorius L. by high-performance liquid chromatography coupled with diode array detector. J Chromatogr A 1216(11):2063–2070. https://doi.org/10.1016/j.chroma.2008.03.046

Fan X-F, Wang X-R et al (2012) Effect of safflower injection on endoplasmic reticulum stress-induced apoptosts in rats with hypoxic pulmonary hypertension. Zhongguo Ying Yong Sheng Li Xue Za Zhi 28(6):561–567

Fu H, Wu R et al (2016) Safflower yellow prevents pulmonary metastasis of breast cancer by inhibiting tumor cell invadopodia. Am J Chin Med 44(07):1491–1506

Guo J, Chen F et al (2020) Impact of safflower seed oil on serum lipid level and immunocompetence of rats. Int J Bioautomation 24(2)

Han X, Zhang Y et al (2016) Hydroxysafflor yellow A improves established monocrotaline-induced pulmonary arterial hypertension in rats. J Int Med Res 44(3):569–584

Han SJ, Lim MJ et al (2021) Safflower seed extract attenuates the development of osteoarthritis by blocking NF-κB signaling. Pharmaceuticals 14(3):258

Hiramatsu M, Takahashi T et al (2009) Antioxidant and neuroprotective activities of Mogami-benibana (safflower, Carthamus tinctorius Linne). Neurochem Res 34(4):795–805

Huang X-X, Yan Z-Y et al (2017) Investigation of chemical constituents of safflower and their tyrosinase inhibitory activity. J Asian Nat Prod Res 21(3):248–256

Hwang E-Y, Yu M-H et al (2016) Defatted safflower seed extract inhibits adipogenesis in 3T3-L1 preadipocytes and improves lipid profiles in C57BL/6J ob/ob mice fed a high-fat diet. Nutr Res 36(9):995–1003

Jo AR, Han H-S et al (2017) Inhibitory effect of moschamine isolated from Carthamus tinctorius on LPS-induced inflammatory mediators via AP-1 and STAT1/3 inactivation in RAW 264.7 macrophages. Bioorg Med Chem Lett 27(23):5245–5251

Ju S, Tan L et al (2020) Safflower yellow a attenuates osteoarthritis via regulating inflammation and cholesterol metabolism. https://doi.org/10.21203/rs.3.rs-112207/v1

Kazuma K, Takahashi T et al (2000) Quinochalcones and flavonoids from fresh florets in different cultivars of Carthamus tinctorius L. Biosci Biotechnol Biochem 64(8):1588–1599

Kim D-H, Moon Y-S et al (2015) Serotonins of safflower seeds play a key role in anti-inflammatory effect in lipopolysaccharide-stimulated RAW 264.7 macrophages. J Plant Biotechnol 42(4):364–369

Koyama N, Suzuki K et al (2008) Effects of safflower seed extract supplementation on oxidation and cardiovascular risk markers in healthy human volunteers. Br J Nutr 101(4):568–575

Lee S-H, Lillehoj HS et al (2007) Immunomodulatory effects of dietary safflower leaf in chickens. Korean J Community Living Sci 18(4):715–724

Lee S-H, Lillehoj HS et al (2008) Immune enhancing properties of safflower leaf (Carthamus tinctorius) on chicken lymphocytes and macrophages. J Poult Sci 45(2):147–151

Lei H, Ren R et al (2020) Neuroprotective effects of safflower flavonoid extract in 6-hydroxydopamine-induced model of Parkinson's disease may be related to its anti-inflammatory action. Molecules 25(21):5206

Li W, Liu J et al (2013) Hydroxysafflor yellow A protects methylglyoxal-induced injury in the cultured human brain microvascular endothelial cells. Neurosci Lett 549:146–150

Li L-J, Li Y-M et al (2015) The value of safflower yellow injection for the treatment of acute cerebral infarction: a randomized controlled trial. Evid Based Complement Alternat Med 2015:478793

Li F, He Z et al (2017) Isocartormin, a novel quinochalcone C-glycoside from Carthamus tinctorius. Acta Pharm Sin B 7(4):527–531

Li X-J, Kang Y et al (2019) The effects of safflower yellow on acute exacerbation of chronic obstructive pulmonary disease: a randomized, controlled clinical trial. Evid Based Complement Alternat Med 2019:5952742

Li X-R, Liu J et al (2021) Polyacetylene glucosides from the florets of Carthamus tinctorius and their anti-inflammatory activity. Phytochemistry 187:112770

Liao H, Li Y et al (2019) Comparison of inhibitory effects of safflower decoction and safflower injection on protein and mRNA expressions of iNOS and IL-1β in LPS-activated RAW264.7 cells. J Immunol Res 2019:1018274

Lin K, Qin Z et al (2021) Hydroxyl safflower yellow B combined with doxorubicin inhibits the proliferation of human breast cancer MCF-7 cells. Oncol Lett 21(5):426

Liu F, Wei Y et al (1992) Hypotensive effects of safflower yellow in spontaneously hypertensive rats and influence on plasma renin activity and angiotensin II level. Yao Xue Xue Bao 27(10): 785–787

Liu R, Zhao T et al (2018) The anti-anaphylactoid effects of hydroxysafflor yellow A on the suppression of mast cell Ca2+ influx and degranulation. Phytomedicine 48:43–50

Lu L-G, Wu S-Q et al (2013) Study on tolerability for safflower peony ointment in clinical trial phase I. Zhongguo Zhong Yao Za Zhi 38(1):123–126

Lu P-H, Kuo C-Y et al (2021) Safflower extract inhibits ADP-induced human platelet aggregation. Plan Theory 10(6):1192

Luo Z, Zeng H et al (2015) Safflower polysaccharide inhibits the proliferation and metastasis of MCF-7 breast cancer cell. Mol Med Rep 11(6):4611–4616

Lv Y, Qian Y et al (2015) Hydroxysafflor yellow A exerts neuroprotective effects in cerebral ischemia reperfusion-injured mice by suppressing the innate immune TLR4-inducing pathway. Eur J Pharmacol 769:324–332

Lyu X, Yan K et al (2022) Intragastric safflower yellow and its main component HSYA improve leptin sensitivity before body weight change in diet-induced obese mice. Naunyn Schmiedeberg's Arch Pharmacol 395(5):579–591

Masterjohn C (2007) The anti-inflammatory properties of safflower oil and coconut oil may be mediated by their respective concentrations of vitamin E. J Am Coll Cardiol 49(17):1825–1826

Menegaes JF, Nunes UR (2020) Safflower: importance, use and economical exploitation. Scientia Agraria Paranaensis 1(1):1–11

Mirhoseini M, Mohamadpour M et al (2012) Toxic effects of Carthamus tinctorius L. (safflower) extract on mouse spermatogenesis. J Assist Reprod Genet 29(5):457–461

Miyazaki M, Takemura N et al (2000) Dietary docosahexaenoic acid ameliorates, but rapeseed oil and safflower oil accelerate renal injury in stroke-prone spontaneously hypertensive rats as compared with soybean oil, which is associated with expression for renal transforming growth factor-beta, fibronectin and renin. Biochim Biophys Acta 1483(1):101–110

Nagatsu A, Zhang H et al (2000) Tyrosinase inhibitory and anti-tumor promoting activities of compounds isolated from safflower (Carthamus tinctorius L.) and cotton (Gossypium hirsutum L.) oil cakes. Nat Prod Lett 14(3):153–158

Nasiri K, Akbari A et al (2021) Safflower seed oil improves steroidogenesis and spermatogenesis in rats with type II diabetes mellitus by modulating the genes expression involved in steroidogenesis, inflammation and oxidative stress. J Ethnopharmacol 275:114139

Nie P-H, Zhang L et al (2012) The effects of hydroxysafflor yellow A on blood pressure and cardiac function. J Ethnopharmacol 139(3):746–750

Norris LE, Collene AL et al (2009) Comparison of dietary conjugated linoleic acid with safflower oil on body composition in obese postmenopausal women with type 2 diabetes mellitus. Am J Clin Nutr 90(3):468–476

Qazi N, Khan RA et al (2014) Effect of Carthamus tinctorius (safflower) on fasting blood glucose and insulin levels in alloxan induced diabetic rabbits. Pak J Pharm Sci 27(2):377–380

Qiong Z, Jian-hua P et al (2005) A clinical study of safflower yellow injection in treating coronary heart disease angina pectoris with Xin-blood stagnation syndrome. Chin J Integr Med 11(3): 222–225

Rakhimov KD, Turgumbayeva AA et al (2018) Study of safety and antimicrobial, anti-inflammatory and anti-inflammatory properties ointment from the safflower extract. Bull Natl Acad Sci Rep Kazakhstan 17
Ren R, Shi C et al (2016) Neuroprotective effects of a standardized flavonoid extract of safflower against neurotoxin-induced cellular and animal models of Parkinson's disease. Sci Rep 6(1): 1–13
Ruyvaran M, Zamani A et al (2022) Safflower (Carthamus tinctorius L.) oil could improve abdominal obesity, blood pressure, and insulin resistance in patients with metabolic syndrome: a randomized, double-blind, placebo-controlled clinical trial. J Ethnopharmacol 282:114590
Santana LF, Dutra TDS et al (2017) Safflower oil (Carthamus tinctorius L.) intake increases total cholesterol and ldl-cholesterol levels in an experimental model of metabolic syndrome. Int J Cardiovasc Sci 30:476–483
Sato M, Yoshida S et al (2000) Superiority of dietary safflower oil over olive oil in lowering serum cholesterol and increasing hepatic mRNAs for the LDL receptor and cholesterol 7-α hydroxylase in exogenously hypercholesterolemic (ExHC) rats. Biosci Biotechnol Biochem 64(6): 1111–1117
Shi X, Ruan D et al (2010) Anti-tumor activity of safflower polysaccharide (SPS) and effect on cytotoxicity of CTL cells, NK cells of T739 lung cancer in mice. Zhongguo Zhong Yao Za Zhi 35(2):215–218
Sun L, Yang L et al (2012) Neuroprotection of hydroxysafflor yellow A in the transient focal ischemia: inhibition of protein oxidation/nitration, 12/15-lipoxygenase and blood-brain barrier disruption. Brain Res 1473:227–235
Sun L-P, Shi F-F et al (2020a) Antioxidant and anti-inflammatory activities of safflower (Carthamus tinctorius L.) honey extract. Foods 9(8):1039
Sun Y, Li SZ et al (2020b) Safflower polysaccharide inhibits hepatocellular carcinoma cells through P38MAPK activation. TMR Cancer 4(2):7
Suzuki K, Tsubaki S et al (2010) Effects of safflower seed extract on arterial stiffness. Vasc Health Risk Manag 6:1007
Tabrett DG, Phillips GD (1982) A clinical evaluation of safflower oil emulsion in total parenteral nutrition. Anaesth Intensive Care 10(3):258–264
Takeuchi H, Matsuo T et al (1995) Diet-induced thermogenesis is lower in rats fed a lard diet than in those fed a high oleic acid safflower oil diet, a safflower oil diet or a linseed oil diet. J Nutr 125(4):920–925
Takimoto T, Suzuki K et al (2011) Effect of N-(p-coumaroyl)serotonin and N-feruloylserotonin, major anti-atherogenic polyphenols in safflower seed, on vasodilation, proliferation and migration of vascular smooth muscle cells. Mol Nutr Food Res 55(10):1561–1571
Wakabayashi T, Hirokawa S et al (1997) Immunomodulating activities of polysaccharide fractions from dried safflower petals. Cytotechnology 25(1):205
Wang C-C, Choy C-S et al (2010) Protective effect of dried safflower petal aqueous extract and its main constituent, carthamus yellow, against lipopolysaccharide-induced inflammation in RAW264. 7 macrophages. J Sci Food Agric 91(2):218–225
Wang C-C, Choy C-S et al (2011) Protective effect of dried safflower petal aqueous extract and its main constituent, carthamus yellow, against lipopolysaccharide-induced inflammation in RAW264. 7 macrophages. J Sci Food Agric 91(2):218–225
Wang Y, Chen P et al (2014) Antinociceptive and anti-inflammatory activities of extract and two isolated flavonoids of Carthamus tinctorius L. J Ethnopharmacol 151(2):944–950
Wang K-H, Li S-F et al (2018) In vitro anticoagulant activity and active components of safflower injection. Molecules 23(1):170
Wang C, Gao Y et al (2020) Safflower yellow alleviates osteoarthritis and prevents inflammation by inhibiting PGE2 release and regulating NF-κB/SIRT1/AMPK signaling pathways. Phytomedicine 78:153305
Wong KH, Deitel M (1981) Studies with a safflower oil emulsion in total parenteral nutrition. Can Med Assoc J 125(12):1328

Wu S, Yue Y et al (2013) Carthamus red from Carthamus tinctorius L. exerts antioxidant and hepatoprotective effect against CCl4-induced liver damage in rats via the Nrf2 pathway. J Ethnopharmacol 148(2):570–578

Xi S-Y, Zhang Q et al (2012) Effects of hydroxy safflower yellow-A on tumor capillary angiogenesis in transplanted human gastric adenocarcinoma BGC-823 tumors in nude mice. J Tradit Chin Med 32(2):243–248

Yan K, Wang X et al (2020a) Safflower yellow and its main component HSYA alleviate diet-induced obesity in mice: possible involvement of the increased antioxidant enzymes in liver and adipose tissue. Front Pharmacol 11:482

Yan K, Wang X et al (2020b) Safflower yellow improves insulin sensitivity in high-fat diet-induced obese mice by promoting peroxisome proliferator-activated receptor-γ2 expression in subcutaneous adipose tissue. J Diabetes Investig 11(6):1457–1469

Yang X-W, Li Y-H et al (2016) Safflower yellow regulates microglial polarization and inhibits inflammatory response in LPS-stimulated Bv2 cells. Int J Immunopathol Pharmacol 29(1): 54–64

Yang T-L, Zhang X-F et al (2017) Safflower injection combined with alprostadil and sildenafil treats the chronic pulmonary heart disease complicated with pulmonary hypertension. Chin Tradl Patent Med 40–46

Yu S-Y, Lee Y-J et al (2013) Phenolic composition, antioxidant activity and anti-adipogenic effect of hot water extract from safflower (Carthamus tinctorius L.) seed. Nutrients 5(12):4894–4907

Yu L, Liu Z et al (2020) Hydroxysafflor yellow A confers neuroprotection from focal cerebral ischemia by modulating the crosstalk between JAK2/STAT3 and SOCS3 signaling pathways. Cell Mol Neurobiol 40(8):1271–1281

Zeng H-H, Dong W et al (2009) Effect of safflower injection on pulmonary hypertension in rat during chronic hypoxia and hypercapnia. Zhongguo Ying Yong Sheng Li Xue Za Zhi 25(1): 36–40

Zhang HL, Nagatsu A et al (1997) Antioxidative compounds isolated from safflower (Carthamus tinctorius L.) oil cake. Chem Pharm Bull 45(12):1910–1914

Zhang Z, Li Q et al (2010) Prevention of diet-induced obesity by safflower oil: insights at the levels of PPAR α, orexin, and ghrelin gene expression of adipocytes in mice. Acta Biochim Biophys Sin 42(3):202–208

Zhang L, Yue YH et al (2012) Study on the type I allergic reaction of safflower injection. Chin J Hosp Pharm 17:1319–1321

Zhang L-L, Tian K et al (2016) Phytochemistry and pharmacology of Carthamus tinctorius L. Am J Chin Med 44(02):197–226

Zhang T, Zhang S et al (2021) Safflower leaf ameliorates cognitive impairment through moderating excessive astrocyte activation in APP/PS1 mice. Food Funct 12(22):11704–11716

Zhou H, Yang J et al (2018) Safflower polysaccharide inhibits the development of tongue squamous cell carcinoma. World J Surg Oncol 16(1):1–7

Chapter 8
Chrysanthemum morifolium Ramat.: A Medicinal Plant with Diverse Traditional Uses, Bioactive Constituents, and Pharmacological Activities

Jitendra Pandey, Tonking Bastola, Bhawana Dhakal, Amrit Poudel, and Hari Prasad Devkota

Abstract *Chrysanthemum morifolium* Ramat. (Family*:* Asteraceae) is a subtropical ornamental perennial herbaceous plant, which acquires a height of about 0.6–1 m. The plant is originated in China and distributed to other Asian countries such as Nepal, Korea, India, Thailand, and Japan. Both commercially and traditionally, the most important part of this plant is the flower, which is the second most valuable flowering crop in the world after the rose. Ethnomedicinally, its flower can be used to cure excessive heat in the body, eye inflammation, cephalalgia, fever, inflammation, migraine, skin infection, vertigo, eye complications (eye itching, tired eyes, blurred vision, loss of vision, redness of the eyes), influenza, angina, dizziness, coronary heart disease, chronic inflammation, colitis, stomatitis, hypertension, and a wide range of biological activities including antioxidant effect, anti-inflammatory activity, anti-obesity effect, cardioprotective effect, neuroprotective effect, anticancer activity, and antidiabetic effect has been explored due to the presence of bioactive phytoconstituents such as flavonoids and their glycosides, other polyphenols, volatiles oils, terpenoids, steroids, and polysaccharides. This chapter is focused on

J. Pandey (✉)
Department of Pharmacy, Crimson College of Technology, Pokhara University, Rupandehi, Nepal

T. Bastola
Institute of Pharmaceutical Research and Development, College of Pharmacy, Wonkwang University, Iksan, South Korea

B. Dhakal
Department of Pharmacy, Kantipur Academy of Health Sciences, Kathmandu, Nepal

A. Poudel
Department of Biodiversity and Bioresources, Satvik Nepal, Pokhara, Nepal

H. P. Devkota
Headquarters for Admissions and Education, Kumamoto University, Kumamoto, Japan

Graduate School of Pharmaceutical Sciences, Kumamoto University, Kumamoto, Japan
e-mail: devkotah@kumamoto-u.ac.jp

H. P. Devkota, T. Aftab (eds.), *Medicinal Plants of the Asteraceae Family*,
https://doi.org/10.1007/978-981-19-6080-2_8

ethnomedicinal uses, commercial importance, health beneficial effects, bioactive molecules, nutritional benefits, biological activities, clinical studies, safety, and reported toxicities of this plant.

Keywords *Chrysanthemum morifolium* Ramat. · Juhuwa · Glycosides · Volatile oils · Flavonoids · Antioxidant

8.1 Introduction

Chrysanthemum morifolium Ramat (Fig. 8.1) is an ornamental perennial herbaceous plant, which acquires a height of about 0.6–1 m (Cockshull 2019). This plant is believed to be originated in China, and it is believed to be cultivated for more than 30 centuries (Liang et al. 2014). Apart from that this plant is also native to other Asian countries such as Korea, India, Thailand, Nepal, and Japan. Commonly, this plant is known as Godawari in Nepali, chrysanthemum in English, Bahupatrika in Sanskrit, Chandramukhi in Hindi, and Juhua in Chinese language. Nowadays, it has been commercially cultivated in North and South America, Australia, South Africa, and Northeast Europe, for ornamental purposes (Cockshull 2019), and *C. morifolium* flower is the second most valuable flowering crop in the world after rose (Kalia

Fig. 8.1 Photographs of whole plants and flowers of *Chrysanthemum morifolium*

2015). The stem of this plant is sparsely enveloped with white-colored threadlike hairs. The leaves are aromatic, olive-green, ovate-lanceolate, or ovate in appearance and are subcordate at the base with an obtuse shape at the apex. The average length and width of the leaves are about 3.5–5 cm and 3–4 cm, respectively, and are subglabrous to weakly pubescent on both surfaces. Furthermore, leaves are pinnatifid to shallowly pinnatipartite in obtuse, short, lateral, and terminal lobes. Flowers have two types of arrangement, i.e., ray florets and disk or tubular florets. Ray florets have a wide range of colors such as yellow, orange, pink, lavender, white, purple, olive-green, rust, bronze, and red in flowers. On other hand, disk florets have yellow to pale greenish-yellow colored flowers with a 5-toothed corolla tube. Similarly, flowers have petals (yellow or white color) in ligulate shape and are located on the axillary or terminal portion of the stem. Furthermore, the central portion of the flower is yellow and has a tubular shape (Hu 2015; Cockshull 2019; Yuan et al. 2020).

The garden chrysanthemum blooms excellently in moist and well-drained fertile soil having moderately acidic conditions with abundant organic components and light exposure. Low exposure to sunlight may weaken the mechanical strength and girth of the plant. Also, the plant may produce very few flowers. Usually, this plant has a shallow type of root and requires regular irrigation during drought and high-temperature conditions (Shahrajabian et al. 2019). Chiefly, the large mass of these flowering plants is extensively grown in Northeastern Europe and East Asia (Yuan et al. 2020; Maddala 2021). Chrysanthemum flowers have diverse colors, mainly, red, yellow, pink, violet, and white (Maddala 2021). This genus has approximately 30 species, which are cosmopolitan in distribution (Kumar et al. 2005). As per Chinese Pharmacopeia, there are four cultivars (Chiju, Hangju, Gongju, and Boju) of Chrysanthemum in China, based on their production area (Hu 2015). In central Asian countries like China, Japan, and Korea, the leaves and flowers of this plant have extensively been used since ancient times for edible purposes. Leaves are being used as a flavoring agent in vinegar and aromatic tea. The petal and flower heads are partially boiled and mixed with soya sauce or vinegar and eaten as a salad with tofu food. Other food items such as pickles, soups, aromatic tea, and tempura are also prepared from *C. morifolium* flowers. In South Korea, flowers are used as a flavoring agent while preparing traditional rice wine (Cockshull 2019). Medicinally, the most useful part of this plant is the flower. The flower head of *C. morifolium* is popularly known as Juhua in China and it has widespread uses in Traditional Chinese Medicine (TCM), herbal tea, nutraceuticals, and food supplements. Furthermore, Juhua has also been described in Chinese Pharmacopeia since 1967 (Chang et al. 2019). Traditionally, Juhua is an effective remedy for inflammatory disease, conjunctivitis, cold, vertigo, headache, etc. (Cockshull 2019; Lu et al. 2016). However, the medicinal utilization of *C. morifolium* flower is limited to only a few Asian countries. Therefore, this chapter will focus on the exploration and dissemination of all the available scientific literature information correlated with bioactive compounds, validated biological activities, beneficial nutritional effects, traditional uses, safety, toxicity, and clinical studies of *C. morifolium* flower.

8.2 Traditional Uses

Diverse ethnomedicinal uses of *C. morifolium* flower have been documented in TCM. In TCM, *Morus alba* leaf, whole plant of *Mentha haplocalyx*, and fruits of *Forsythia suspense* are mixed with *C. morifolium* flower to prepare a decoction, which is very much effective for the eradication of exogenous pathological symptoms caused due to excessive heat and wind (fever, cough, and headache). Also, the decoction prepared by the combination of *Cassia obtusifolia* seeds, *Gentiana manshurica* rhizome, *Morus alba* leaf, and *C. morifolium* flower is a popular remedy for the treatment of pain and redness of the eyes caused due to excessive wind and liver fire. Besides that, a polyherbal Chinese formulation containing *Haliotis diversicolor* shells, *Paeonia lactiflora* roots, *C. morifolium* flower, and claws of Chinese Cats have been extensively used to overcome the problems of headache and dizziness, created due to overexpression of Liver-Yang component. Fresh juice of the chrysanthemum flower can be consumed orally to cure furunculosis and furuncle (Hu 2015; Sun et al. 2010). In TCM, there is a belief that the chrysanthemum flower can induce modest cold energy in our body which can pass through the channels that connect to several visceral organs such as the spleen, lungs, kidney, and liver. Thus, this plant is worthwhile to mitigate the initial stage of several diseases that infect the upper part of the respiratory tract, nose, throat, ears, skin, and eye (Shahrajabian et al. 2019). Consumption of *C. morifolium* petals as a salad is believed to enhance the longevity of human life (Kalia 2015). The Japanese pharmacopeia has described *C. morifolium* flower as a remedy for the treatment of eye inflammation, cephalalgia, fever, inflammation, migraine, skin infection, vertigo, and eye complications (eye itching, tired eyes, blurred vision, loss of vision, and redness of the eyes) (Kitano et al. 2011; Wang et al. 2015; Youssef et al. 2020). *C. morifolium* flowers are also consumed as a refreshing infusion which helps to promote blood circulation, alleviate headaches, boost vision, combat infections like influenza, and cure angina, dizziness, coronary heart disease, chronic inflammation, colitis, stomatitis, and hypertension. The flower decoction is efficacious to promote menstruation, acts as a washing solution for carbuncles and cancerous sores, as a liniment for swollen glands, as an enema, and as a detoxifying agent and overcomes the problems of a stomachache. The flowers are immersed with wine and used to rejuvenate the digestive and nervous systems (Kim et al. 2013; Kitano et al. 2011; Luyen et al. 2015). In South Korea, tea or infusion of *C. morifolium* flowers is taken as a stimulant which makes people awaken and alert. In Western herbal medicine, this tea is a beneficial remedy for varicose veins, nausea, immune suppression, and atherosclerosis (Sharma et al. 2011; Sinha et al. 2013; Zheng et al. 2015). Furthermore, *C. morifolium* flower heads are commonly utilized as a parasiticide and also for the treatment of Parkinson's disease, type 2 diabetes, night blindness, and tinnitus (Maddala 2021; Shahat et al. 2001). Moreover, ethnomedicinal uses of different parts of *C. morifolium* are depicted in Table 8.1.

Table 8.1 Traditional uses of different parts of *C. morifolium*

S. N.	Different parts of plant	Traditional uses
1	*C. morifolium* flower heads	Parasiticides, insect repellent, mosquito repellents, and insecticide (Maddala 2021), as a food and beverage, hypolipidemic, and cardio protective (Wang et al. 2013; Yu et al. 2013)
2	*C. morifolium* whole plant	For the treatment of early stages of ear-, throat-, eye-, and nose-related disease (Maddala 2021)
3	*C. morifolium* flower heads tea	Suppress the mental pressure and fear, protect from possible inflammation, bone depletion, and oxidative stress, act as an immune booster, and alleviate neuronal pain, heat rashes, and blurred vision (Shahrajabian et al. 2019). Tea is also effective to cure Alzheimer's disease and diabetes (Yu et al. 2021)
4	*C. morifolium* whole plant extract	Protect from skin infection (Marongiu et al. 2009), diminishes indoor air pollution (Wolverton et al. 1984)
5	*C. morifolium* roots	To treat headache (Maddala 2021)
6	*C. morifolium* petals	Used as a salad (Maddala 2021)
7	*C. morifolium* leaves	Used as a juice and applied on external wounds (Maddala 2021)
8	*C. morifolium* corolla	To mitigate eye related complications (Maddala 2021)
9	*C. morifolium* inflorescence	As an immune booster and anti-inflammatory agent (Cheng et al. 2005)
10	*Decoction of C. morifolium* flowers and leaves	To relieve stomachache (Maddala 2021)
11	*C. morifolium* flowers soaked in wine	To restore the function of nervous, digestive, and circulatory system (Maddala 2021)

8.3 Chemical Constituents

Diverse classes of bioactive compounds have been isolated from the roots, stem barks, aerial parts, and flower of the *C. morifolium.* Among them, extensive studies have been reported about the photochemistry of flowers. In this section we have reported information about different phytochemicals isolated from different parts of the *C. morifolium.*

8.3.1 Chemical Constituents from C. morifolium *Flowers*

Flavonoids

Different subclasses of flavonoids, namely flavanone (eriodictyol and eriodictyol-7-*O*-glucoside), flavones (eupatorin, apigenin, diosmetin, acacetin, and luteolin), flavonols (kaempferol, quercetin, kaempferide, chrysosphenol D, isorhamnetin, and

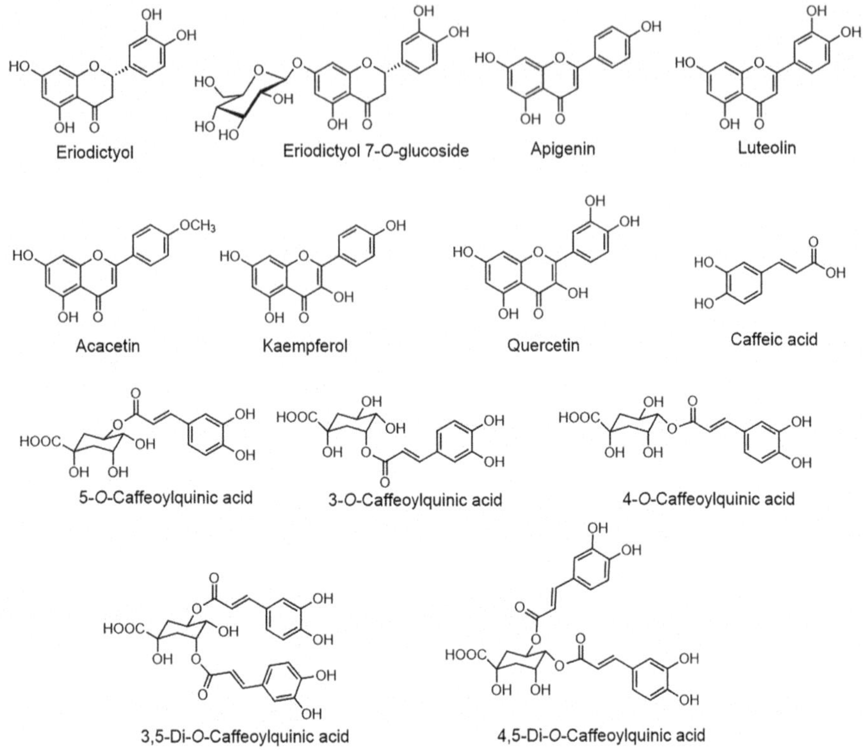

Fig. 8.2 Chemical structures of major compounds *Chrysanthemum morifolium*

chrysosphenol C), as well as their acetyl glycosides and glycosides derivatives have been identified from *C. morifolium* flowers. Luteolin is the most abundant flavonoid which is usually utilized as a marker compound in quality control analysis of its flower (Hu 2015; Lai et al. 2007; Lin and Harnly 2010; Sun et al. 2010; Wang et al. 2013). Similarly, Yuwan et al (Yuan et al. 2020) isolated a total of 60 different flavonoids, and their structure was elucidated by using NMR spectroscopy. Some of those compounds were morifonoside A, apigenin, quercitrin, diosmin, scolimoside, eupatilin, isorhamnetin, spinacetin, axillarin, bonanzin, cirsilol, artemetin, hespertin, hesperetin, naringenin, etc. The majority of those compounds were isolated in the C-7 glycosidic form of eriodictyol, acacetin, luteolin, diosmetin, and apigenin. In addition to that similar types of flavonoids were also identified in a study conducted by Chen et al. and Peng et al. (Chen et al. 2021; Peng et al. 2020). In another finding, other bioactive flavonoids, scolimoside, diosmetin-7-*O*- β -D- glucopyranoside, *vitexin-2-O*-rhamnoside, and *myricetin* were obtained from the *C. morifolium* ethanolic flower extract (Sun et al. 2010; Xie et al. 2009). Three glycosidic flavonoids, apigenin-7-*O*-beta-D-glucoside, luteolin-7-*O*-beta-D-glucoside, and acacetin-7-*O*-beta-D-glucoside, were also reported from the *C. morifolium* flower extract (Liu et al. 2001). The structures of main compounds are presented in Fig. 8.2.

Polyphenols

A total of 14 polyphenol compounds such as 4-*O*-caffeoylquinic acid, 3-*O*-caffeoylquinic acid, caffeic acid, 5-*O*-caffeoylquinic acid, 3,5-di-*O*-caffeoylquinic acid, and 4,5-di-*O*-caffeoylquinic acid were isolated from the hydromethanolic extract of *C. morifolium* flowers (Gong et al. 2019). Other studies confirmed that the flowers also contain macranthoin F, and 1,3-dicaffeoyl-epi-quinic acid, 1,4-di-caffeoylquinic acid, 1,5-di-caffeoylquinic acid, and 3,4-di-caffeoylquinic acid, 3-methyoxaloyl-1,5-di-caffeoylquinic acid, 3,4,5-tricaffeoylquinic acid, 4,5-di-caffeoylquinic acid, 4-caffeoyl-5-feruloylquinic acid (Chen et al. 2021; Hu 2015; Xie et al. 2009). Generally, the concentration of caffeoylquinic acid is high, when 70% of ray florets and 50% of tubular florets become open (Hu 2015). Furthermore, four new types of phenolic glycosides (2,6-dimethoxyl-4-hydroxymethyl-phenol-1-*O*-(6-*O*-caffeoyl)-β-D-glucopyranoside, (2S)-propane-1,2-diol 1-*O*-(6-*O*-caffeoyl)-β-D-glucopyranoside, ethylene glycol 1-*O*-(6-*O*-caffeoyl)-β-D-glucopyranoside, and butane-2,3-diol 2-*O*-(6-*O*-caffeoyl)-β-D-glucopyranoside) (Yang et al. 2019).

Terpenoids

Hu and Chen (Hu and Chen 1997) isolated four sesquiterpenes, chrysanthediacetate B, β-dictyopterol, chrysanthediacetate C, and chrysanthediol A, from the *C. morifolium* alcoholic flower extract. A new type of endoperoxysesquiterpene lactone (10a-hydroxy-1a,4a-endoperoxy-guaia-2-en-12,6a-olide) was also reported identified from the methanolic extract of its flower, which was very much effective against lipase and α-glucosidase enzymes (Luyen et al. 2013). Similarly, four other sesquiterpenes, chrysartemin A, chrysandiol, chrysartemin B, and chlorochrymorin were also discovered from its flower (Tang and Eisenbrand 1992). From several studies, a total of 68 triterpenoids such as brein, arnidiol, coflodiol, calenduladiol, 3-epicabraleadiol, dammarenediol II, faradio, erythrodiol, (24R)-saringosterol, maniladiol, faradiol R-epoxide, heliantriol A1, longispinogenin, heliantriol C, and 22-α-methoxyfaradiol) were isolated from *C. morifolium* flowers by using nonpolar solvents. Terpenoids isolated were of oleanane type (9), taraxastane-type (20), lupane-type (8), taraxerane-type (1), tirucallane-type (3), dammarane-type (10) cycloartane-type (11), ursane type (6) (Hu 2015; Ukiya et al. 2001; Yuan et al. 2020)

Volatile Components

In a previous study, a GC-MS total ion chromatogram was used to characterize the volatile components present in *C. morifolium* flowers. The study enabled the identification of 36 different volatile components such as benzene, nonane, xylene, α-pinene, camphene, camphor, D-limonene, eucalyptol, carveol, thymol, verbenol, caryophyllene, cis-ocimine, endo-borneol, r-elemene, and bornyl acetate (Yang et al. 2017). In other studies, GC-MC analysis of C. morifolium flowers confirmed the presence of similar types of volatile compounds. Some new volatile chemicals identified in that study were camphene, sabinene, 1,8-cineole, 1-phellandrene, pinocarvone, γ-curcumene, myrtenol, zingiberine, levomenol, tetracosene, valencene, hexadecanoic acid, tricosnae, trans-cadinol, β-guaiene, trans-cadinol, α-Cedrene, ionene, α and β-farnesne, β-bisabolene, 7-α-selinene, n-eicosane, β-sesquiphellendrene, n-heneicosane, α-fenchene, ortho-cymene, cineole, terpinene, α-thujone, terpinoline, β-thujone, verbenol, myrtenol, α-terpinol, β-farnesen, β-elemene, copaene, safranal, morillo, chrysanthenone, β-cubebene, cubedol, nerolidol, globulol, β-guaiene, cadinol, β-humulene, cedrol, longifolenaldehyde, germacrene, spathulenol, α-farnesene, umbellulone, cuminal, α-ionene, (+)-carvotanaacetone, carvone, berbenone, para-cymen-8-ol, methyl salicylate, linalool, fillifolone, sulcatone, eucalyptol, pseudocumene, α-phellandrene, isoborneol, para-cresol, neric acid, piperitenone, trans-α-bergamotene, cis-p-mentha-2,8-dien-1-ol, ocimene, etc. (Boukhebti et al. 2020; Kuang et al. 2018; Lawal et al. 2014; Peng et al. 2020; Sun et al. 2010; Woo et al. 2008; Youssef et al. 2020; Zhang et al. 2020a, b).

Other Chemical Compounds

In a previous study, a large complex polysaccharide (Zheng et al. 2004) and two acidic polysaccharides were reported from the aqueous extract of *C. morifolium* flowers (Zheng et al. 2006). These polysaccharides were composed of arabinose, galactose, glucose, mannitose, rhamnose, and xylose (Liang et al. 2021; Zheng et al. 2004). Also, two glycosidic derivative dendranacetylene A and 8E-decaene-4,6-diyn-1-*O*-β-D-glucopyranosyl-(1″ →2″)- -β--D-glucopyranoside were also isolated from the flowers of this plant (Li et al. 2021). Another study confirmed that its flower also contains carotenes, α-carotene, β-carotene, lutein, and β-cryptoxanthin (Chen et al. 2021). A numbing principle having chemical formula of [N-isobutyl-6-2-thienyl -2E,4E- hexadienamide] (Shahat et al. 2001), new polyacetylene glucoside dendranacetylene A (Li et al. 2021), phenylpropanoid types of two neuro-protective arylnaphthalene (chrysanthelignanoside A and B) and other twelve known compounds ((7R,7′R,8S,8′S)-neo-olivil-4-*O*-β-D-glucopyranoside, urolignoside, (7R,7′R,8S,8′S)-neo-olivil-9-*O*-β-D-glucopyranoside, rosin, secoisolariciresinol-4-*O*-β-D-glucopyranoside, sinapoyl-4-*O*-β-D-glucopyranoside, 5-*O*-β-D-glucopyranosyl-2-hydroxy benzoic acid methyl ester, butane-2,3-diol 2-*O*-(6-*O*-caffeoyl)-β-D-glucopyranoside, ethylene glycol 1-*O*-(6-*O*-caffeoyl)-β-D-glucopyranoside, (2S)-propane-1,2-diol 1-*O*-(6-*O*-caffeoyl)-β-D-glucopyranoside, 2,6-dimethoxyl-4-hydroxymethyl-phenol 1-*O*-(6-*O*-caffeoyl)-β-D-glucopyranoside, and viridoside) were also reported from its ethanolic flower extract (Yang et al. 2019; Yuan et al. 2020). Besides, sufficient amount of essential metals (zinc, calcium, and iron) and ascorbic acid have also been reported from the flower extract (Cheng and

Yuhua 2007). Moreover, other less abundantly present bioactive compounds in its flower include alkaloids, phytosterols (β-sitosterol, campesterol, β-amyrin, stigmasterol, and α-amyrin), coumarins, neolignans, lignans, and bisabolol (Hu 2015; Lu et al. 2016) along with physcin, (Z)-B-ring-homotonghaosu, edomin, (E)-B-ring-homotonghaosu, chrysophanol, edomin, pviridoside, anthraquinones, chrysandiol, (3R,7R,9R)-3,9-dihydroxygermacra-4(15),10(14),11(12)-triene, cyperusol, jinsidajuol A and B, β-dictyopterol C, hrysanthediacetate C, 10α-hydroxy-1α,4-α-endoperoxy-guaia-2-en-12,6α-olide, chrysanthediol A, and chrysanthediol B (Yuan et al. 2020).

8.3.2 Chemical Constituents from C. morifolium *Leaves*

A flavanone (eriodictyol 7-O-glucuronide) and two polyphenol compounds (chlorogenic acid and 3,5-dicaffeoylquinic acid), having insecticidal properties, were isolated from the methanolic leaves extract of *C. morifolium* (Beninger et al. 2004).

8.3.3 Chemical Constituents from C. morifolium *Stem*

Research about the phytochemicals of *C. morifolium* stem is scanty. However, in a study conducted by Qu et al. (Qu et al. 2017), four new molecules (heterophyllol-1-*O*-β-D-glucopyranoside and morineoliganosides A, B, and C were isolated along with 27 known compounds. The known compounds were derivatives of caffeoylquinic acids (such as chlorogenic acid methyl ester, 1,5-di-*O*-caffeoylquinic acid, 3-*O*-caffeoylquinic acid, methyl 3,4-di-*O*-caffeoylquinate, and 1,4-di-*O*-caffeoylquinic acid) and flavonoids(such as luteolin, luteolin-7-*O*-rutinoside, acacetin 7-*O*-β-D-glucopyranoside, luteolin-7-*O*-β-D-glucopyranoside, crisilineol, acacetin 7-*O*-rutinoside, (2S)-hesperetin, apigenin-6,8-di-C-β-D-xylopyranoside, and (2S)-eriodictyol), similar to compounds isolated from its flower.

8.4 Pharmacological Activities

8.4.1 Antioxidant Activity

Total flavones extracted from *C. morifolium* showed potential antioxidant action in rat brain with ischemia and reperfusion injury, where it significantly attenuated the reactive oxygen species (ROS) level and also reversed the decrease of superoxide dismutase (SOD) activity and increase of (malondialdehyde) MDA content (Lin et al. 2010). Total flavonoids from *C. morifolium* increased the antioxidant enzyme levels such as glutathione (GSH), glutathione peroxidase (GSH-Px), SOD, and (carbonic anhydrase) CA in the brain, liver, and kidney of mice with lead-induced oxidative injury (Xia et al. 2008). Flower extract of *C. morifolium in* different solvents showed potential antioxidant activity determined by DPPH free radical scavenging activity and reducing power assay (Yang et al. 2011). Dicaffeoylquinic acid derivatives isolated from *C. morifolium* showed strong superoxide anion radical scavenging activity and DPPH radical scavenging activity (Kim and Lee 2005). Pretreatment with *C. morifolium* extract in 1, 10, and 100 μg/mL concentration attenuated the 1 mM MPP+ induced oxidative stress in human neuroblastoma SH-SY5Y cells (Kim et al. 2009). Aqueous and ethanolic extract of flower of *C. morifolium* inhibited the production of cellular ROS induced by oxidized LDL in human umbilical vein endothelial cells (Lii et al. 2010). Flower extract of *C. morifolium* showed antioxidant activity evaluated using DPPH and FRAP assay (Yang et al. 2017).

8.4.2 Anti-inflammatory Activity

Ethanolic flower extract of *C. morifolium* treated in 50, 100, and 200 mg/Kg for 7 days in C57BL/6 J male mice with LPS (3 mg/Kg)-induced acute lung injury showed protective and anti-inflammatory activity such as inhibition of an increased number of white blood cells, lymphocytes neutrophils, and levels of TNF-α and IL-6 (Liu et al. 2020). In another study, inflammation in the ear and paw of rats was induced by using xylene and carrageenan, respectively. Oral administration of different concentrations of C. *morifolium* flower extracts (5–10 g/kg) for 1 week displayed a significant reduction in inflammation by diminishing the prostaglandin E2 (PGE_2) release and inflammatory action of cyclooxygenase (COX) enzymes (Yuan et al. 2020). A new polyacetylene glucoside dendranacetylene A, isolated from *C. morifolium* flower (in 50% acetone), prominently suppressed nitric oxide (NO) production in the lipopolysaccharide-induced murine macrophages cells (RAW 264.7) (Li et al. 2021). Another study confirmed that C. *morifolium* flower extract treatment on macrophage cells (*RAW264.7*) of the mouse can suppress the synthesis of prostaglandin E2 (PGE_2), induced by lipopolysaccharide (IC_{50}, 0.6 mg/mL). In the same cell line, luteolin and its glycoside form isolated from its flower performed a significant reduction in inflammation by diminishing the prostaglandin E2 (PGE_2) release and inflammatory action of cyclooxygenase enzymes (Hu 2015).

8.4.3 Antimicrobial Activity

Bioassay of root and leaf samples of *C. morifolium* showed antifungal activity against three species of pathogenic fungi, *Fusarium oxysporum*, *Magnaporthe oryzae*, and *Verticillium dahlia* (Xue et al. 2019; Zhang et al. 2020a, b). Silver nanoparticles by using in clinical ultrasound gel prepared by treating sliver ion with flower extract of *C. morifolium* showed potential bactericidal activity against *Staphylococcus aureus* and *Escherichia coli* (He et al. 2013). A study on intestinal bacteria from humans and rats revealed that flower extract of *C. morifolium* significantly inhibited pathogenic bacteria such as *Enterobacter*, *Enterococcus*, *Clostridium*, and *Bacteroides* meanwhile promoting the commensal probiotics such as *Lactobacillus* and *Bifidobacterium* (Tao et al. 2016). A flavonoid acetin-7-galactopyranoside, isolated from the *C. morifolium* flower extract, exhibited a profound inhibitory effect on HIV, in the H0 cell model. In a structure-activity relationship study, it was discovered that the presence of the C_2-C_3 olefin bond along with hydroxyl moiety at carbon numbers 5 and 7 is very much crucial for the suppression of HIV growth (Hu 2015).

8.4.4 Anticancer and Anti-tumorigenic Activity

In vitro study revealed that polyhydroxylated flavonoids, luteolin, and diosmetin isolated from *C. morifolium* flower induced cytotoxicities against human colon cancer cells (Colon 205) with IC_{50} values 96.9 and 82.9 μM, respectively, showing its therapeutic potential for the treatment of colon cancer (Xie et al. 2009). In the MTT assays, polysaccharides and flavonoids present in *C. morifolium* flowers were found to be effective against cancerous cells (LO2 and PANC-1 cells) (Fan 2013) and tumor cells (MKN45 cells) (Liu et al. 2018), respectively. Aqueous extract of *C. morifolium* had shown significant antitumor potency by inhibiting 68.85% tumor formation, in an experiment performed by using potato disk tumor assay (PDTA) (Kalia et al. 2016). Furthermore, bioassay-guided isolation of *C.* morifolium flower extract demonstrated that cycloartane-3,24,25-triol and acacetin remarkably inhibited the growth of various prostate cancerous cells, including PC-3, LNCap, and DU145. Also, acacetin reduced the progression of in vivo DU145 tumor cells (Kim et al. 2014; Singh et al. 2005). Moreover, two triterpenoids, faradiol and taraxasterol, isolated from the *C. morifolium* flower extract, at the concentration of 2 μmol/mice, effectively suppressed tumor cells promotion in the skin induced by 12-*O*-tetradecanoylphorbol-13-acetate (TPA) (Yuan et al. 2020).

8.4.5 Cardiovascular Protection

C. morifolium showed a cardioprotective effect on isolated hearts of rats and ventricular myocytes during ischemia/anoxia and reperfusion/reoxygenation (Jiang et al. 2004). Treatment with *C. morifolium* extract showed a cardiovascular protective effect in oxidized LDL-induced atherogenic effects in human umbilical vein endothelial cells (Lii et al. 2010). In another experiment, incubation of *C. morifolium* flower extract with thoracic aorta of rat (in vitro) resulted in induction of both endothelium-independent and dependent relaxation (Jiang et al. 1986). Also, the extract exhibited an attribution on K^+ channel, NO, and Ca^{2+} regulation. In addition, the extract increased the activity of superoxide dismutase to reverse the diminished contraction of the isolated heart, and it also reduced the extent of ischemia, induced by cardiomyocytes (Xu et al. 1989).

8.4.6 Anti-osteoporotic Activity

Water extract of *C. morifolium* inhibited the osteoclast differentiation induced by recombinant murine-soluble RANKL in bone marrow-derived macrophages isolated from the C57BL/6 J mice, which signify its therapeutic potential against osteoporosis, rheumatoid arthritis, and periodontitis (Jang et al. 2021).

8.4.7 Antidiabetic Activity

The flower extract of *C. morifolium* showed potential α-glucosidase inhibitory activity, which signifies its importance in the treatment of postprandial hyperglycemia (Yang et al. 2011). Also, a bioactive compound, 10α-hydroxyl-1α,4-α-endoperoxyl-guaia-2-en[1]12,6α-olide present in *C. morifolium* flower extract was efficacious in inhibiting lipase and α-glucosidase enzymes, with IC_{50} values of 161.0 and 229.3 μM, respectively. Similarly, the other two compounds, acacetin-7-*O*-a-L-rhamnoside and acacetin-7-glucoside, were successful to inhibit both α-amylase and α-glucosidase enzymes whereas eriodictyol only suppressed only the activity of α-glucosidase (Luyen et al. 2013). Nine-week-old male KK-Ay mice (obese type 2 diabetic mice) treated with hot water extract of edible *C. morifolium* for 5 weeks showed antidiabetic effects such as improved blood glucose levels, insulin resistance, and increased adiponectin level in adipose tissue and plasma, with decreased expression of pro-inflammatory adipocytokines (Yamamoto et al. 2015). In a study conducted by Shang (Shang et al. 2017), *C. morifolium* flower extract (300 mg/kg) was fed to both diabetic and normal mice for 42 and 45 days, respectively. The extract evinced a potent hypoglycemic effect by inducing the recovery of partially damaged islets β-cells and elevating the expression of hepatic proteins, glutamine synthase (GS), and peroxisome proliferator-activated receptor α (PPARα) and glucose transporter 2 (Glut-2).

8.4.8 Neuroprotective Activity

Pretreatment with *C. morifolium* flower extract showed a neuroprotective effect in rat brains with ischemia and reperfusion injury (Lin et al. 2010). In an in vitro study of *C. morifolium* extract revealed its neuroprotective activity against MPP + -induced a cytotoxic effect in human neuroblastoma SH-SY5Y cells (Kim et al. 2009).

8.4.9 Antimutagenic Activity

Four flavonoid compounds, luteolin, acacetin, quercetin, and apigenin, isolated from the ethyl acetate fraction of *C. morifolium* methanolic flower extract were successful to suppress the SOS response of umu gene expression in a mutant bacterium *Salmonella typhimurium* TA1535WpSK1002. All the compounds were treated at the concentration of 0.7 μmol/mL, and furylfuramide was utilized as a mutagen. Their inhibitory dose50 (ID_{50}) were reported to be 0.44, 0.62, 0.59, and 0.55 μmol/mL. Apart from that these molecules were also effective to suppress the SOS response of umu gene expression in response to other liver metabolizing enzymes not requiring mutagens (N-methyl[1]N′-nitro-N-nitrosoguanidine and 4-nitroquinolin 1-oxide) as well as those mutagens which need UV IR radiation and liver-metabolizing enzymes (3-amino-1,4-dimethyl-5H-pyrido[4,3-b]indole and aflatoxin B1) (Miyazawa and Hisama 2003).

8.4.10 Anti-obesity Activity

The presence of the major amount of chlorogenic acid, apigenin, and luteolin in *C. morifolium* potentiates its anti-obesity effects by suppressing weight gain (Lin and Harnly 2010; Lee and Kim 2020). Treatment with hot water extract of *C. morifolium* flower inhibited the adipocyte lipid accumulation along with inhibition of expression of genes related to adipogenesis and lipogenesis in 3 T3-L1 cells (Lee and Kim 2020). In an in vivo study of obesity, the ethanol extract of *C. morifolium* leaves inhibited the high fat diet-induced obesity in C57BL/6 J mice when fed for 16 weeks along with a high-fat diet (Ryu et al. 2019). It prevented not only the obesity but also the complications related to obesity such as attenuated hepatic steatosis, hepatic lipogenesis, insulin resistance, dyslipidemia, adipocyte inflammation, and lipotoxicity of peripheral tissue (Ryu et al. 2019).

8.5 Clinical Studies

Very few scientific studies have been reported regarding clinical trials of *C. morifolium*. While performing a randomized controlled clinical trial for around 2 months, by orally ingesting 150 and 50 mg *C. morifolium* flower extract every day, it was manifested to be very safe. However, the major clinical aim of improving effect on skin conditions was not achieved (Yagi et al. 2012). In another human clinical trial of traditional Chinese herbal formulation (*C. morifolium* flower extract was the major component), which was administered twice a day, it was proven that the herbal formulation is effective to control stage 1 and stage 2 hypertension, as it diminished blood pressure from 159.2/100.2 to 136.9/86.2 mmHg in 1 month (Jiang et al. 2002). Furthermore, the study of *C. morifolium* in humans demonstrated that luteolin-rich chrysanthemum flower extract (LCE) administration in the form of a capsule containing 100 mg of LCE (10 mg of luteolin) for 4 weeks suppressed the fasting serum uric acid level in Japanese men with mild hyperuricemia (Hirano et al. 2017).

8.6 Studies Related to Safety and Toxicity

Safety and nontoxicity are prerequisites in all traditional and herbal medicine. Traditionally, *C. morifolium* was considered a safe medicinal plant, and recently many studies have demonstrated that it is safe and possesses nontoxic properties in in vivo and in vitro studies (Yuan et al. 2020). In China, the ministry of health has categorized the flowers of this plant as a safe food and medicine (Hu 2015). The study on the acute and long-term toxicity of ethanolic extract of *C. morifolium* in rats showed that the extract was safe at a limited dose level. Rats were administrated with extract, at a single dose of 15 g/kg (which contains 5.19% apigenin and 7% luteolin) for acute toxicity and up to 1280 mg/kg/day for 26 weeks for long-term toxicity, and the result showed no treatment-related death, no change in biochemical parameters, no toxicological changes in body weight, and other physiological parameters (Lii et al. 2010). In another investigation, pregnant rats were subjected to a toxicity test by feeding *C. morifolium* flower extract at the dose of 6.09 g/kg, regularly for 10 days. As per Chinese Pharmacopeia, this dose is considered to be 300 times greater than the dose recommended for human consumption. However, no statistically significant difference was observed between placebo and test groups, in terms of an average live fetus, length of the fetus, bodyweight of the fetus, weight gain in pregnant rats, and the ratio of the absorptive fetus, ensuring that no possibility of teratogenic effect and embryonic toxicity (Yagi et al. 2012). In contrast, there is evidence of occurrences of contact dermatitis by different parts of *C. morifolium* in the order of stem < entire plant< leaf< flower (Sharma et al. 1989).

8.7 Conclusion and Future Recommendations

In summary, *C. morifolium* flowers have a wide range of health advantages, nutritional value, commercial importance, and traditional values. From the above information, it is pretty evident that medicinal investigations on this ornamental plant are growing exponentially. Despite its worldwide cultivation for commercial purposes, emphasis on its utilization for modern herbal formulation and natural tonic preparations is not flourished yet. Therefore, precedence should be given to its clinical effect and safety evaluation in human samples. Moreover, spreading awareness and training about its cultivation in low-income families definitely would help to strengthen their socioeconomic status.

References

Beninger CW, Abou-Zaid MM, Kistner AL, Hallett RH, Iqbal MJ, Grodzinski B, Hall JC (2004) A flavanone and two phenolic acids from *Chrysanthemum morifolium* with phytotoxic and insect growth regulating activity. J Chem Ecol 30(3):589–606

Boukhebti H, Demirtas I, Omar L, Chaker AN (2020) Chemical composition, antibacterial activity of essential oil and anatomical study of *Chrysanthemum morifolium*. J Drug Deliv Ther 10(2-s):7–13

Chang X, Wei D, Chen D, Chen D, Yan H, Sun X, Zhu W, Duan J (2019) Historical origin and development of medicinal and tea *Chrysanthemum morifolium* resources. Mod Chin Med 21: 116–123

Chen S, Liu J, Dong G, Zhang X, Liu Y, Sun W, Liu A (2021) Flavonoids and caffeoylquinic acids in Chrysanthemum morifolium Ramat flowers: A potentially rich source of bioactive compounds. Food Chem 344:128733

Cheng H, Yuhua CAO (2007) Determination of hydroxyl radical in CuSO4-vitamin C reaction system using capillary electrophoresis with electrochemical detection and its application in the determination of the scavenging activities of chrysanthemum. Chin J Chromatogr 25(5): 681–685

Cheng W, Li J, You T, Hu C (2005) Anti-inflammatory and immunomodulatory activities of the extracts from the inflorescence of *Chrysanthemum indicum* Linne. J Ethnopharmacol 101(1–3): 334–337

Cockshull KE (2019) Chrysanthemum morifolium. In: CRC handbook of flowering. CRC Press, Boca Raton, FL, pp 238–257

Fan LJ (2013) Structure features of polysaccharides from *Chrysanthemum morifolium* and their activities against tumor cells and NF-κB. Chin Tradit Herb Drug 2364–2371

Gong J, Chu B, Gong L, Fang Z, Zhang X, Qiu S, Wang J, Xiang Y, Xiao G, Yuan H, Zheng F (2019) Comparison of phenolic compounds and the antioxidant activities of fifteen *Chrysanthemum morifolium* Ramat cv.'Hangbaiju'in China. Antioxidants 8(8):325

He Y, Du Z, Lv H, Jia Q, Tang Z, Zheng X, Zhang K, Zhao F (2013) Green synthesis of silver nanoparticles by *Chrysanthemum morifolium* Ramat. extract and their application in clinical ultrasound gel. Int J Nanomedicine 8:1809–1815

Hirano M, Takeda S, Hitoe S, Shimoda H (2017) Luteolin-rich chrysanthemum flower extract suppresses baseline serum uric acid in Japanese subjects with mild hyperuricemia. Integr Mol Med 4(2):1–5

Hu C (2015) *Chrysanthemum morifolium* Ramat 菊花 (Juhua, florists chrysanthemum). In: Dietary Chinese herbs. Springer, Vienna, pp 681–691

Hu L, Chen Z (1997) Sesquiterpenoid alcohols from Chrysanthemum morifolium. Phytochemistry 44(7):1287–1290

Jang HY, Lee HS, Noh EM, Kim JM, You YO, Lee G et al (2021) Aqueous extract of *Chrysanthemum morifolium* Ramat. inhibits RANKL-induced osteoclast differentiation by suppressing the c-fos/NFATc1 pathway. Arch Oral Biol 122:105029

Jiang H, Wang L, Zhou X, Xia Q (1986) Vasorelaxant effect and underlying mechanism of EtOAc extract from *Chrysanthemum morifolium* in rat thoracic aorta. Chin J Pathophysiol 21(2): 334–338

Jiang W, Zhang WG, Ma XS (2002) Clinical and experimental study on jiangzhi tiaoya granule in treating essential hypertension and protecting function of vascular endothelium. Zhongguo Zhong Xi Yi Jie He Za Zhi 22(1):18–20

Jiang H, Xia Q, Xu W, Zheng M (2004) *Chrysanthemum morifolium* attenuated the reduction of contraction of isolated rat heart and cardiomyocytes induced by ischemia/reperfusion. Pharmazie 59(7):565–567

Kalia R (2015) Effect of different concentrations of auxins on the regeneration of *Chrysanthemum morifolium* plantlets. Int J Tech Res Appl 3:106–107

Kalia R, Katnoria JK, kaur Nagpal, A. (2016) Antitumor activity of aqueous leaf extracts of different cultivars of *Chrysanthemum morifolium* R. using potato disc tumor assay. J Pharm Sci Res 8(11):1262

Kim HJ, Lee YS (2005) Identification of new dicaffeoylquinic acids from *Chrysanthemum morifolium* and their antioxidant activities. Planta Med 71(09):871–876

Kim IS, Koppula S, Park PJ, Kim EH, Kim CG, Choi WS, Lee KH, Choi DK (2009) *Chrysanthemum morifolium* Ramat (CM) extract protects human neuroblastoma SH-SY5Y cells against MPP+-induced cytotoxicity. J Ethnopharmacol 126(3):447–454

Kim C, Kim MC, Kim SM, Nam D, Choi SH, Kim SH, Ahn KS, Lee EH, Jung SH, Ahn KS (2013) *Chrysanthemum indicum* L. extract induces apoptosis through suppression of constitutive STAT3 activation in human prostate cancer DU145 cells. Phytother Res 27(1):30–38

Kim HR, Park CG, Jung JY (2014) Acacetin (5, 7-dihydroxy-4′-methoxyflavone) exhibits in vitro and in vivo anticancer activity through the suppression of NF-κB/Akt signaling in prostate cancer cells. Int J Mol Med 33(2):317–324

Kitano H, Yagi M, Nomoto K, Shono S, Yonei Y, Hara H, Yamaji A (2011) Research on the inhibitory effect of edible purple chrysanthemum on generation of advanced glycation end products (AGEs). New Food Industry 53:1–10

Kuang CL, Lv D, Shen GH, Li SS, Luo QY, Zhang ZQ (2018) Chemical composition and antimicrobial activities of volatile oil extracted from *Chrysanthemum morifolium* Ramat. J Food Sci Technol 55(7):2786–2794

Kumar A, Singh SP, Bhakuni RS (2005) Secondary metabolites of Chrysanthemum genus and their biological activities. Curr Sci 89:1489–1501

Lai JP, Lim YH, Su J, Shen HM, Ong CN (2007) Identification and characterization of major flavonoids and caffeoylquinic acids in three Compositae plants by LC/DAD-APCI/MS. J Chromatogr B 848(2):215–225

Lawal OA, Ogunwande IA, Olorunloba OF, Opoku AR (2014) The essential oils of *Chrysanthemum morifolium* Ramat. from Nigeria. Am J Essential Oils Nat Prod 2(1):63–66

Lee MS, Kim Y (2020) *Chrysanthemum morifolium* flower extract inhibits adipogenesis of 3T3-L1 cells via AMPK/SIRT1 pathway activation. Nutrients 12(9):2726

Li M, Zeng M, Zhang J, Shi J, Lv J, Tang Y, Zheng XK, Feng WS (2021) Anti-inflammatory Dendranacetylene A, a new polyacetylene glucoside from the flower of *Chrysanthemum morifolium* Ramat. Nat Prod Res 35(24):5692–5698

Liang X, Wu H, Su W (2014) A rapid UPLC-PAD fingerprint analysis of Chrysanthemum morifolium Ramat combined with chemometrics methods. Food Anal Methods 7(1):197–204

Liang WL, Gong D, Zhang WK (2021) The composition of chrysanthemum extracts and their pharmacological functions. STEMedicine 2(5):e69

Lii CK, Lei YP, Yao HT, Hsieh YS, Tsai CW, Liu KL, Chen HW (2010) *Chrysanthemum morifolium* Ramat. reduces the oxidized LDL-induced expression of intercellular adhesion molecule-1 and E-selectin in human umbilical vein endothelial cells. J Ethnopharmacol 128(1):213–220

Lin LZ, Harnly JM (2010) Identification of the phenolic components of chrysanthemum flower (*Chrysanthemum morifolium* Ramat). Food Chem 120(1):319–326

Lin GH, Lin L, Liang HW, Ma X, Wang JY, Wu LP, Xian HD, Bruce IC, Xia Q (2010) Antioxidant action of a *Chrysanthemum morifolium* extract protects rat brain against ischemia and reperfusion injury. J Med Food 13(2):306–311

Liu JQ, Shen QQ, Liu JS, Wu DL, Wang JT (2001) Studies on the chemical constituents from *Chrysanthemum morifolium* Ramat. Zhongguo Zhong Yao Za Zhi 26(8):547–548

Liu YH, Mou X, Zhou DY, Zhou DY, Shou CM (2018) Extraction of flavonoids from *Chrysanthemum morifolium* and antitumor activity in vitro. Exp Ther Med 15(2):1203–1210

Liu G, Zheng Q, Pan K, Xu X (2020) Protective effect of *Chrysanthemum morifolium* Ramat. ethanol extract on lipopolysaccharide induced acute lung injury in mice. BMC Complement Med Ther 20(1):1–11

Lu B, Li M, Yin R (2016) Phytochemical content, health benefits, and toxicology of common edible flowers: a review (2000–2015). Crit Rev Food Sci Nutr 56(sup1):S130–S148

Luyen NT, Hanh TTH, Binh PT, Dang NH, Van Minh C, Dat NT (2013) Inhibitors of α-glucosidase, α-amylase and lipase from *Chrysanthemum morifolium*. Phytochem Lett 6(3): 322–325

Luyen BTT, Tai BH, Thao NP, Cha JY, Lee HY, Lee YM, Kim YH (2015) Anti-inflammatory components of *Chrysanthemum indicum* flowers. Bioorg Med Chem Lett 25(2):266–269

Maddala VKS (2021) Chrysanthemum traditional medicine and its role in biosorption. Ann Rom Soc Cell Biol 20256–20263

Marongiu B, Piras A, Porcedda S, Tuveri E, Laconi S, Deidda D, Maxia A (2009) Chemical and biological comparisons on supercritical extracts of Tanacetum cinerariifolium (Trevir) Sch. Bip. with three related species of chrysanthemums of Sardinia (Italy). Nat Prod Res 23(2):190–199

Miyazawa M, Hisama M (2003) Antimutagenic activity of flavonoids from *Chrysanthemum morifolium*. Biosci Biotechnol Biochem 67(10):2091–2099

Peng A, Lin L, Zhao M (2020) Screening of key flavonoids and monoterpenoids for xanthine oxidase inhibitory activity-oriented quality control of *Chrysanthemum morifolium* Ramat.'Boju' based on spectrum-effect relationship coupled with UPLC-TOF-MS and HS-SPME-GC/MS. Food Res Int 137:109448

Qu L, Ruan JY, Jin LJ, Shi WZ, Li XX, Han LF, Zhang Y, Wang T (2017) Xanthine oxidase inhibitory effects of the constituents of *Chrysanthemum morifolium* stems. Phytochem Lett 19: 39–45

Ryu R, Kwon EY, Choi JY, Shon JC, Liu KH, Choi MS (2019) Chrysanthemum leaf ethanol extract prevents obesity and metabolic disease in diet-induced obese mice via lipid mobilization in white adipose tissue. Nutrients 11(6):1347

Shahat AA, Apers S, Pieters L, Vlietinck AJ (2001) Isolation and complete NMR assignment of the numbing principle from *Chrysanthemum morifolium*. Fitoterapia 72(1):89–91

Shahrajabian MH, Sun W, Zandi P, Cheng Q (2019) A review of Chrysanthemum, the eastern queen in traditional Chinese medicine with healing power in modern pharmaceutical sciences. Appl Ecol Environ Res 17(6):13355–13369

Shang X, Zhu ZY, Wang F, Liu JC, Liu JY, Xie ML (2017) Hypoglycemic effect of *Chrysanthemum morifolium* extract on alloxan-induced diabetic mice is associated with peroxisome proliferator-activated receptor α/γ-mediated hepatic glycogen synthesis. J Appl Biomed 15(1): 81–86

Sharma SC, Tanwar RC, Kaur S (1989) Contact dermatitis from chrysanthemums in India. Contact Dermatitis 21(2):69–71

Sharma Y, Hegde RV, Venugopal CK (2011) Health and nutrition from ornamentals. Int J Res Ayurveda Pharm 2(2):375–382

Singh RP, Agrawal P, Yim D, Agarwal C, Agarwal R (2005) Acacetin inhibits cell growth and cell cycle progression, and induces apoptosis in human prostate cancer cells: structure–activity relationship with linarin and linarin acetate. Carcinogenesis 26(4):845–854

Sinha S, Mishra RK, Sinam G, Mallick S, Gupta AK (2013) Comparative evaluation of metal phytoremediation potential of trees, grasses, and flowering plants from tannery-wastewater-contaminated soil in relation with physicochemical properties. Soil Sediment Contam Int J 22(8):958–983

Sun QL, Hua S, Ye JH, Zheng XQ, Liang YR (2010) Flavonoids and volatiles in *Chrysanthemum morifolium* Ramat flower from Tongxiang County in China. Afr J Biotechnol 9(25):3817–3821

Tang W, Eisenbrand G (1992) *Chrysanthemum indicum* L. and *C. morifolium* Ramat. In: Chinese drugs of plant origin. Springer, Berlin, pp 309–313

Tao JH, Duan JA, Qian YY, Qian DW, Guo JM (2016) Investigation of the interactions between *Chrysanthemum morifolium* flowers extract and intestinal bacteria from human and rat. Biomed Chromatogr 30(11):1807–1819

Ukiya M, Akihisa T, Yasukawa K, Kasahara Y, Kimura Y, Koike K, Nikaido T, Takido M (2001) Constituents of compositae plants. 2. Triterpene diols, triols, and their 3-o-fatty acid esters from edible chrysanthemum flower extract and their anti-inflammatory effects. J Agric Food Chem 49:3187–3197

Wang T, Zhu Z, Guo Q, Mao P (2013) Variation in major flavonoids glycosides and caffeoylquinic acids during florescence of three *Chrysanthemum morifolium* Ramat cv.'Hangju'genotypes. Biochem Syst Ecol 47:74–79

Wang S, Hao LJ, Zhu JJ, Wang ZM, Zhang X, Song XM (2015) Comparative evaluation of chrysanthemum Flos from different origins by HPLC-DAD-MS n and relative response factors. Food Anal Methods 8(1):40–51

Wolverton BC, McDonald RC, Watkins E (1984) Foliage plants for removing indoor air pollutants from energy-efficient homes. Econ Bot 38(2):224–228

Woo KS, Yu JS, Hwang IG, Lee YR, Lee CH, Yoon HS, Jeong HS (2008) Antioxidative activity of volatile compounds in flower of *Chrysanthemum indicum, C. morifolium*, and C. zawadskii. J Korean Soc Food Sci Nutr 37(6):805–809

Xia DZ, Lv GY, Yu XF, Wang HM, Yang Q (2008) Antagonism of total flavonoids from *Chrysanthemum morifolium* against lead induced oxidative injury in mice. Zhongguo Zhong Yao Za Zhi 33(23):2803–2808

Xie YY, Yuan D, Yang JY, Wang LH, Wu CF (2009) Cytotoxic activity of flavonoids from the flowers of *Chrysanthemum morifolium* on human colon cancer Colon205 cells. J Asian Nat Prod Res 11(9):771–778

Xu W, Cao C, Xia Q, Jiang H, YE, Z. (1989) *Dendranthema morifolium* attenuated the reduction of contraction of isolated heart and cardiomyocytes induced by ischemia/reperfusion. Chin J Pathophysiol 20(5):822–825

Xue H, Jiang Y, Zhao H, Köllner TG, Chen S, Chen F, Chen F (2019) Characterization of composition and antifungal properties of leaf secondary metabolites from thirteen cultivars of *Chrysanthemum morifolium* Ramat. Molecules 24(23):4202

Yagi M, Nomoto K, Hori M, Kitano T, Yabukita H, Ogura M, Hamada U, Yonei Y (2012) The effect of edible purple chrysanthemum extract on advanced glycation end products generation in skin: a randomized controlled clinical trial and in vitro study. Anti-Aging Med 9(2):61–74

Yamamoto J, Tadaishi M, Yamane T, Oishi Y, Shimizu M, Kobayashi-Hattori K (2015) Hot water extracts of edible *Chrysanthemum morifolium* Ramat. Exert antidiabetic effects in obese diabetic KK-ay mice. Biosci Biotechnol Biochem 79(7):1147–1154

Yang J, Kim JS, Jeong HJ, Kang HH, Cho JC, Yeom HM, Kim MJ (2011) Determination of antioxidant and α-glucosidase inhibitory activities and luteolin contents of *Chrysanthemum morifolium* Ramat extracts. Afr J Biotechnol 10(82):19197–19202

Yang L, Nuerbiye A, Cheng P, Wang JH, Li H (2017) Analysis of floral volatile components and antioxidant activity of different varieties of *Chrysanthemum morifolium*. Molecules 22(10): 1790

Yang PF, Yang YN, Feng ZM, Jiang JS, Zhang PC (2019) Six new compounds from the flowers of *Chrysanthemum morifolium* and their biological activities. Bioorg Chem 82:139–144

Youssef FS, Eid SY, Alshammari E, Ashour ML, Wink M, El-Readi MZ (2020) *Chrysanthemum indicum* and *Chrysanthemum morifolium*: chemical composition of their essential oils and their potential use as natural preservatives with antimicrobial and antioxidant activities. Foods 9(10): 1460

Yu Y, Zhu C, Wang S, Song W, Yang Y, Shi J (2013) Homosecoiridoid alkaloids with amino acid units from the flower buds of *Lonicera japonica*. J Nat Prod 76(12):2226–2233

Yu Q, Chen W, Zhong J, Qing D, Yan C (2021) Structural elucidation of three novel oligosaccharides from Kunlun Chrysanthemum flower tea and their bioactivities. Food Chem Toxicol 149: 112032

Yuan H, Jiang S, Liu Y, Daniyal M, Jian Y, Peng C, Shen J, Liu S, Wang W (2020) The flower head of *Chrysanthemum morifolium* Ramat.(Juhua): A paradigm of flowers serving as Chinese dietary herbal medicine. J Ethnopharmacol 261:113043

Zhang K, Jiang Y, Zhao H, Köllner TG, Chen S, Chen F, Chen F (2020a) Diverse terpenoids and their associated antifungal properties from roots of different cultivars of *Chrysanthemum morifolium* Ramat. Molecules 25(9):2083

Zhang X, Yu X, Shi Y, Zhao X, Xing M, Tian C, Guo L, Xia D (2020b) *Chrysanthemum morifolium* cv. Hang-ju leaves: an abundant source of preservatives for food industry. Eur Food Res Technol 246(5):939–946

Zheng Y, Liu LIU, Fang JN (2004) A novel polysaccharide from Chrysanthemum morifolium. Acta Bot Sin 46(8):997–1001

Zheng Y, Wang XS, Fang J (2006) Two acidic polysaccharides from the flowers of Chrysanthemum morifolium. J Asian Nat Prod Res 8(3):217–222

Zheng C, Dong Q, Du Z, Wang P, Ding K (2015) Structural elucidation of a polysaccharide from *Chrysanthemum morifolium* flowers with anti-angiogenic activity. Int J Biol Macromol 79:674–680

Chapter 9
Crassocephalum crepidioides (Benth.) S. Moore: Traditional Uses, Chemical Constituents, and Biological Activities

Sila Gurung, Prakash Poudel, Namuna Adhikari, Gopal Lamichhane, and Rashmi Thapa

Abstract Medicinal plants belonging to the Asteraceae family are reported to have long history in traditional medicines which are widely distributed throughout the world in several ecological habitats. The genus *Crassocephalum* consists of 24 accepted species in which species *C. crepidioides* is one of them. It is commonly known as "fireweed ragleaf" in English and "Anikale jhar" in Nepali. There are several traditional uses associated with the plant which have been proven by several scientific studies. *C. crepidioides* is consumed as leafy vegetables in Africa. Several pharmacological activities including antibacterial, wound healing, antidiabetic, anti-inflammatory, and antioxidant properties are reported in the plant. Secondary metabolites such as alkaloids, flavonoids, and phenolic compounds are the main constituents present in *C. crepidioides*. In this chapter, the traditional uses, pharmacological activities, and phytochemical constituents of C. *crepidioides* are presented along with the safety and toxicity profile.

Keywords *Crassocephalum crepidioides* · Pharmacological activity · Redflower ragweed · Phytochemical constituents

S. Gurung
Department of Biodiversity and Bioresources, Satvik Nepal, Pokhara, Nepal

P. Poudel (✉) · N. Adhikari
Pharmacy Program, Gandaki University, Pokhara, Nepal
e-mail: prakash.poudel@gandakiuniversity.edu.np

G. Lamichhane
Department of Oriental Pharmacy, Wonkwang University, Iksan, Republic of Korea

R. Thapa
School of Health and Allied Sciences, Faculty of Health Sciences, Pokhara University, Pokhara, Nepal
e-mail: rashmithapa@pu.edu.np

H. P. Devkota, T. Aftab (eds.), *Medicinal Plants of the Asteraceae Family*,
https://doi.org/10.1007/978-981-19-6080-2_9

Fig. 9.1 Photograph of the mature plant (**a**), flowerhead with red florets (**b**), flower with silky white hairs (**c**), leaves (**d**), roots (**e**), and fruits and seeds (**f**) of *Crassocehalum crepidioides* (Benth.) S. Moore

9.1 Introduction

Crassocephalum crepidioides (Benth.) S. Moore (Fig. 9.1) is distributed throughout the world (Can and Thao 2020; Joshi 2014). It is commonly known as "fireweed ragleaf", "thickhead", and "redflower ragweed" in English, "anikale jhar" and "salaha ko jhar" in Nepali, "Kaiklai" in Chepang, and "lodomran" in Tamang (NP Manandhar 2002). It is an annual terrestrial herb which is edible, erect, and sparingly branched and grows up to the height of 40–100 cm tall. Stem is rounded or fluted cross-section, solid, and hairy. Stipules are absent. Fruit is achene with pappus (Ayodele et al. 2020a, b; https://portal.wiktrop.org/species/show/79; https://www.cabi.org/isc/datasheet/15870). Seeds are composed of floating balls that possess several silky white hairs (Ayodele et al. 2020a, b).

Crassocephalum has 24 types of accepted species (http://www.theplantlist.org/browse/A/Compositae/Crassocephalum/). Species *crepidioides* is distributed up to an altitude of 2500 m above sea level and grows well on well-drained soil and shaded region. The main aim of the chapter is to document the traditional and pharmacological activities of *Crassocephalum crepidioides* along with the phytochemistry of the plant.

9.2 Traditional Uses

The leaves of *C. crepidioides* are reported to be used for the treatment of indigestion and stomachache. It is also used to rejuvenate the spleen. Likewise, young leaves are edible and are used as vegetable (Can and Thao 2020). The people of Africa use the succulent leaves and stems as vegetables in soups and stews. Similarly, leaves are also used to relieve helminth's diseases, malaria, colds, intestinal worms, liver disorders, hepatic insufficiency, and lowering blood pressure. Moreover, the plant is also used for the treatment of nose bleeding, fever, inflammation, and edema (Ayodele et al. 2020a, b; Nguemfo et al. 2020). In Cameroon, *C. crepidioides* is used to treat intestinal worms (Hung et al. 2019). In Nepal, juice of the plant is used to treat diarrhea and also applied topically for healing cuts and wounds (Manandhar 2002). It is given to pregnant women to prevent anemia. In addition, breastfeeding mothers consume the plant to increase milk production (Adjatin et al. 2012).

9.3 Chemical Constituents

The major phytochemicals reported in *C. crepidioides* are flavonoids, phenolic compounds, alkaloids, glycosides, tannins, saponins, steroids, essential oils, mucilage, coumarins, and ascorbic acid (Can and Thao 2020; Ayodele et al. 2020a, b; Dansi et al. 2013). Phenolic and flavonoid compounds including gallic acid, catechin, chlorogenic acid, caffeic acid, ellagic acid, phenol glycoside, rutin, isoquercitrin, quercetin, and kaempferol (Fig. 9.2) are reported to be present in the

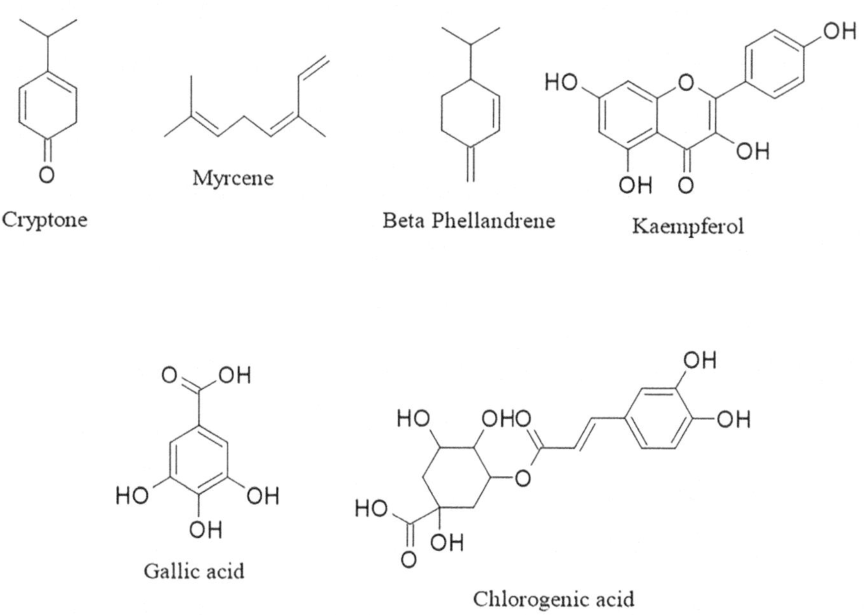

Fig. 9.2 Chemical structures of some of the compounds present in *C. crepidioides*

leaves of *C. crepidioides* according to the study performed by Adefegha et al. and Can and Thao (Adefegha et al. 2014; Can and Thao 2020). Likewise, aerial parts of the plant contain pyrrolizidine alkaloids such as jacobine and jacoline (Can and Thao 2020; Joshi 2014; Schramm et al. 2021).

9.4 Pharmacological Activities

Several studies have reported the antibacterial, hypoglycemic, antioxidant, anti-inflammatory, antitumor, and antidiabetic properties of *C. crepidioides*.

9.4.1 Wound Healing Activity

According to the study performed by Can and Thao, the ethanol extract of leaf of *C. crepidioides* exhibited wound healing activity in albino mice model when the extract was administered at a dose of 50 mg/kg/day in comparison to animal model treated with vehicle. In the study, it has been reported that the wound healing activity of ethanolic leaf extract may be enhanced due to its anti-inflammatory activity. Moreover, it has been reported that treatment of the mice model with plant extract significantly enhanced the fibroblast density at wound sites. The topical application of plant extract at a dose of 50 mg/kg/d showed significant wound closure by 28% as compared to control group in which the wound closure was only 15.4%. In addition, epithelialization also occurred 3.5 days faster in plant extract-treated groups as compared to control group. In the histopathological study, plant extract decreased the inflammatory cell density. On the other hand, fibroblast density and blood vessel number were increased in extract treated group (Can and Thao 2020).

9.4.2 Anti-inflammatory and Antioxidant Activity

An in vitro study of leaf extract of *C. crepidioides* performed by Can and Thao reported the anti-inflammatory activity of the plant. The in vitro study was performed on macrophage cell RAW 264.7 isolated from the mouse model. The leaf extract showed anti-inflammatory effect by decreasing the nitric oxide production in macrophage cell of the mouse. The effect was dose-dependent in which the highest inhibition of nitric oxide was observed when the cells were treated with leaf extract of 125 μg/mL concentration. The leaf extract also reported the antioxidant potential with IC_{50} value 48.0 μg/L against DPPH free radicals (Can and Thao 2020).

9.4.3 Antidiabetic Property

In the research study performed by Bahar et al., *C. crepidioides* extracts reported significant inhibition of α-amylase enzyme of 66.37% at 200 mg/mL concentration which in turn delays carbohydrate digestion resulting into overall increase in digestion time. The extract reported α-amylase enzyme inhibition with IC_{50} value 126.85 ± 2.10 mg/mL. As a result, blood glucose level was found to be significantly reduced in Wister albino rats thus indicating the antidiabetic potential of the plant (Bahar et al. 2017).

9.4.4 Antitumor Property

According to Tomimori et al., *C. crepidioides* extract reported the inhibition of tumor growth in nude mice model that was inoculated with murine S-180-cells which indicate the antitumor property of the plant. In the study, mice bearing tumor cells were treated with plant extract at a dose of 5 g/kg in which the tumor volume was significantly lower as compared to control group. The extract was treated to the test groups for 14 days. However, when the plant extract was treated to S-180 cells, significant apoptosis of the cells was observed. Therefore, the study indicates that *C. crepidioides* exhibited in vivo antitumor property (Tomimori et al. 2012).

9.4.5 Anticoagulant Activity

According to the clinical research study performed by Ayodele et al., methanol leaf extract of *C. crepidioides* increased the clotting time and prothrombin time. The extract also activated the partial thromboplastin times of healthy human blood. Hexane and aqueous fractions of the plant extract showed highest clotting times at 10 mg/mL concentration. Similarly, methanol extract along with hexane, ethyl acetate, and butanol fractions increased the prothrombin time at 10 mg/mL concentration. Among all the fractions, hexane fraction showed longer prothrombin time as compared to aqueous fraction and methanol extract. Thus, *C. crepidioides* indicated the anticoagulant activity which may be utilized for the treatment of blood coagulation disorders (Ayodele et al. 2019).

9.4.6 Hepatoprotective Effect

Aniya et al. performed the research study in which male Sprague-Dawley rats were used for in vivo study. Hepatotoxicity was induced in the test groups by carbon tetrachloride (CCl_4) and galactosamine (GalN) plus LPS. *C. crepidioides* extract significantly reduced the lipid peroxide in liver homogenate. The extract also reduced the activities of AST and ALT in serum. Therefore, the plant extract significantly reduced the free radicals and thereby indicating the hepatoprotective property (Aniya et al. 2005).

9.4.7 Anticholinesterase Activity

According to Owokotomo et al., thymol is the major constituent present in stem of C. crepidioides. Likewise, α-caryophyllene and β-cubebene are the major constituents of leaf of the plant. All these essential oils present in stem and leaf of the plant exhibited the significant inhibition of acetyl-cholinesterase in which the IC_{50} of stem and leaf were 0.82 and 0.83, respectively, which was as high as 80% of galantamine which is the standard acetyl-cholinesterase inhibitor. Thus, the study exhibited the in vitro acetylcholinesterase inhibiting potential of *C. crepidioides* (Owokotomo et al. 2015).

9.5 Studies Related to Safety and Toxicity

According to the research study performed by Nguemfo et al., acute and subacute toxicity study of aqueous leaf extract of *C. crepidioides* was performed in Wister albino rats. For acute toxicity study, single dose of leaf extract, i.e., 5000 mg/kg body weight was orally administered to the test group whereas control group were fed with vehicle. Likewise, in subacute toxicity study model, control groups were treated with vehicle and test groups were treated with 250 mg/kg, 500 mg/kg, and 1000 mg/kg body weight of plant extract. There was no any mortality of animals in the acute toxicity study of aqueous leaf extract. Moreover, lethal dose of the leaf extract was found to be greater than that of administered dose (5000 mg/kg body weight) in the study which is considered to be safe for animals and humans. However, the risk of hepatotoxicity and nephrotoxicity was higher when the extract was repeatedly administered at a dose of up to 500 mg/kg body weight. Therefore, acute toxicity study exhibited safety profile of *C. crepidioides* leaf extract (Nguemfo et al. 2020).

According to Dansi et al., toxicity of leaf extract of *C. crepidioides* was performed on Brine Shrimp nauplii in which different dose levels of extract was tested. In the study, lethal concentration (LC50) of *C. crepidioides* was found to be between 0.781 and 1.562 mg/mL. Therefore, the extract was not toxic to shrimp larvae since the lethal concentration of the extract was found to be greater than leaf extract concentration of 0.1 mg/mL thus revealing the safety profile of *C. crepidioides* leaves (Dansi et al. 2013).

9.6 Conclusions and Future Recommendations

Crassocephalum crepidioides is a succulent herb native to Tropical Africa and is distributed throughout the world. It has been reported to possess traditional medicinal values since ages with pharmacological benefits. Leaves, stem, flower, roots, and aerial parts of the plant possess antibacterial, hypoglycemic, antioxidant, anti-inflammatory, antitumor, and antidiabetic properties. Despite having innumerable pharmacological benefits, *C. crepidioides* also accumulates pyrrolizidine alkaloids such as jacobine and jacoline which are hepatotoxic and tumorigenic in nature. Therefore, further scientific investigations regarding its domestication and safe use are crucial.

References

Adefegha SA, Oboh G, Molehin OR, Saliu JA, Athayde ML, Boligon AA (2014) Chromatographic fingerprint analysis, acetylcholinesterase inhibitory properties and antioxidant activities of Redflower ragleaf (*Crassocephalum crepidioides*) extract. J Food Biochem 40:109–119. https://doi.org/10.1111/jfbc.12200

Adjatin A, Dansi A, Eze CS, Assogba P, Dossou-Aminon I, Akpagana K, Akoegninou A, Sanni A (2012) Ethnobotanical investigation and diversity of Gbolo (*Crassocephalum rubens* (Juss. Ex Jacq.) S. Moore and *Crassocephalum crepidioides* (Benth.) S. Moore), a traditional leafy vegetable under domestication in Benin. Genet Resour Crop Evol 59(8):1867–1881

Aniya Y, Koyama T, Miyagi C, Miyahira M, Inomata C, Kinoshita S, Ichiba T (2005) Free radical scavenging and Hepatoprotective actions of the medicinal herb, *Crassocephalum crepidioides* from the Okinawa Islands. Biol Pharm Bull 28(1):19–23. https://pubmed.ncbi.nlm.nih.gov/15635156/

Ayodele OO, Onajobi FD, Osoniyi O (2019) In vitro anticoagulant effect of *Crassocephalum crepidioides* leaf methanol extract and fractions on human blood. J Exp Pharmacol 11:99–107. https://doi.org/10.2147/JEP.S218261

Ayodele OO, Onajobi FD, Osoniyi OR (2020a) Chemical profiling of the hexane fraction of *Crassocephalum crepidioides* Benth S. Moore leaves by GC-MS. Afr J Pure Appl Chem 14(1):1–8. https://doi.org/10.5897/AJPAC2019.0815

Ayodele O, Osonlyl RO, Onajobi FD (2020b) Phytochemical profiling of the hexane fraction of *Crassocephalum crepidioides* Benth S. Moore leaves by GC-MS. Afr J Pure Appl Chem 14(1): 1–8. https://www.researchgate.net/publication/340310897

Bahar E, Akter KM, Lee GH, Lee HY, Rashid HO, Choi MK, Bhattarai KR, Hossain MMM, Ara J, Mazumder K, Raihan O, Chae HJ, Yoon H (2017) β-Cell protection and antidiabetic activities of *Crassocephalum crepidioides* (Asteraceae) Benth. S. Moore extract against alloxan-induced oxidative stress via regulation of apoptosis and reactive oxygen species (ROS). BMC Complement Altern Med 17:179. https://doi.org/10.1186/s12906-017-1697-0

Can NM, Thao DTP (2020) Wound healing activity of *Crassocephalum crepidioides* (Benth.) S. Moore. Leaf hydroethanolic extract. Oxid Med Cell Longev 2020:2483187. https://doi.org/10.1155/2020/2483187

Dansi A, Adjatin A, Badoussi E, Loko YL, Dansi M, Gbaguidi F, Ajokpota P, Ahissou H, Akoegninou A, Akpagana K, Sanni A (2013) Phytochemical screening and toxicity studies of *Crassocephalum rubens* (Juss. Ex Jacq.) S. Moore and *Crassocephalum crepidioides* (Benth.) S. Moore consumed as vegetable in Benin. J Chem Pharm Res 5(6):160–167

Hung NH, Satyal P, Dai DN, Tai TA, Huong LT, Hong Chuong NT, Hieu HV, Tuan PA, Vuong PV, Setzer WN (2019) Chemical compositions of *Crassocephalum crepidioides* essential oils and larvicidal activities against Aedes aegypti, Aedes albopictus, and Culex quinquefasciatus. Nat Prod Commun 14(6):1–5. https://doi.org/10.1177/1934578X19850033

Joshi RK (2014) Study of essential oil composition of the roots of *Crassocephalum crepidioides* (Benth.) S. Moore. J Chil Chem Soc 59:1. https://doi.org/10.4067/S0717-97072014000100025

Manandhar NP (2002) Plants and people of Nepal. Timber Press, Inc., Portland

Nguemfo EL, Mbock AJ, Zangueu Bogning C, Magne Fongang AL, Belle Ebanda Kedi P, Dongmo AB (2020) Acute and sub-acute toxicity assessment of aqueous leaves extract of *Crassocephalum crepidioides* (Asteraceae) in Wister rats. J Complement Integr Med 18(2): 295–302. https://doi.org/10.1515/jcim-2020-0018

Owokotomo IA, Ekundayo O, Abayomi TG, Chukwuka AV (2015) In-vitro anti-cholinesterase activity of essential oil from four tropical medicinal plants. Toxicol Rep 2:850–857. https://pubmed.ncbi.nlm.nih.gov/28962420/

Schramm S, Rozhon W, Adedeji-Badmus AN, Liang Y, Nayem S, Winkelmann T, Poppenberger B (2021) The orphan crop *Crassocephalum crepidioides* accumulates the pyrrolizidine alkaloid jacobine in response to nitrogen starvation. *Frontiers*. Plant Sci 12:702985. https://doi.org/10.3389/fpls.2021.702985

Tomimori K, Nakama S, Kimura R, Tamaki K, Ishikawa C, Mori N (2012) Antitumor activity and macrophage nitric oxide producing action of medicinal herb *Crassocephalum crepidioides*. BMC Complement Altern Med 12:78. https://doi.org/10.1186/1472-6882-12-78

Chapter 10
Dolomiaea costus (Falc.) Kasana and A. K. Pandey: Ethnomedicinal Uses, Bioactive Chemical Constituents, and Pharmacological Activities

Abhay Prakash Mishra, Manisha Nigam, Hari Prasad Devkota, and Motlalepula Gilbert Matsabisa

Abstract *Dolomiaea costus* (Syn. *Saussurea costus*), commonly known as the costus, Indian costus, kut, or putschuk, is a species of thistle native to South Asia and a member of the *Dolomiaea* genus. Studies conducted on *Dolomiaea* species around the globe have demonstrated the existence of bioactive secondary metabolites including polyphenols, flavonoids, lignans, sesquiterpenoids, and lactones. The goal of this chapter is to compile all the scientific data on *D. costus*' traditional usage, bioactive chemical components, and pharmacological properties. In summation, our investigation shows that D. costus has a wide variety of bioactive substances contributing its anti-inflammatory, antibacterial, antioxidant, and anticancer activities. To provide greater scientific support for their traditional applications, more in vivo and clinical research for mechanism-based pharmacological evaluation should be conducted in the future.

Keywords Saussurea · *Aucklandia costus* · Sesquiterpene lactones · Costunolide · Santamarine · Anticancer · Clinical studies

---

A. P. Mishra · M. G. Matsabisa
Department of Pharmacology, Faculty of Health Sciences, School of Clinical Medicine, University of the Free State, Bloemfontein, South Africa
e-mail: Mishra.ap@ufs.ac.za; MatsabisaMG@ufs.ac.za

M. Nigam (✉)
Department of Biochemistry, Hemvati Nandan Bahuguna Garhwal University, Srinagar Garhwal, Uttarakhand, India
e-mail: m.nigam@hnbgu.ac.in

H. P. Devkota
Headquarters for Admissions and Education, Kumamoto University, Kumamoto, Japan
e-mail: devkotah@kumamoto-u.ac.jp

H. P. Devkota, T. Aftab (eds.), *Medicinal Plants of the Asteraceae Family*,
https://doi.org/10.1007/978-981-19-6080-2_10

10.1 Introduction

The genus *Dolomiaea* DC (Asteraceae: Cardueae) is represented by 21 species worldwide and is endemic to the alpine regions of China, India, Myanmar, Nepal, and Pakistan. It is one of the well-established genera in subtribe Saussureinae N. Garcia and Susanna (earlier informally known as the *Saussurea* or *Saussurea—Jurinea* group) and occurs up to 5000 m elevation. *Dolomiaea* has been of great interest for systematists and has remained an integral part of the *Saussurea—Jurinea* group (Kasana et al. 2020a, b).

Dolomiaea costus (Syn: *Saussurea costus*) is a species of *Dolomiaea* genus indigenous to China and South Asia. It is also referred to as costus, Indian costus, Kuth, or Putchuk. Plants belonging to this species have therapeutic benefits and, as a result, commercial value. In addition, they are employed in religious rites. These plants are especially abundant in edibles, dyes, flavoring substances, oil, rubber, insecticides, and other products. Several species are also produced for their ornamental qualities (Mishra et al. 2021).

The presence of remarkable bioactive secondary metabolites with antioxidant, anti-inflammatory, anticancer, and hepatoprotective properties has been investigated on the basis of phytochemical and bioactivity studies on *Dolomiaea* (*Saussurea*) species throughout the world. These compounds include phenolics, flavonoids, lignans, sesquiterpenes, and lactones (Mishra et al. 2018). In India, Kashmiri Rishi mystics have a long history of using the plant *Dolomiaea costus*. Since ancient times, essential oils extracted from the root have already been consumed in folk medicine and perfumes. According to Google Arts & Culture, *Dolomiaea costus*'s root is called Costus. The plant's root is the most important portion for medicinal or homoeopathic uses. The root is also known as radix aucklandiae in Latin (Costus 2021; Waly 2009).

According to the Plants of the World Online, *Dolomiaea costus* has six synonyms as *Aucklandia costus* Falc, *Aplotaxis lappa* Decne, *Saussurea costus* (Falc.) Lipsch, *Aucklandia lappa* (Decne.) Decne, *Saussurea lappa* (Decne.) Sch. Bip, and *Theodorea costus* (Falc.) Kuntze (Plants of the World Online 2021). The goal of this chapter is to compile all of the scientific evidence that is currently accessible regarding the pharmacological properties, bioactive chemical components, and traditional usage of Dolomiaea costus.

10.2 Traditional Uses

Traditional Korean, Chinese, and Japanese medicine have employed the dried root of *Dolomiaea costus* to alleviate stomach pain for ages (Baz 2009). In addition bronchitis, coughing, and promoting urination are all treated with *D costus*. Women, on the other hand, can use it to alleviate menstrual issues. *Dactylorhiza hatagirea*, *Picrorhiza kurrooa*, *Bunium persicum*, *Dolomiaea costus*, and *Aconitum heterophyllum* are given to the mother if the kid refuses milk. *Bunium persicum* and *Dolomiaea costus*, taken combined, also prevent pregnancy hemorrhage (Kumari et al. 2012). The hot water extract of the *D. costus* has been used in traditional Chinese medicine for generations as an arthritic pain reliever (Pandey et al. 2007). According to Baz (2009), an oral combined drug made from *C. mukul* and *D. costus* root in a dosage of 2 g twice daily significantly reduced stiffness, pain, swelling, and tenderness in patients with osteoarthritis within a 3-week period (Baz 2009).

10.3 Chemical Constituents

It has been discovered that *D. costus* roots contained lignans, triterpenes, sesquiterpenoids, monoterpenes, glycosides, flavonoids, alkaloids, resinoids, organic acids, peptides, etc. (Ansari et al. 2021b; Liu et al. 2021). Sesquiterpene lactones are found in a variety of medicinal plants and have traditionally been used to treat inflammatory illnesses (Cho et al. 2000). Sesquiterpenes and sesquiterpene lactones (Choi et al. 2012) are the main active components found in *D. costus* root, particularly the high boiling sesquiterpene lactones (Figs. 10.1 and 10.2). Dehydrocostus lactone's potential as an anti-inflammatory, anticancer, and antioxidant substance is being discovered in a growing number of investigations (Li et al. 2018, 2019; Lee et al. 2019; Yang et al. 2016; Wu et al. 2021).

By using GCMS analysis, Omer et al. (2019) reported 18 and 37 components in the aqueous and ethanol extracts of *D. costus* roots, respectively. Among them are cyclodecacyclotetradecene,14,15-didehydro, 2-(hydroxymethyl)-2-nitro, 1,3-propanediol, bufa-20,22-dienolide,14,15-epoxy-3,11-dihydroxy, 2-methylene-5-(1-methylvinyl)-8-, bicycle (5.3.0) decane, isosteviol methyl ester, 2(3H)-benzofuranone, and 6-ethyenlhexahydro-6-methylene-7-(dehydrosaussurea lactone), 4,7,10,13,16,19-docosahexaenoic acid, methyl ester, and androstan-17-one, 3-ethyl-3-hydroxy-(5), cholest-7-en-3-ol,4-methyl-(3, 4), −guaiene, 3-oxatricyclo (20.8.0.0 (7,16)) triaconta-1, 9,12,15-octadecatrienoic acid, (z, z, z), and cyclohexane, 1,2-diethenyl-4-(1-methylethylidene), cis- was identified with large peak zones in both extracts (Omer et al. 2019). Liu et al. (2021) mentioned eight compounds from *A. lappa* (*D. costus*) as dehydrocostus lactone (DHL), costunolide, arbusculin A, betulin, 5-hydroxymethyl-2-furaldehyde, 3,5-dimethoxy-4-hydroxybenzaldehyde, n-butyl-bD-fructopyranoside, and 1-oleoylglycerol (Liu et al. 2021; Duan et al. 2010). Numerous investigations have established that the *D. costus* plant contains sesquiterpenes of the guaiane class, namely cynaropicrin (Baz 2009); lappalone; zaluzanin C; saussureamine A, B, C, D, and E; dihydrodehydrocostus lactone; DHL; isodehydrocostus lactone; 3-epizaluzanin C; 11,13-dihydro-3-epizaluzanin C; 11,13-epoxyisozaluzanin C; 11,13-epoxy-3-ketodehydrocostus lactone; 11,13-epoxydehydrocostus lactone; lappadilactone; 4,7-dimethoxydehydrocostuslactone; 13-sulfodihydrosantamarine; 13-sulfodihydro reynosin; saussurealdehyde; and others with sulfonic acid, 11,13-dihydrolucozanin-C, and 12-methoxydihydrodehydrocostus lactone (Cárdenas et al. 2017; Matsuda

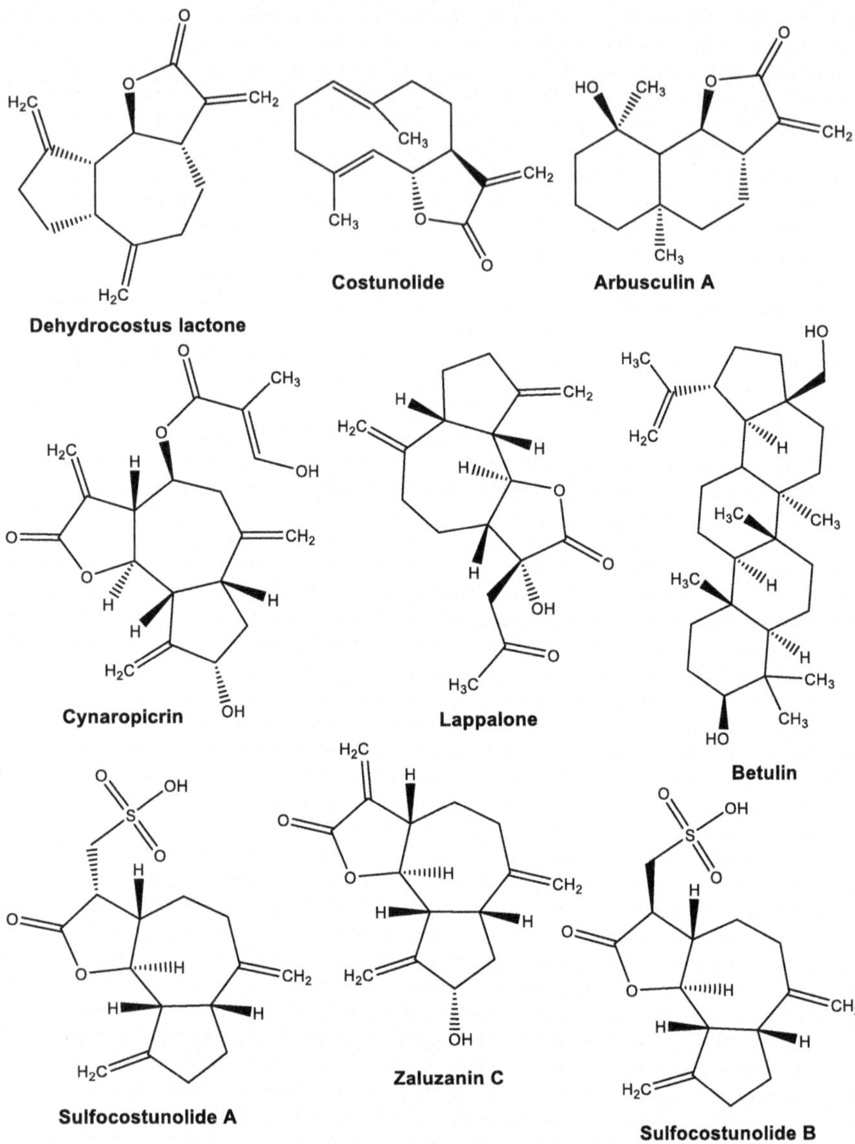

Fig. 10.1 Chemical structures of chemical constituents isolated from *D. costus*

et al. 2000; Wang et al. 2008, 2010; Yin et al. 2005; Hassan and Masoodi 2020). Duan et al. (2010) isolated sesquiterpenes, including santamarine, 10-hydroxyl-artemisinic acid, β-cyclocostunolide and 4-hydroxy-4-methyldihydrocostol from the roots of *D. costus*, as well as other compounds such as β-sitosterol, daucosterol, 5-hydroxymethyl-furaldehyde, and trans-syingin (Duan et al. 2010).

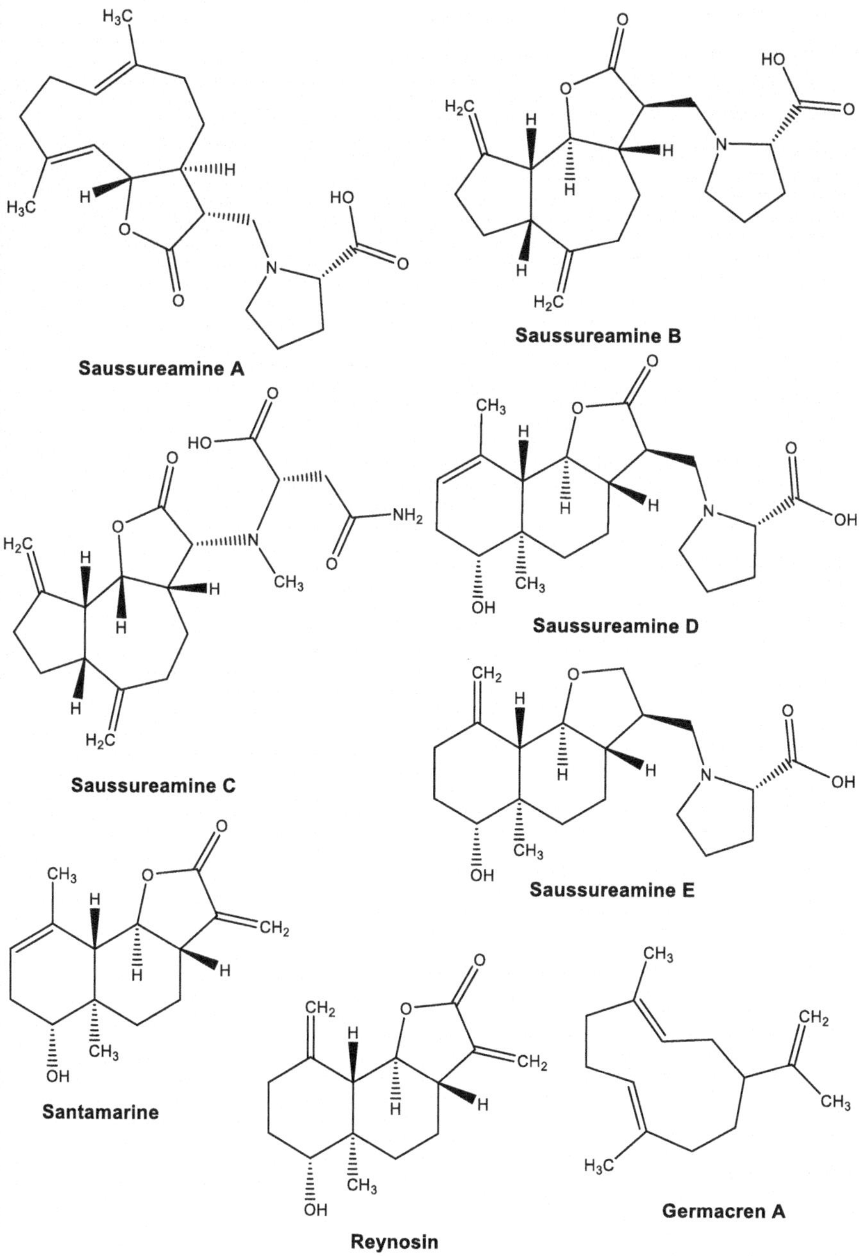

Fig. 10.2 Chemical structures of chemical constituents isolated from *D. costus*

Apart from that, 13-methoxydihydrodehydrocostus lactone and isodehydrocostus lactone 4′-methoxydehydrocostus lactone, 15-aldehyde, and D-isodehydrocostus lactone saussureal, 14-aldehyde, 13-sulfodihydrosantamarine, saussureamine D, saussureamine E 11-, 13-dihydroreynosin, 13-sulfodihydroreynosin, reynosin, 1,6-dihydroxycostic acid ethyl ester, cyclocostunolide, magnolialide, -costic acid, -isocostic acid, -costol, arbusculin A, colaritin, santamarine, 4-hydroxyeudesm-11 (13)-en-12-al, 4-hydroxy-4-methyldihydrcostol, and 4-hydroxy-4-hydroxyeudesmanes are other eudesmanes found in *D. costus* (Hassan and Masoodi 2020; Yin et al. 2005; Wang et al. 2010; Matsuda et al. 2000). Numerous germacranes, including germacrene A, germacra-1, 4,11 (13)-trienoic acid, germacra-1, 4,11 (13)-trien-12-ol, and germacra-1, 4,11, are reported in *D. costus* (13) costunolide, costunolide 15-O-D-glucopyranoside, and costunolide 12 methoxydihydrocostunolide (de Kraker et al. 2001; Matsuda et al. 2000; Hassan and Masoodi 2020). Forty flavone glycosides with antifungal action have already been found in the roots of *D. costus*. Among them are 3′-(3R-acetoxy-5,5-dimethylcyclopent-1-ene)-4′-omethylscutellarein 7-O-(6″-oacetyl-ß-D-glucopyranosyl-(1 → 3)-[a-L-rhamnopyranosyl-(1 → 2)]-β-D-glucopyranoside, kaempferol-3-O-a-L-(2a′,3a′-E-di-*p*-coumaroyl)-rhamnoside-7-O-(6″-O-acetyl-β-D-glucopyranosyl-(1 → 3)-[a-L-rhamnopyranosyl-(1 → 2)]-β-D glucopyranoside, kaempferol-3-O-β-D-glucopyranosyl (1 → 2)-β-D-(6a′-O-caffeoyl)-glucopyranoside-7-O-(6″-O-acetyl-β-D-glucopyranosyl-(1 → 3)-[a-L-rhamno-pyranosyl-(1 → 2)]-β-D-glucopyranoside, and kaempferol-3-O-β-D glucopyransoyl-(1 → 4)-a-L-rhamnopyranosyl-(1 → 6)-β-D-galactopyranoside 7-O-(6″-O-acetyl-β-D-glucopyranosyl-(1 → 3)-[a-L-rhamnopyranosyl-(1 → 2)]-β-D-glucopyranoside (Rao et al. 2007; Hassan and Masoodi 2020). Kumar et al. (2014) quantified isoalantolactone and alantolactone by the UPLC/MS/MS method in *D. costus* root collected from the Western Himalayan area of Himachal Pradesh. A UPLC–DAD study of sesquiterpene lactones in ethanolic extract revealed that alantolactone concentration was greatest in root extract compared to isoalantolactone (Kumar et al. 2014).

10.4 Pharmacological Activities

D. costus is still used according to its traditional uses in modern science. Numerous studies have examined the biological activities of *D. costus* using several in vitro and in vivo test models. It has been found that the plant's extracts have anticancer, anti-ulcer, anti-inflammatory, immunomodulatory, hepatoprotective, pesticidal, and many other properties (Baz 2009).

10.4.1 Anti-Inflammatory Activity

In mice and rats, 50, 100, and 200 mg/kg of body weight of *D. costus* ethanol extract were tested for effects on acute and chronic inflammation. It was observed that an oral dosage of 50–200 mg/kg of body weight greatly diminished paw oedema and prohibited inflammatory cells from aggregating (Gokhale et al. 2002; Hassan and Masoodi 2020). The sesquiterpene lactone santamarin, discovered from *D. costus* inhibited iNOS protein, significantly reduced nitric oxide (NO) derived from iNOS, suppressed COX-2 protein, and decreased COX-derived PGE2 generation in lipopolysaccharide (LPS)-stimulated RAW264.7 cell lines and murine peritoneal macrophage. Similar to this, santamarin reduced the production of interleukin-1 and tumor necrosis factor (TNF) (IL-1). Santamarin also inhibited the LPS-induced phosphorylation, degradation, and nuclear translocation of IB- and p65 in RAW264.7 cell lines. Additionally, santamarin enhanced heme oxygenase (HO)-1 protein and mRNA levels, which play a cytoprotective impact against inflammation (Choi et al. 2012).

The anti-inflammatory potential of the *D. costus* ethanol extract substantially reduced inducible nitric oxide synthase (iNOS) expression, COX-2, interleukin-6 (IL-6), IL-1, and tumor necrosis factor (TNF) in lipopolysaccharide (LPS)-stimulated RAW 264.7 cell lines. It was discovered that the decrease of LPS-induced nuclear translocation of p65, followed by the degradation of IB and suppression of phosphorylation of mitogen-activated protein kinases, was the mechanism behind *D. costus*' anti-inflammatory effects (Lim et al. 2020).

Recently, it was reported that DHL isolated from *D. costus* exhibits strong anti-inflammatory activity at intraperitoneal dosages of 2.5 or 5 mg/kg on the activation of macrophages by gram-positive bacteria and acute lung injury (ALI) in a mouse model (Wu et al. 2021). In gram-positive bacterium-challenged macrophages, DHL dramatically decreased M1 polarization and increased M2 polarization via blocking the p38 MAPK/NF-B pathway and activating AMPK/Nrf2 signaling (Wu et al. 2021). The outcomes of Lammari et al. (2021) evidenced that activity of metalloprotease MMP-9 and transcription of inflammatory cytokines TNF-, GM-CSF, and IL1 were reduced in nano-capsules containing *D. costus* essential oil (Lammari et al. 2021).

10.4.2 Anticancer/Antitumor Effect

It was investigated the root component costunolide, derived from *D. costus*, promotes apoptosis in human leukemia cells. Costunolide induces apoptosis by increasing the potential of the mitochondrial membrane and intensifies its effect by generating reactive oxygen species (ROS), changing the permeability of the mitochondria, and causing cytochrome C to be released into the cytoplasm. N-acetylcysteine (NAC) reduced the production of ROS, the breakdown of mitochondria, and apoptotic cell death as a result of costunolide therapy (Lee et al. 2001). The extracts induced apoptosis by upregulating pro-apoptotic molecules and downregulating anti-apoptotic molecules. After 3 days of treatment with *D. costus* extract, cell viability was assessed. The cell viability decreased dose-dependently, with the majority of cells no longer viable at 500 mg/mL of extract (Ko et al. 2005). Thara and Zuhra (2012) investigated the antiproliferative capacity of *D. costus*, via MTT assay against DLA cell lines and revealed an IC_{50} value of 70 mg/mL. The methanol extract of *D. costus* showed 85% viability (100 g/mL), whereas the chloroform extract exhibited 58.5% viability (100 g/mL) (Thara and Zuhra 2012). Jeong et al. (2002) examined the effect of costunolide on the growth of human umbilical vein endothelial cells (HUVECs). Costunolide considerably decreased HUVEC proliferation but had less than significant inhibitory action against KB3–1 cell growth. The same bioactive substance was also employed to treat activated endothelial cells by vascular endothelial growth factor (VEGF). Costunolide at noncytotoxic doses reduced HUVEC proliferation driven by VEGF. As a consequence of the findings, that compound appears to be a powerful angiogenesis inhibitor to be used as a new anticancer drug (Jeong et al. 2002).

On three soft tissue sarcoma cells of varied origins, the effects of costunolide and DHL, derived from *D. costus*, were explored (Kretschmer et al. 2012). Both medicines ($IC_{50} = 6.2–9.8$ g/mL) reduced cell viability in a dose- and time-dependent manner. The cell cycle was unaffected by costunolide, and it had little caspase 3/7 activity and cleaved caspase-3 levels. DHL reduced the number of G1 cells while boosting the S and G2/M cell numbers. Additionally, it enhanced caspase 3/7 activity, cleaved caspase-3, and PARP to cause apoptosis. Thus, the evolution of drug-resistant tumor therapeutics might benefit from this molecule as a potential lead candidate.

A popular molecular target for the treatment of cancer is the TNF-induced NF-κB pathway. *D. costus* blocks NF-κB-dependent MMP-9 expression, and one of its derivatives, costunolide, also blocks TNF-induced NF-κB transcription and MMP-9 expression. Without causing weight loss, costunolide also diminished in vivo tumor development and metastasis. We can infer from this that *D. costus* and costunolide, one of its components, may be useful in the management of highly metastatic cancer progression and metastases (Choi et al. 2013).

Similarly, in MCF-7 cells also, ethanolic extract of *D. costus* (2 or 4 M) is a significant inhibitor of 12-O-tetradecanoylphorbol-13-acetate (TPA)-induced MMP-9 production and strongly inhibits the NF-B signaling pathway (Kim et al. 2014). This establishes that suppressing NF-B pathways in MCF-7 cells inhibits MMP-9 production, which in turn lowers TPA-stimulated cancer cell invasion.

The effects and mechanisms of action of *D. costus* dried roots methanol extract (30 g/mL) on oral cancer were already studied (Moon et al. 2013). With an IC_{50} value of 30 g/mL, treatment of KB cells with *D. costus* extract reduced cell viability considerably. Starting with the 24 h treatment, the DNA ladder was seen. The proteolytic processing of caspase-3, 9, and poly (ADP-ribose) polymerase was also increased with a considerable increase in Bax and a marked drop in Bcl-2. Caspases-3/−7 activation was also confirmed in KB cells. Thus, *D costus* suppressed cell proliferation in KB human oral cancer cell lines by inducing apoptosis. The cytotoxicity of isoalantolactone and alantolactone extracted from *D. costus* roots was higher against A549 (human lung carcinoma) and C-6 (rat glioma) cells (Kumar et al. 2014).

According to research, *D. costus* root extract has cytotoxic and growth-inhibitory properties on human neuroblastoma cells (SH-SY5Y) is accomplished via mitochondria-mediated pathway (Rahman et al. 2015). The sesquiterpene lactones DHL and costunolide, both derived from *D. costus*, were studied for their potential anticancer effects in neuroblastoma cell lines (Tabata et al. 2015). The neuroblastoma cells IMR-32, NB-39, SK-N-SH, and LA-N-1 were utilized to evaluate both compounds' cytotoxicity. Following treatment with these compounds, hallmarks of cellular death such as nuclear condensation and membrane inversion were found. Caspase-7 activation and PARP cleavage were induced by both drugs. Furthermore, the sesquiterpene lactones inhibited neuroblastoma cell invasion and migration. These findings imply that these chemicals could be effective against neuroblastoma.

According to Dong et al., DCH and costunolide possess anti-colorectal cancer profile via blocking Wnt/catenin pathway (Dong et al. 2015). Numerous investigations have shown that costunolide has a variety of modes of action. The effects of costunolide against human lung squamous carcinoma cells SK-MES-1 were studied (Hua et al. 2016). It was shown that costunolide, in a dose-dependent way, induced apoptosis and cell cycle arrest at the G1/S phase in SK-MES-1 cells, which reduced their survival. Additionally, it causes downregulation of Bcl-2 and activation of caspase-3, as well as the overexpression of p53 and Bax. Furthermore, it was suggested that costunolide could induce apoptosis in SK-MES-1 cells via a mitochondria-dependent mechanism due to the substantial decline in mitochondrial membrane potential. Similar to this, DHC impaired the activity of thioredoxin reductase 1 (TrxR1) in HeLa cells, which led to an increase in ROS, a breakdown of the intracellular redox equilibrium and ultimately apoptosis. TrxR1, which is ubiquitous and highly generated in cancer cells, is an important protein needed for cellular redox control, tumor growth, apoptosis, and metastasis (Yang et al. 2016).

A study found that SMMC-7721 xenografts significantly improved therapeutically while experiencing no adverse effects from the volatile oil from *D. costus* root (Lin et al. 2016). By arresting the cell cycle at the S and G2/M phases and inducing death by activating the caspase-3 pathway, it reduced hepatocellular carcinoma cell (HCC) proliferation in vitro. MMP-9 inhibition also reduced the ability of HCC cells to penetrate and migrate. Additionally, a mechanistic analysis discovered that it can act as an inhibitor of the EGFR, inhibiting the downstream MEK/P38 and PI3-K/Akt pathways as well as EGFR activation. The results indicated that a possible anti-HCC medication may be made from the volatile oil derived from the root of *D. costus*.

Additionally, *D. costus* extract inhibits the growth of prostate cancer cells via controlling both apoptosis and autophagy (Tian et al. 2017). According to studies, the combination of costunolide and dehydrocostuslactone was more effective against breast cancer and had less negative side effects than either costunolide or DCH by itself. Proteomics and Western blot affirmation were used to evaluate the basic mechanisms of costunolide and DCH combination therapy (CD) on breast cancer cell lines (Peng et al. 2017). CD affects protein kinase A (PKA) signaling as well as 14-3-3 and c-Myc-associated apoptotic signaling. While the expressions of c-Myc, p-AKT, and p-BID were noticeably downregulated, the expressions of p53 and p-14-3-3 were drastically increased. The ratio of BAX/BCL-2 was also noticeably improved in breast cancer cells following CD therapy. The effectiveness of *D. costus* in preventing the proliferation of HeLa, T-47D, and Hep G-2 cell lines were examined as compared to the HCC1937 BL non-cancer cells (Hasson et al. 2018). As shown by the dose-dependent suppression of LDH and NO production as well as death receptors triggering in T-47D and HeLa cells but not the Hep G-2 cells, treatment of T-47D, HeLa, and HepG-2 cells with *D. costus* extract decreased cell viability, growth, and proliferation by inducing either extrinsic or intrinsic apoptotic signaling. Interestingly, lyophilized *D. costus* considerably decreased the growth of HEp-2 and T-47D cells after 48 h of treatment, whereas 24 h of treatment significantly hindered the growth of T-47D, Hep G-2, and HeLa cells. It has been suggested that the *D. costus* extract might be a source of bioactive compounds that could be utilized to develop an anticancer medication to treat liver, colon, and breast cancers (Shati 2020).

Costunolide (COS) and dehydrocostuslactone's (DCH) effects on HepG2 cell autophagy regulation have recently been studied (Okubo et al. 2020). Costunolide and DCH inhibit autophagy, resulting in the build-up of SQSTM1/p62 and microtubule-associated protein 1 light chain 3 (LC3). CL and DCH suppress autophagy by hindering autophagic flux, promoting the accumulation of LC3-II and p62. These findings call for more research into COS and DCH as possible autophagy inhibitors in the treatment of liver cancer. Interestingly, cytotoxic findings against Chinese hamster ovary (CHO) cells using *D. costus* rhizome extract-based zinc oxide nanoparticles revealed that 5 μg/mL of the ZnO particles showed cytotoxicity with IC_{50} value 3.164 ± 0.8956 μg/mL (Kolahalam et al. 2021). It has been shown that *D. costus*'s anti-oncogenic action is mediated through the bcl-2-mediated apoptotic pathway and mitochondrial cytochrome C release in the cancer cell line hepg2, suggesting that it may be effective in treating liver cancer (Alotaibi et al. 2021).

By controlling inflammation and apoptotic imbalance, *D. costus* (20, 40 mg/kg) may help to prevent and treat benign prostatic hyperplasia (Choi et al. 2021). It reduces prostate index, prostate weight, and hormone regulatory variables significantly. *D. costus* also lowered the amounts of inflammatory cytokines while increasing the levels of apoptotic marker proteins. Additionally, it raised BCL-2 accompanying X protein (BAX) expression in the prostate while diminishing B-cell lymphoma 2 (BCL-2) expression.

10.4.3 Nematocidal Activity

A nematocidal test of *D. costus* (rhizome) was conducted against pre-juvenile nematodes *Cephalobus litoralis* and *Helicotylenchus indicus* to find out their possible toxicity. The control for the mortality test was diluted water containing nematode larvae. The results from in vitro tests indicated that methanol extract from *D. costus* caused a significant portion of the mortality of second stage juveniles of both *H. indicus* and *C. litoralis*. In comparison to 0.5%, 2, and 1%, concentrations were found to be more effective. Mortality rates also increased with time with a higher percentage after 72 h (Zia-UL-Haq et al. 2012).

10.4.4 Antimicrobial Activity

The root displayed antibacterial activity against several bacteria, including *Neisseria gonorrhoea*, *Pseudomonas aeruginosa*, *Klebsiella pneumonia*, *Escherichia coli*, and *Enterobacter aerogenes*, with a minimum inhibitory concentration (MIC) of 250–400 μg/mL (El-Sawi et al. 2010).

Chitosan (a biopolymer) and *D. costus* root extract were combined to create nanoconjugates that were then studied for their potential to suppress pathogenic *Candida* species, particularly fungicide-resistant strains of *C. glabrata* and *C. albicans* (Alshubaily 2019). The antimycotic activities of the synthesized nanoconjugates from fungal chitosan/costus extract (NCt/CE) and their nanocomposite against common and antibiotic-resistant strains of *Candida albicans* and *Candida glabrata* were reviewed. It was noticed that all of the components had substantial antimycotic potentiality against all strains tested; however, the NCt/CE nanoconjugates had noticeably stronger antimicrobial action. Significant structural and morphological changes were seen in resistant strains after exposure to NCt/CE using scanning microscope imaging, pointing NCt/CE nanoconjugates as a biodegradable (natural) substance and effective antimycotic drug for managing resistant pathogenic strains of yeast.

The antifungal property of *D. costus* root extracts was investigated against a number of potent phytopathogenic fungi using a variety of solvent extracts (water, methanol, ethanol, ethyl acetate, and dichloromethane), including *Fusarium solani*, *Fusarium moniliforme*, *Fusarium oxysporum*, *Helminthosporium sativum*, and *Macrophomina phaseolina* (Al Otibi et al. 2019). All of the examined isolates were found to be very susceptible to the antifungal effects of methanol, ethanol, and dichloromethane extracts. With a 5% methanol extract of *D. costus*, M. phaseolina exhibited the highest inhibition (89%), and the morphology and cellular structure were severely damaged. Additionally, the 5% dichloromethane extract entirely halted conidial growth.

The antifungal action of *D. costus* ethanol extract against *Candida albicans* was shown to be due to a decrease in chitin synthesis or assembly and (1,3)-D-glucan production (Lee and Kim 2020). The extract's overall MIC against the tested *Candida* species ranged from 98 to 780 μg/mL. The *D. costus* ethanol extract had MICs of 780 μg/mL and 98 μg/mL for both *Candida albicans* and *Candida glabrata*, respectively. DHL, a naturally occurring sesquiterpene lactone derived from the traditional Chinese herb *D. costus*, was mentioned in the section above as well.

Researchers have demonstrated that it diminishes methicillin-resistant *Staphylococcus aureus* (MRSA)-induced ALI, suggesting that it may be an effective method of treating Gram-positive bacterial inflammation (Wu et al. 2021). *D. costus* hydroalcoholic roots extract has been documented to exhibit antimicrobic action against several microorganisms such as *Klebsiella pneumoniae*, *Staphylococcus aureus*, *Escherichia coli*, *Salmonella typhi*, *Pseudomonas aeruginosa*, *Aspergillus flavus*, and *Candida albicans* showing MIC value of 100, 50, 25, 50, 100, 25, and 50 μg/disk, respectively (Ansari et al. 2021b).

Due to its enormous potential for application in medicine, this plant species has been exploited to make nanoparticles. The biomasses of *D. costus* roots were used to synthesize different morphological MgO nanoparticles (MgONPs) (Amina et al. 2020). The green synthesis of MgONPs employed the biomass of two *D. costus* species, *Qustal hindi* and *Qustal bahri*. The antibacterial property of the produced MgONPs was tested against 6 pathogenic strains namely *E. coli*, *P. aeruginosa*, *C. tropicalis*, *C. glabrata*, *S. aureus* and *B. subtilis*. *Qustal bahri* biomass nanoparticles had inhibition zones of 12, 8, and 17 mm against *B. subtilis*, *E. coli*, and *C. tropicalis*, as opposed to *Qustal hindi* nanoparticles, which had inhibition zones of 15, 16, 17, 18, 14, and 10 mm against *E. coli*, *P. aeruginosa*, *C. glabrata*, *C. tropicalis*, *S. aureus*, and *B. subtilis* respectively. Interestingly, a notable antimicrobial effect of the ZnO nanoparticles (at 170 ppm concentration), prepared using *D. costus* methanol roots extract indicated that the gram-negative strains exhibited comparatively significant inhibitory action than gram-positive strains (Kolahalam et al. 2021). Additionally, at 170 ppm concentration, *Aspergillus niger* and *flavus*, *Fusarium oxysporum*, and *Rhizopus oryzae* all demonstrated substantial inhibitory action.

10.4.5 Antioxidant Activity

An aqueous extract of *D. costus* root exhibits its antioxidant profile by reducing the oxidative cardiac damage that isoproterenol (85 mg/kg) induces in rats. The aqueous extract of *S. lappa* (AESL) was administered orally to the rats in three doses (100, 200, and 300 mg/kg) (Mohamed Saleem et al. 2013). Long-term oral therapy of AESL in three distinct doses dramatically regained cardiac levels of creatinine kinase (CK), lactate dehydrogenase (LDH), thiobarbituric acid reactive substances (TBARS), glutathione (GSH) and aspartate transaminase (AST).

The aqueous extract of *D.* costus has been documented to be high in phenolic compounds (80 mg/g dry weight), possessing antioxidant and anti-scavenging properties (Alnahdi et al. 2017). As a result, it can be used in stable phytoformulations for cosmetic purposes. Because of the inclusion of kaempferol, gallic acid, caffeic acid, and other important phenolics, topical administration of a cosmetic cream containing *D. costus* extract has considerable anti-aging effects (Adnan et al. 2017). Interestingly, essential oils from *D. costus* collected at different harvesting periods also differs in chemical composition and antioxidant activity (Benedetto et al. 2018). The antioxidant properties of *D. costus* extract may explain their protective effects against paracetamol-induced toxicity, resulting in improved effects on the liver and reproductive system (Kadhem and Kadhum 2019). The stabilization of the Nrf2 pathway increases the synthesis of heme oxygenase (HO-1), which is another explanation for *D. costus*' antioxidant action (Lim et al. 2020). In a study, the ability of *D. costus* to prevent thorium deposition in the parts of the brain of male albino rats, which results in oxidative stress, was examined. It was reported that because of its antioxidant and chelating qualities, costus extract helps to mitigate some of the harmful effects of thorium that accumulates in different parts of the brain, causing oxidative stress and disrupting ionic content and neurotransmitter levels (Abdel-Rahman et al. 2020).

By reducing oxidative stress, as shown by restored antioxidant enzyme activity and suppressed lipid peroxidation, the ethanolic extract of *D. costus* also lessens triamcinolone acetonide-triggered splenic and pulmonary tissue injury in mice (Abdel-Rahman et al. 2020). Recent research on a human hepatoma cells to evaluate the anticancer, antioxidant, and hepatoprotective effects of the *D. costus* root revealed that the ethanolic, aqueous, and hydroethanolic extracts, all exhibited outstanding antioxidant properties (Ansari et al. 2021a, b). Thiamethoxam (TMX, neonicotinoid insecticide) reproductive toxicity on male rats was investigated, as well as the capacity of *D. costus* (costus roots) to relieve TMX toxicity (Habotta et al. 2021). The administration of costus roots (200 mg/kg) significantly reduced TMX-induced alterations in testicular glutathione levels and antioxidant activities of SOD and CAT, demonstrating that these plant components have potent antioxidant properties.

10.4.6 Anticonvulsant Activity

To test their anticonvulsant potency against pentylenetetrazole (PTZ), picrotoxin-induced convulsion, and maximum electric shock (MES) in mice, numerous extracts of *Saussurea lappa* (SL) roots were prepared from a range of solvents, including petroleum ether, methanol, and water. Out of these, petroleum ether extract at dosages of 100 and 300 mg/kg (intraperitoneal route) enhanced latency to clonic convulsions and decreased mortality in mice treated with pentylenetetrazole (PTZ) and picrotoxin (Ambavade et al. 2009). Maximum electric shock-triggered convulsions and pentylenetetrazol-triggered clonic convulsions were used to test the anticonvulsant efficacy of an ethanolic extract of SL. The extract showed strong anticonvulsant activity in both assays (Gupta Pushpraj et al. 2009).

10.5 Clinical Studies

Very few clinical studies are there on therapeutic uses of *D. costus*. In order to treat chronic hepatitis B, a study intended to assess the antiviral property of *D. costus* against the hepatitis B virus (HBV) (CHB) (Ansari et al. 2018). At week 12, HBsAg loss was detected in 35.71% ($n = 14$) of patients in the HBeAg positive ($P < 0.05$) and negative ($P = 0.10$) groups, respectively (after treatment). At week 12, HBeAg loss was detected in 71.42% of individuals ($P < 0.001$). At week 12, cVR was accomplished in 57.14% ($P < 0.01$) and 37.5% ($P < 0.05$) of patients in the HBeAg positive and negative groups, respectively, with ALT normalization in 85.71% and 81.25% of patients ($P < 0.001$).

It was shown that *D. costus* exhibit antiviral properties against HBV in CHB patients, as well as being quite safe during a 12-week treatment period. Patients with persistent superficial gastritis were administered a decoction of the *D. costus*, and their serum gastrin, stomach acidity, and plasma somatostatin levels were measured (Chen et al. 1994). A decoction of SL was also administered to five healthy participants. It was discovered that stomach emptying time and endogenous motilin discharge both increase. Nevertheless, there were no alterations in the levels of serum gastrin, acidity output, or plasma somatostatin.

10.6 Studies Related to Safety and Toxicity

About the toxicity and safety of *D. costus*, little information is available. In general, the root of *D. costus* is safe to consume; however, it may contain an ingredient called aristolochic acid that has nephrotoxic and carcinogenic properties (Mujammami 2020). A study suggests that ethephon-induced renal toxicity can be reduced through consumption of *D. costus* root extract. In addition, by improving sperm abnormalities, testicular tissue damage and DNA damage, as well as P53 protein expression, *D costus* has shown protective effects against ethephon induced reproductive toxicity in rats. In many allergic disorders, dehydrocostus lactone suppresses inflammation in the airways by binding to dimerized tumor proteins (Mujammami 2020). According to Abd El-Rahman et al. (2020), *D. costus* oral administration was found to effectively reverse triamcinolone acetonide-induced immunosuppression in the combined biochemical, histopathological, and immunohistochemical studies. The anti-inflammatory effects of triamcinolone acetonide may also be moderately enhanced by *D. costus*. Moreover, *D. costus* could combat the oxidative stress caused by triamcinolone acetonide in lung tissues and protecting them from apoptosis. Additionally, how *D. costus* is supplemented has a significant impact on how promptly triamcinolone acetonide side effects are recovered (Abd El-Rahman et al. 2020). Similar to this, researchers examined thyroid gland activity, antioxidant effect, and thorium toxicity in adult rats using *D. costus* root extract. To counteract the effects of thorium, *D. costus* extract stimulated the body's antioxidant system.

The sesquiterpene lactones present in *D. costus* may produce an allergic reaction in those with a history of allergic reactions to any *Saussurea* species. There have been reports of those exposed to sesquiterpene lactones developing contact dermatitis as well as genotoxic, embryotoxic, and mutagenic effects (Seca et al. 2017).

10.7 Conclusions and Future Recommendations

D. costus is widely utilized in traditional medicine around the world for a variety of well-known medicinal uses; Ayurvedic, Unani, and Chinese traditional medicine are just a few of the traditional medical systems that use them. Different compounds isolated are costunolide, isodihydrocostunolide, cynaropicrin, and DHL that display these various biological actions. *D. costus* has a lot of potential for human health, and its therapeutic benefits should be studied more thoroughly. Preclinical and clinical studies on the utilization of these plants are urgently needed, as are further in-depth investigations into all bioactive chemicals and their biomolecular mechanisms at the cellular and tissue levels.

However, the majority of these studies concentrated on in vitro assessments of biological activities, with only a few studies focusing on comprehensive mechanism of action in vivo models. Similarly, investigations that focus on bioactivity-guided separation and identification of active components are limited when compared to studies that focus on biological activity screening. Future research should be designed to fill in these research gaps.

References

Abd El-Rahman GI, Behairy A, Elseddawy NM, Batiha GE-S, Hozzein WN, Khodeer DM, Abd-Elhakim MY (2020) Saussurea lappa ethanolic extract attenuates triamcinolone acetonide-induced pulmonary and splenic tissue damage in rats via modulation of oxidative stress, inflammation, and apoptosis. Antioxidants (Basel) 9(5):396. https://doi.org/10.3390/antiox9050396

Abdel-Rahman M, Rezk MM, Ahmed-Farid OA, Essam S, Abdel Moneim AE (2020) Saussurea lappa root extract ameliorates the hazards effect of thorium induced oxidative stress and neuroendocrine alterations in adult male rats. Environ Sci Pollut Res 27(12):13237–13246. https://doi.org/10.1007/s11356-020-07917-y

Adnan Q, Akhtar N, Khan BA (2017) Phytoformulation of Sassurea lappa plant extract: a single blind, noninvasive and split face study of cream on various skin parameters. Pak J Pharm Sci 30(5(Supplementary)):1981–1986

Al Otibi F, Rizwana H, Alharbi RI, Alshaikh N, Albasher G (2019) Antifungal effect of Saussurea lappa roots against phytopathogenic fungi and resulting morphological and ultrastructural changes. Gesunde Pflanz 72(1):57–67. https://doi.org/10.1007/s10343-019-00483-5

Alnahdi HS, Danial EN, Elhalwagy MEAE, Ayaz NO (2017) Phytochemical studies, antioxidant properties and antimicrobial activities of herbal medicinal plants costus and cidir used in Saudi Arabia. Int J Pharm 13(5):481–487. https://doi.org/10.3923/ijp.2017.481.487

Alotaibi AA, Bepari A, Assiri RA, Niazi SK, Nayaka S, Rudrappa M, Nagaraja SK, Bhat MP (2021) Saussurea lappa exhibits anti-oncogenic effect in hepatocellular carcinoma, HepG2 cancer cell line by Bcl-2 mediated apoptotic pathway and mitochondrial cytochrome C release. Curr Issues Mol Biol 43(2):1114–1132. https://doi.org/10.3390/cimb43020079

Alshubaily FA (2019) Enhanced antimycotic activity of nanoconjugates from fungal chitosan and Saussurea costus extract against resistant pathogenic Candida strains. Int J Biol Macromol 141: 499–503. https://doi.org/10.1016/j.ijbiomac.2019.09.022

Ambavade SD, Mhetre NA, Muthal AP, Bodhankar SL (2009) Pharmacological evaluation of anticonvulsant activity of root extract of Saussurea lappa in mice. Eur J Integr Med 1(3): 131–137. https://doi.org/10.1016/j.eujim.2009.08.159

Amina M, Al Musayeib NM, Alarfaj NA, El-Tohamy MF, Oraby HF, Al Hamoud GA, Bukhari SI, Moubayed NMS (2020) Biogenic green synthesis of MgO nanoparticles using Saussurea costus biomasses for a comprehensive detection of their antimicrobial, cytotoxicity against MCF-7 breast cancer cells and photocatalysis potentials. PLoS One 15(8):e0237567. https://doi.org/10.1371/journal.pone.0237567

Ansari S, Siddiqui MA, Maaz M (2018) Hepatocurative effect of Saussurea lappa, C.B Clarke and Artemisia absinthium, Linn in chronic hepatitis B. J Young Pharm 10(3):354–357. https://doi.org/10.5530/jyp.2018.10.78

Ansari S, Hasan K, Bhat S (2021a) Anticancer, antioxidant, and hepatoprotective activity of Saussurea lappa, C.B. clarke (qust) on human hepatoma cell line. J Cancer Res Ther 17(2): 499. https://doi.org/10.4103/jcrt.JCRT_571_19

Ansari S, Maaz M, Ahmad I, Syed Kazim H, Sajad Ahmad B, Syed Kazim N, Husain M (2021b) Quality control, HPTLC analysis, antioxidant and antimicrobial activity of hydroalcoholic extract of roots of qust (Saussurea lappa, C.B Clarke). Drug Metabol Drug Interact 36(2): 145–153. https://doi.org/10.1515/dmpt-2020-0159

Baz LA (2009) Assessment of therapeutic value of black Costus (Saussurea lappa) using several parameters. King Abdul Aziz University, Jeddah, Saudi Arabia

Benedetto C, D'Auria M, Mecca M, Prasad P, Singh P, Singh S, Sinisgalli C, Milella L (2018) Chemical and biological evaluation of essential oil from Saussurea costus (Falc.) Lipsch. from Garhwal himalaya collected at different harvesting periods. Nat Prod Res 33(16):2355–2358. https://doi.org/10.1080/14786419.2018.1440219

Cárdenas DM, Cala A, Molinillo JMG, Macías FA (2017) Preparation and phytotoxicity study of lappalone from dehydrocostuslactone. Phytochem Lett 20:66–72. https://doi.org/10.1016/j.phytol.2017.04.017

Chen S, Li Y, He F (1994) Effect of Saussurea lappa on gastric functions. Zhongguo Zhong Xi yi jie he za zhi Zhongguo Zhongxiyi jiehe zazhi 14(7):406–408

Cho JY, Baik KU, Jung JH, Park MH (2000) In vitro anti-inflammatory effects of cynaropicrin, a sesquiterpene lactone, from Saussurea lappa. Eur J Pharmacol 398(3):399–407

Choi HG, Lee DS, Li B, Choi YH, Lee SH, Kim YC (2012) Santamarin, a sesquiterpene lactone isolated from Saussurea lappa, represses LPS-induced inflammatory responses via expression of heme oxygenase-1 in murine macrophage cells. Int Immunopharmacol 13(3):271–279. https://doi.org/10.1016/j.intimp.2012.04.016

Choi YK, Cho S-G, Woo S-M, Yun YJ, Jo J, Kim W, Shin YC, Ko S-G (2013) Saussurea lappaClarke-derived costunolide prevents TNFα-induced breast cancer cell migration and invasion by inhibiting NF-κB activity. Evid Based Complement Alternat Med 2013:1–10. https://doi.org/10.1155/2013/936257

Choi D-H, Kim J-Y, An J-H, Sung S-H, Kong H-S (2021) Effects of Saussurea costus on apoptosis imbalance and inflammation in benign prostatic hyperplasia. J Ethnopharmacol 279:114349. https://doi.org/10.1016/j.jep.2021.114349

Costus (2021) Google Arts & Culture. https://artsandculture.google.com/entity/costus/m03nr7p1?hl=en. Accessed 20 Oct 2021 de Kraker J-W, Franssen MCR, de Groot A, Shibata T, Bouwmeester HJ (2001) Germacrenes from fresh costus roots. Phytochemistry 58(3):481–487. https://doi.org/10.1016/s0031-9422(01)00291-6

Dong G-z, Shim A-R, Hyeon JS, Lee HJ, Ryu J-H (2015) Inhibition of Wnt/β-catenin pathway by Dehydrocostus lactone and Costunolide in colon cancer cells. Phytother Res 29(5):680–686. https://doi.org/10.1002/ptr.5299

Duan J-A, Hou P, Tang Y, Liu P, Su S, Liu H (2010) A new sesquiterpene and other constituents from Saussurea lappa root. Nat Prod Commun 5(10):1934578X1000501002

El-Sawi NM, Backer W, Aly MM, Baz L (2010) Assessment of therapeutic value of black costus (Saussurea lappa) using several parameters. Int J Environ Sci 5(5):832–841

Gokhale AB, Damre AS, Kulkarni KR, Saraf MN (2002) Preliminary evaluation of anti-inflammatory and anti-arthritic activity of S. lappa, A. speciosa and A. aspera. Phytomedicine 9(5):433–437. https://doi.org/10.1078/09447110260571689

Gupta Pushpraj S, Jadhav S, Ghaisas M, Deshpande A (2009) Anticonvulsant activity of Saussurea lappa. Pharmacology 3:809–814

Habotta OA, Ateya A, Saleh RM, El-Ashry ES (2021) Thiamethoxam-induced oxidative stress, lipid peroxidation, and disturbance of steroidogenic genes in male rats: palliative role of Saussurea lappa and Silybum marianum. Environ Toxicol 36(10):2051–2061. https://doi.org/10.1002/tox.23322

Hassan R, Masoodi MH (2020) Saussurea lappa: a comprehensive review on its pharmacological activity and phytochemistry. Curr Tradit Med 6(1):13–23. https://doi.org/10.2174/2215083805666190626144909

Hasson SS, HA-S AS, Al-Busaidi JZ, Al-Balushi MS, Hakkim FL, Rashan L, Aleemallah GM, Al-Jabri AA (2018) Potential of Aucklandia lappa decne ethanolic extract to trigger apoptosis of human T47D and hela cells. Asian Pac J Cancer Prev 19(7):1917–1925. https://doi.org/10.22034/APJCP.2018.19.7.1917

Hua P, Zhang G, Zhang Y, Sun MEI, Cui R, Li XIN, Li B, Zhang X (2016) Costunolide induces G1/S phase arrest and activates mitochondrial-mediated apoptotic pathways in SK-MES 1 human lung squamous carcinoma cells. Oncol Lett 11(4):2780–2786. https://doi.org/10.3892/ol.2016.4295

Jeong S-J, Itokawa T, Shibuya M, Kuwano M, Ono M, Higuchi R, Miyamoto T (2002) Costunolide, a sesquiterpene lactone from Saussurea lappa, inhibits the VEGFR KDR/Flk-1 signaling pathway. Cancer Lett 187(1–2):129–133. https://doi.org/10.1016/s0304-3835(02)00361-0

Kadhem MA, Kadhum SA (2019) Protective effect of ethanolic extract of Saussurea Lappa on paracetamol induced toxicity in female rabbits. J Phys Conf Ser 6:062043

Kasana S, Dwivedi MD, Uniyal PL, Pandey AK (2020a) An updated circumscription of Saussurea (Cardueae, Asteraceae) and allied genera based on morphological and molecular data. Phytotaxa 450(2):173–187. https://doi.org/10.11646/phytotaxa.450.2.3

Kasana S, Uniyal PL, Pandey AK (2020b) A taxonomic revision of the genus Dolomiaea (Asteraceae: Cardueae) in India. J Indian Assoc Angiosperm Taxonomy 30(1):270–277

Kim H-R, Kim J-M, Kim M-S, Hwang J-K, Park Y-J, Yang S-H, Kim H-J, Ryu D-G, Lee D-S, Oh H, Kim Y-C, Rhee Y-J, Moon B-S, Yun J-M, Kwon K-B, Lee Y-R (2014) Saussurea lappa extract suppresses TPA-induced cell invasion via inhibition of NF-κB-dependent MMP-9 expression in MCF-7 breast cancer cells. BMC Complement Altern Med 14(1):170. https://doi.org/10.1186/1472-6882-14-170

Ko SG, Kim H-P, Jin D-H, Bae H-S, Kim SH, Park C-H, Lee JW (2005) Saussurea lappa induces G2-growth arrest and apoptosis in AGS gastric cancer cells. Cancer Lett 220(1):11–19. https://doi.org/10.1016/j.canlet.2004.06.026

Kolahalam LA, Prasad KRS, Murali Krishna P, Supraja N (2021) Saussurea lappa plant rhizome extract-based zinc oxide nanoparticles: synthesis, characterization and its antibacterial, antifungal activities and cytotoxic studies against Chinese hamster ovary (CHO) cell lines. Heliyon 7(6):e07265. https://doi.org/10.1016/j.heliyon.2021.e07265

Kretschmer N, Rinner B, Stuendl N, Kaltenegger H, Wolf E, Kunert O, Boechzelt H, Leithner A, Bauer R, Lohberger B (2012) Effect of costunolide and dehydrocostus lactone on cell cycle, apoptosis, and ABC transporter expression in human soft tissue sarcoma cells. Planta Med 78(16):1749–1756. https://doi.org/10.1055/s-0032-1315385

Kumar A, Kumar S, Kumar D, Agnihotri VK (2014) UPLC/MS/MS method for quantification and cytotoxic activity of sesquiterpene lactones isolated from Saussurea lappa. J Ethnopharmacol 155(2):1393–1397. https://doi.org/10.1016/j.jep.2014.07.037

Kumari A, Sharma D, Bawa R, Kaushal P (2012) Non timber forest products used for mother and child health care in tribal and remote areas of Himachal Pradesh. Indian For 138(12):1151–1159

Lammari N, Demautis T, Louaer O, Meniai AH, Casabianca H, Bensouici C, Devouassoux G, Fessi H, Bentaher A, Elaissari A (2021) Nanocapsules containing Saussurea lappa essential oil: formulation, characterization, antidiabetic, anti-cholinesterase and anti-inflammatory potentials. Int J Pharm 593:120138. https://doi.org/10.1016/j.ijpharm.2020.120138

Lee H-S, Kim Y (2020) Aucklandia lappa causes cell wall damage in Candida albicans by reducing chitin and (1,3)-β-D-glucan. J Microbiol Biotechnol 30(7):967–973. https://doi.org/10.4014/jmb.2002.02025

Lee M-G, Lee K-T, Chi S-G, Park J-H (2001) Costunolide induces apoptosis by ROS-mediated mitochondrial permeability transition and cytochrome C release. Biol Pharm Bull 24(3): 303–306. https://doi.org/10.1248/bpb.24.303

Lee HI, Lee J, Hwang D, Lee GR, Kim N, Kwon M, Lee H, Piao D, Kim HJ, Kim NY, Kim HS, Seo EK, Kang D, Jeong W (2019) Dehydrocostus lactone suppresses osteoclast differentiation by regulating NFATc1 and inhibits osteoclast activation through modulating migration and lysosome function. FASEB J 33(8):9685–9694. https://doi.org/10.1096/fj.201900862R

Li W, Ma YB, Mao YQ, Lin T (2018) Dehydrocostus lactone suppresses cell growth and induces apoptosis in recombinant human papilloma virus-18 HaCaT cells via the PI3K/Akt signaling pathway. Mol Med Rep 17(6):7925–7930. https://doi.org/10.3892/mmr.2018.8805

Li Z, Yuan G, Lin X, Liu Q, Xu J, Lian Z, Song F, Zheng J, Xie D, Chen L, Wang X, Feng H, Zhou M, Yao G (2019) Dehydrocostus lactone (DHC) suppresses estrogen deficiency-induced osteoporosis. Biochem Pharmacol 163:279–289. https://doi.org/10.1016/j.bcp.2019.02.002

Lim JS, Lee SH, Lee SR, Lim H-J, Roh Y-S, Won EJ, Cho N, Chun C, Cho Y-C (2020) Inhibitory effects of Aucklandia lappa Decne. Extract on inflammatory and oxidative responses in LPS-treated macrophages. Molecules 25(6):1336. https://doi.org/10.3390/molecules25061336

Lin X, Peng Z, Fu X, Liu C, Xu Y, Ji W, Fan J, Chen L, Fang L, Huang Y, Su C (2016) Volatile oil from Saussurea lappa exerts antitumor efficacy by inhibiting epithelial growth factor receptor tyrosine kinase-mediated signaling pathway in hepatocellular carcinoma. Oncotarget 7(48): 79761–79773. https://doi.org/10.18632/oncotarget.12962

Liu X-n, Li H-m, Wang S-p, Zhang J-z, Liu D-l (2021) Sesquiterpene lactones of Aucklandia lappa: pharmacology, pharmacokinetics, toxicity, and structure–activity relationship. Chin Herb Med 13(2):167–176. https://doi.org/10.1016/j.chmed.2020.11.005

Matsuda H, Kageura T, Inoue Y, Morikawa T, Yoshikawa M (2000) Absolute stereostructures and syntheses of saussureamines A, B, C, D and E, amino acid–sesquiterpene conjugates with gastroprotective effect, from the roots of Saussurea lappa. Tetrahedron 56(39):7763–7777. https://doi.org/10.1016/s0040-4020(00)00696-7

Mishra AP, Saklani S, Sharifi-Rad M, Iriti M, Salehi B, Maurya VK, Rauf A, Milella L, Rajabi S, Baghalpour N, Sharifi-Rad J (2018) Antibacterial potential of Saussurea obvallata petroleum ether extract: a spiritually revered medicinal plant. Cell Mol Biol (Noisy-le-Grand) 64(8):65–70

Mishra AP, Saklani S, Parcha V, Nigam M, Coutinho HDM (2021) Antibacterial activity and phytochemical characterisation of Saussurea gossypiphora D. Arch Microbiol 203(8): 5055–5065. https://doi.org/10.1007/s00203-021-02494-1

Mohamed Saleem TS, Prasanthi A, Vishnu MN, Lokanath N, Madhavi M, Mallika G (2013) Aqueous extract of Saussurea lappa root ameliorate oxidative myocardial injury induced by isoproterenol in rats. J Adv Pharm Technol Res 4(2):94. https://doi.org/10.4103/2231-4040.111525

Moon S-M, Yun SJ, Kook J-K, Kim H-J, Choi MS, Park BR, Kim S-G, Kim B-O, Lee S-Y, Ahn H, Chun HS, Kim DK, Kim CS (2013) Anticancer activity of Saussurea lappa extract by apoptotic pathway in KB human oral cancer cells. Pharm Biol 51(11):1372–1377. https://doi.org/10.3109/13880209.2013.792847

Mujammami M (2020) Clinical significance of Saussurea costus in thyroid treatment. Saudi Med J 41(10):1047–1053. https://doi.org/10.15537/smj.2020.10.25416

Okubo S, Ohta T, Fujita H, Shoyama Y, Uto T (2020) Costunolide and dehydrocostuslactone from Saussurea lappa root inhibit autophagy in hepatocellular carcinoma cells. J Nat Med 75(1): 240–245. https://doi.org/10.1007/s11418-020-01462-1

Omer REE, Koua FHM, Abdelhag IM, Ismail AM (2019) Gas chromatography/mass spectrometry profiling of the costus plant Saussurea lappa (Decne.) CB clarke root extracts and their anti-bacterial activity. J Appl Pharm Sci 9(05):073–081

Pandey MM, Rastogi S, Rawat AKS (2007) Saussurea costus: botanical, chemical and pharmacological review of an ayurvedic medicinal plant. J Ethnopharmacol 110(3):379–390. https://doi.org/10.1016/j.jep.2006.12.033

Peng Z, Wang Y, Fan J, Lin X, Liu C, Xu Y, Ji W, Yan C, Su C (2017) Costunolide and dehydrocostuslactone combination treatment inhibit breast cancer by inducing cell cycle arrest and apoptosis through c-Myc/p53 and AKT/14-3-3 pathway. Sci Rep 7(1):41254. https://doi.org/10.1038/srep41254

Plants of the World Online (2021) Dolomiaea costus (Falc.) Kasana & A. K. Pandey. Plants of the World Online, Board of Trustees of the Royal Botanic Gardens, Kew. http://powo.science.kew.org/taxon/urn:lsid:ipni.org:names:77210782-1. Accessed 20 Oct 2021

Rahman MA, Hong J-S, Huh S-O (2015) Antiproliferative properties of Saussurea lappa clarke root extract in SH-SY5Y neuroblastoma cells via intrinsic apoptotic pathway. Anim Cell Syst 19(2): 119–126. https://doi.org/10.1080/19768354.2015.1008041

Rao K, Babu G, Ramnareddy Y (2007) Acylated flavone glycosides from the roots of Saussurea lappa and their antifungal activity. Molecules 12(3):328–344. https://doi.org/10.3390/12030328

Seca AM, Silva AM, Pinto DC (2017) Parthenolide and parthenolide-like sesquiterpene lactones as multiple targets drugs: current knowledge and new developments. Stud Nat Prod Chem 52:337–372

Shati AA, Alkahtani MA, Alfaifi MY, Elbehairi SEI, Elsaid FG, Prasanna R, Mir MA (2020) Secondary metabolites of saussurea costus leaf extract induce apoptosis in breast, liver, and colon cancer cells by caspase-3-dependent intrinsic pathway. BioMed Res Int

Tabata K, Nishimura Y, Takeda T, Kurita M, Uchiyama T, Suzuki T (2015) Sesquiterpene lactones derived from Saussurea lappa induce apoptosis and inhibit invasion and migration in neuroblastoma cells. J Pharmacol Sci 127(4):397–403. https://doi.org/10.1016/j.jphs.2015.01.002

Thara K, Zuhra K (2012) Comprehensive in-vitro pharmacological activities of different extracts of Saussurea lappa. Eur J Exp Biol 2(2):417–420

Tian X, Song HS, Cho YM, Park B, Song Y-J, Jang S, Kang SC (2017) Anticancer effect of Saussurea lappa extract via dual control of apoptosis and autophagy in prostate cancer cells. Medicine 96(30):e7606. https://doi.org/10.1097/md.0000000000007606

Waly N (2009) Verifying the scientific name of costus [Saussurea lappa ((Decne.)C.B.Clarke.)—Asteraceae]. Science 21(2):327–334. https://doi.org/10.4197/Sci.21-2.10

Wang F, Xie Z-H, Gao Y, Xu Y, Cheng X-L, Liu J-K (2008) Sulfonated Guaianolides from Saussurea lappa. Chem Pharm Bull 56(6):864–865. https://doi.org/10.1248/cpb.56.864

Wang Y-F, Ni Z-Y, Dong M, Cong B, Shi Q-W, Gu Y-C, Kiyota H (2010) Secondary metabolites of plants from the genus Saussurea: chemistry and biological activity. Chem Biodivers 7(11): 2623–2659. https://doi.org/10.1002/cbdv.200900406

Wu Y-X, Jiang F-J, Liu G, Wang Y-Y, Gao Z-Q, Jin S-H, Nie Y-J, Chen D, Chen J-L, Pang Q-F (2021) Dehydrocostus lactone attenuates methicillin-resistant Staphylococcus aureus-induced inflammation and acute lung injury via modulating macrophage polarization. Int J Mol Sci 22(18):9754. https://doi.org/10.3390/ijms22189754
Yang M, Zhang J, Li Y, Han X, Gao K, Fang J (2016) Bioassay-guided isolation of dehydrocostus lactone from Saussurea lappa: a new targeted cytosolic thioredoxin reductase anticancer agent. Arch Biochem Biophys 607:20–26. https://doi.org/10.1016/j.abb.2016.08.008
Yin H-q, Fu H-w, Hua H-m, Qi X-l, Li W, Sha Y, Pei Y-h (2005) Two new sesquiterpene lactones with the sulfonic acid group from Saussurea lappa. Chem Pharm Bull 53(7):841–842. https://doi.org/10.1248/cpb.53.841
Zia-UL-Haq M, Shahid SA, Khan BA, Imran I, Qayum M, Akhter M, Khan Z, Muhammed S (2012) Nematicidal potential of selected flora of Pakistan. J Med Plant Res 6(24):4087–4090

Chapter 11
Eclipta prostrata (L.) L.: Traditional Use, Phytochemistry, and Pharmacology

Rupa Sanyal, Saheli Nandi, Sujata Mandal, Saikat Dewanjee, Abdel Rahman Al-Tawaha, Ercan Bursal, Protha Biswas, Manoj Kumar, Radha, Samapika Nandy, Abilash Valsala Gopalakrishnan, Md. Habibur Rahman, Mahipal S. Shekhawat, Devendra Kumar Pandey, Tabarak Malik, and Abhijit Dey

Abstract Herbs had been exploited by different cultures throughout history. In public wellness programs of the countries, drugs of natural origin play a crucial role. *Eclipta prostrata* (L.) L. also known by the name of *Eclipta alba* (L.) Hassk is generally called bhringraj in Indian traditional medicine and false daisy in English. It is a common folk medicinal plant falling under the family Asteraceae. It is an annual, herbaceous, small, branched plant occurring mainly in the tropical and subtropical areas. The plant performs a historic contribution in the pharmaceuticals and has a reassuring cosmetic application in addition to therapeutic applications. The plant

R. Sanyal · S. Nandi
Department of Botany, Bhairab Ganguly College, Kolkata, West Bengal, India

S. Mandal · P. Biswas · S. Nandy · A. Dey (✉)
Department of Life Sciences, Presidency University, Kolkata, West Bengal, India
e-mail: abhijit.dbs@presiuniv.ac.in

S. Dewanjee
Advanced Pharmacognosy Research Laboratory, Department of Pharmaceutical Technology, Jadavpur University, Kolkata, West Bengal, India

A. R. Al-Tawaha
Department of Biological Sciences, Al-Hussein Bin Talal University, Ma'an, Jordon

E. Bursal
Department of Biochemistry, Mus Alparslan University, Mus, Turkey

M. Kumar
Chemical and Biochemical Processing Division, ICAR—Central Institute for Research on Cotton Technology, Mumbai, Maharashtra, India

Radha
School of Biological and Environmental Sciences, Shoolini University of Biotechnology and Management Sciences, Solan, Himachal Pradesh, India

A. V. Gopalakrishnan
Department of Biomedical Sciences, School of Bio Sciences and Technology, Vellore Institute of Technology (VIT), Vellore, Tamil Nadu, India

H. P. Devkota, T. Aftab (eds.), *Medicinal Plants of the Asteraceae Family*,
https://doi.org/10.1007/978-981-19-6080-2_11

possesses a number of bioactive compounds including coumestans, flavonoids, volatile oils, terpenoids and their glycosides, steroids, polyacetylenes, and polypeptides, etc. which are responsible for some of the important pharmacological activities including antimicrobial, analgesic, anti-nociceptive, anti-inflammatory, hepatoprotective, antiviral, immunomodulatory activity, etc. In ayurveda, the extract of the leaf is contemplated as a rejuvenative especially for black hairs and a potential tonic for liver. Several researchers evaluated the acute toxicity of the plant concluding that the LD_{50} (Lethal dose 50) of *E. prostrata* was more than 2.0 g/kg in mice and rats, hence considered a moderately safe drug. This book chapter highlights the information on taxonomic description, photochemical constituents, traditional and folk uses, pharmacology and safety, and toxicity of *E. prostrata* in an aim to furnish recommendations for future research and possibilities for a good application of the herb. Particular attention is given to hepatoprotective, antidiabetic effect, hair growth and alopecia, analgesic, anti-inflammatory, antioxidant, antimicrobial, immunomodulatory, anticancer, and antitumor effects of the plant so that the possible utilization of the plant can be evaluated in pharmaceutics as well as in agricultural resource.

Keywords *Eclipta prostrata* (L.) L. · Pharmacological activities · Bhringraj · Safety and toxicity · Coumestans · Traditional use

11.1 Introduction

E. prostrata (L.) L. is a member of the Asteraceae family. It is called Bhringraj in Ayurveda, Ecliptae Herba in China and false daisy in English (Puri 2003). It possesses a sharp, bitter, and dry taste and exploited in the Ayurvedic system of medication in India (Mukhopadhyay et al. 2018). The plant thrives in subtropical, tropical, and warm temperate regions across the globe. It usually occurs in wet and poorly drained regions; along the dikes of paddy fields, streams and often seen in grassland and in uplands exhibiting precipitation of about 1200 mm or above. The herb is found across the World but is generally native to Asia and extensively disseminate all around India, Brazil, China, and Thailand (Shekokar and Nayak 2017). *E. Prostrata* has multiple therapeutic uses among which Bhringaraj oil

M. H. Rahman
Department of Global Medical Science, Wonju College of Medicine, Yonsei University, Wonju, Gangwon-do, South Korea

M. S. Shekhawat
Plant Biotechnology Unit, KM Government Institute for Postgraduate Studies and Research, Puducherry, India

D. K. Pandey
Department of Biotechnology, Lovely Professional University, Phagwara, Punjab, India

T. Malik
Institute of Health, Jimma University, Jimma, Ethiopia

extracted from the plant, often called "king of the hair," serves as a popular hair oil to support hair growth and retain black hairs. All the parts of the plant contain some phytochemicals. The major phytochemical compounds include coumestans, polyacetylenes, polypeptides, steroids, derivatives of thiophene, flavonoids, and triterpenes (Chauhan et al. 2012). The presence of these phytochemicals is responsible for a number of pharmacological activities which includes analgesic, anticancer, antimyotoxic, antileprotic, antioxidant, antihemorrhagic, antiviral, antihepatotoxic, hypotensive, antibacterial, ovicidal, spasmogenic, and hepatoprotective. Hence, the plant extract exhibits a crucial role in pharmaceutical and cosmetic applications (Mukhopadhyay et al. 2018). The herb has been given much attention by several researchers and investigators. Multiple investigations had been conducted on the pharmacological activities and biomolecules of the plant and found that this herb is rich in steroids, phenolic compounds, polysaccharides, alkaloids, and polyacetate.

Three types of *E. prostrata* have been found in the Asian subcontinent. The first type is white flowered, the second type is yellow flowered, and the third type is black fruit color (Bhalerao et al. 2013). The herb is annual, erect, or prostrate and roughly haired. Stem is herbaceous, cylindrical, much branched, having distinct nodes with often rooting at lower nodes, green in color sometimes brown, and having uneven texture due to the presence of white, small hairs. Leaves usually opposite, oblong, lanceolate, acute to subacute, with dense hairs on both surfaces, sessile to subsessile, length varies from 2.2 cm to 8.5 cm, and width varies from 1.2 cm to 2.3 cm (Yadav et al. 2011). The stalks of the flowers emerge from the leaf axis, and the color of the flowers ranges from white to yellow. On the axillary peduncle lies the solitary flower head, with roughly eight ovated involucre, herbaceous, bracts which are strigose with dense hairs. Each flower head consists of small, white, or yellow ray and disk flowers. Ray flowers are ligulate, not serrated and about the length of bracts. The herb possesses tubular disc flowers with four-toothed corolla, pappus minute, or absent. Stamens are epipetalous, five in number, and filament free, anthers forming a tube, bicarpellary pistil, ovary unicellular and inferior consisting of single basal ovule. Fruit brown in color and single seeded with a slender wing. Two carpels unit together to form the dry fruit, each has a single seed and which do not break open. The seed is hairy, dark brown, non-endospermic, 0.2–0.25 cm and 0.1 cm length and breadth, respectively. Root system well established, shallow taproot, cylindrical, about diameter of 7 mm, grayish in color, fibrous with secondary branches emerging from the primary root (Chung et al. 2017; Marble et al. 2015).

The book chapter furnishes a comprehensive summary of photochemistry, traditional uses, pharmacological profile, safety and toxicity study, respective in vitro and in vivo analysis and other knowledge regarding *E. prostrate* which would serve as a source for future investigation.

11.2 Traditional Uses

E. prostrata commonly known as bhringraj in ayurveda and unani system is a popular traditionally commended medicinal herb around the world especially in subtropical and tropical countries. Being an effective remedy to treat various hair concerns for centuries, this plant is called the "King of hairs" (Kirtikar and Basu 2005). Traditionally this herb has been utilized to stimulate hair growth and check loss of hairs. Plant juice is externally administered over the scalp to blacken the hair (Datta et al. 2009). The herb has good recognition as an antiaging agent. It is recommended as a general tonic in case of physical weakness and tiredness. Externally it is applied in case of burn, inflammation, cut, and the extracted decoction of the fresh leaf is considered to be exceptionally powerful to stop bleeding. Honey together with the leaf decoction is given to the child suffering from upper respiratory tract infections. It is also applied in case of ear as well as eye contagions (Mithun et al. 2011). In Ayurveda, the herb is considered as a powerful medicine for the liver and is often employed as a deobstruent for assisting bile flow and protecting the liver. It is known for its pharmaceutical value including antiseptic, analgesic, antiviral, antioxidant, antibacterial, anti-hyperglycemic, and antihemorrhagic. The extracted juice is given to infants for catarrh after mixing with honey. The chloroform extracts of the herb also exhibit remarkable antidiabetic effect. Immunomodulatory activity is also exhibited by the plant, therefore, utilized as a potential memory modulator (Indian Herbal Pharmacopoeia 1998; The Indian Pharmacopoeia 2010). The herb has traditional external application on dermatitis, eczema, athlete foot, and scalp surface to arrest loss of hairs. The leaves have been extensively utilized as an antivenom agent against scorpion stings and snake bites in Brazil and China (Shekokar and Nayak 2017). It is used as an extremely powerful tonic in the treatment of jaundice, dermatitis, hepatitis, ulcers, hemorrhoids, liver cirrhosis, night blindness, blood pressure, knee and joint aches, hematuria, tuberculosis, and diarrhea with blood stools. Additionally, it is useful in conventional Asian medication for the treatment of different gynecological issues like endometrial cancer, menstrual pain, and irregular menstruation. It also strengthens teeth and gums and clears tongue when gargled with leaf juice. It is believed that consumption of five fresh leaves everyday aids digestion, heals constipation, and restores appetite (Chung et al. 2017). As stated by the Chinese Pharmacopoeia (2015), the herb tastes sweet and sour and serves cold properties. The main purpose is to nurture the kidney and liver, cool blood, and check bleeding. On that account, the herb is conventionally used for treating hepatitis, weakness and pain of joints and knees, gray hairs, renal disease, dizziness, hemorrhages, tinnitus (China Pharmacopoeia Commission 2015). The use of this plant is almost similar in India and Bangladesh, *E. prostrata* has great medicinal value in both the countries and used for skin disease (e.g., burns and wounds), jaundice, respiratory disorders, diabetes, hair fall, fever, and fatigue (Rahmatullah et al. 2010; Khan and Khan 2008). In Nepal the whole plant extract is administered on wounds and cuts (Panthi and Singh 2013). In Thailand, the leaf has been exploited for treating different skin infections and the stem as a blood tonic for boils, anemia, itching, amoebiasis, asthma, piles, and tuberculosis. On the other hand, the root is utilized as a hepatoprotectant, antibacterial agent, and tonic (Tewtrakul et al. 2011). In addition to these, in Brazilian conventional medication, the herb is extensively utilized for treating syphilis, snake bites, and leprosy (Leal et al. 2000). Table 11.1 gives a briefing of the conventional use of the plant in various countries throughout the world.

11.3 Chemical Constituents

E. prostrata has a legendary reputation as a medicinal herb and widely used in various health care systems throughout the world since ancient times. The herb possesses an outstanding pharmaceutical significance responsible for the phytochemical constituents. Numerous phytochemical analysis of the herb had reported the occurrence of several chemical constituents including coumestans, triterpenoid, saponins, glycosides, wedelolactone, flavonoids, thiophene derivatives, alkaloids, phytosterol, alkenynes, steroids, lipids, and polyacetylene compounds (Figs. 11.1 and 11.2) (Datta et al. 2009; Gani and Devi 2015). The aerial portion of the plant consists of wedelolactone, β-amyrin, phytosterol, β-glucoside, and luteolin-7-glucoside. The herb consists of polypeptides and when isolated yield some amino acids upon hydrolysis which include cysteine, tyrosine, glutamic acid, methionine, and phenylalanine. The plant was also found to contain nicotinic acid and nicotine (Bhalerao et al. 2013). The leaves possess wedelolactone, demethylwedelolactone-7-glucoside, stigmasterol, demethylwedelolactone and β-terthienylmethanol (Chopra et al. 1955). The root comprises hentriacontanol, heptacosanol, and polyacetylene replaced thiophene (Chopra et al. 1969). The saponins which are reported in the root include eclalbatin together with dasyscyphin C (Khanna and Kannabiran 2008; Tewtrakul et al. 2007). Some markers of *E. prostrata* as identified are caffeic acid, wedelolactone, and stigmasterol (Chan et al. 2014).

11.3.1 Coumestans

Coumestans are organic compounds derived from coumarin. The chief coumestan identified in *E. prostrata* comprises wedelolactone and desmethyl wedelolactone (Kaushik-Basu et al. 2008). The other types of coumestan isolated from this plant are strigolactone, isodemethylewedelolactone, and demethylwedelolactone-7-glucoside (Zhang and Guo 2001).

Table 11.1 Traditional uses of different parts of *E. prostrata* around the world

Country/ region	Plant part	Traditional use	Formulation	Reference
Africa	Whole plant	Fetal development	Extract of the plant is taken to facilitate childbirth and fetus development	Malan and Neuba (2011)
Bangladesh	Leaf	Diabetes	The leaf of white flowered plant is boiled in water containing earthen vessels along with the leaf of *Scoparia dulcis* and *Cynodon dactylon*. The water after straining through cloth is administered orally to diabetic patients on empty stomachs in the morning and evening	Rahmatullah et al. (2009)
China	Leaf, root, stem	Diphtheritis	Leaf, stem, and root is crushed to obtain juice which is added to equivalent honey and taken orally	State Administration of Traditional Chinese Medicine of the People's Republic of China (1999)
China	Whole plant	Dysentery	200 g of the plant and 50 g sugar are soaked in water and taken 3–4 times to cure the disease	State Administration of Traditional Chinese Medicine of the People's Republic of China (1999)
China	Whole plant	Hemorrhagic disease	Taken in a combination with herbs exhibiting hemostatic and blood cooling properties	State Administration of Traditional Chinese Medicine of the People's Republic of China (1999)
India	Whole plant	Jaundice and fever	Herb is crushed along with black pepper and formulated into small pills. Two pills two times a day are given to children in case of jaundice and fever	Sahu et al. (2013)
India	Whole plant	Scorpion sting, snake bite	Decoction prepared from the whole plant	Panghal et al. (2010) and Khan et al. (2014)
India, Thailand	Whole plant	Asthma	Mixture of plant ash and honey is administered orally 3 times a day for 3 months	Khan and Khan (2008) and Tewtrakul et al. 2007
India	Leaf	Skin infections and diseases	Paste of the leaf is employed externally to treat eczema and sores for 15 days and also administered orally two times daily	Khan and Khan (2008)

(continued)

Table 11.1 (continued)

Country/ region	Plant part	Traditional use	Formulation	Reference
India	Whole plant	Acidity	The decoction is taken with cow's milk, 3 times regularly for 15 days before each meal	Khan and Khan (2008)
India	Leaf	Cut and wound	The crushed leaf is administered over cuts and wounds	Agnihotri and Gupta (2013) and Tewari et al. (2013)
India	Leaf	Hair problems	The crushed leaf paste is massaged on hair and scalp to rid of premature graying, hair fall, dandruff, and promote hair growth	Tewari et al. (2013), Yesodharan and Sujana (2007) and Kumar et al. (2012)
Nepal	Whole plant	Cut and wound	The whole plant is crushed and applied on the affected areas	Panthi and Singh (2013)
Nepal	Whole plant	Catarrhal problems	Plant decoction in combination with essential oil	Neeraja and Margaret (2012)
Pakistan	Leaf	Ringworm, allergies and athlete's foot	Paste of the leaf is applied on the surface of the affected parts	Hussain et al. (2010)
Philippines	Whole plant	Hemoptysis of pulmonary tuberculosis	Decoction prepared from dried plant or tincture	Neeraja and Margaret (2012)

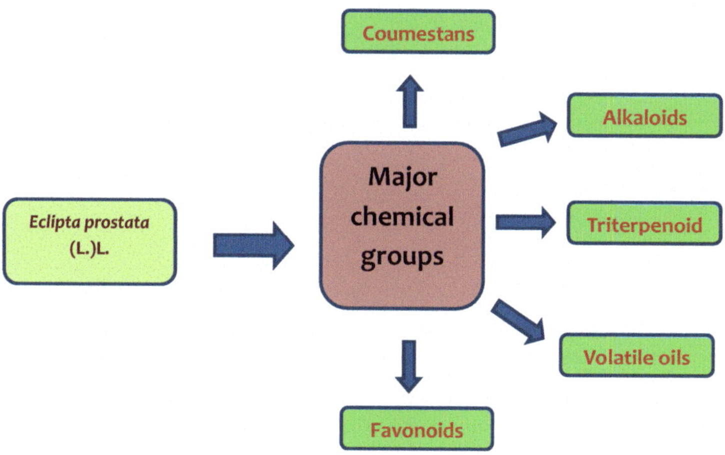

Fig. 11.1 Major chemical groups of *Eclipta prostate* (L.)

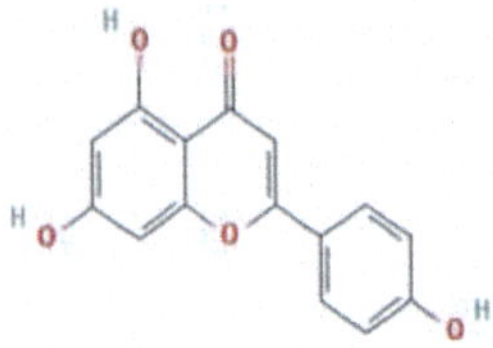

Apigenin

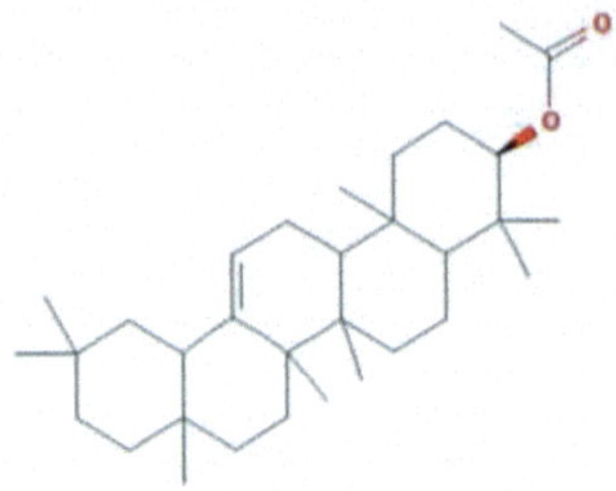

β-amyrin

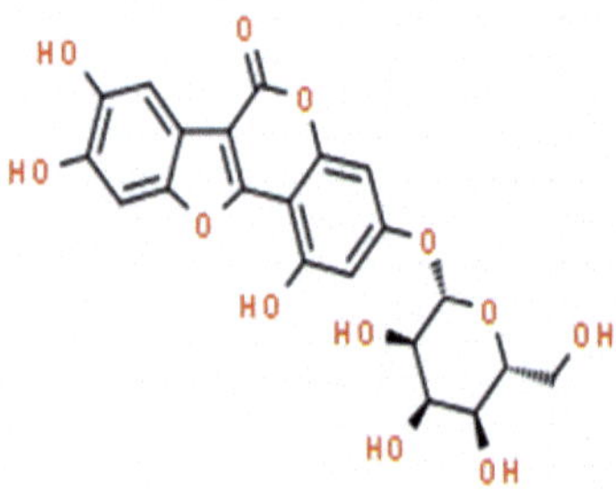

Demethylwedelolactone-7-glucoside

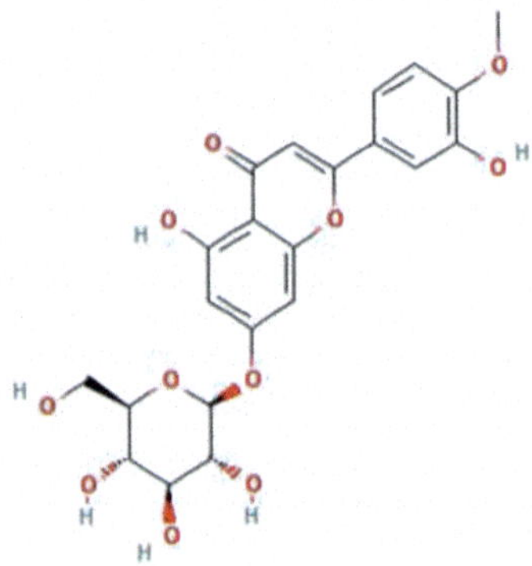

Diosmetin

Fig. 11.2 Chemical constituents of *Eclipta prostrata* (L.) L

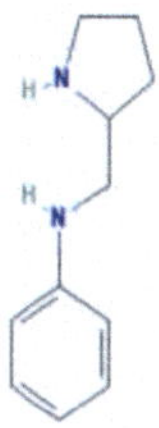

echinocysticacid

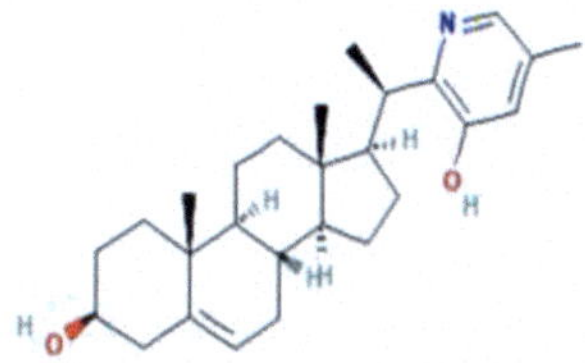

Ecliptalbine

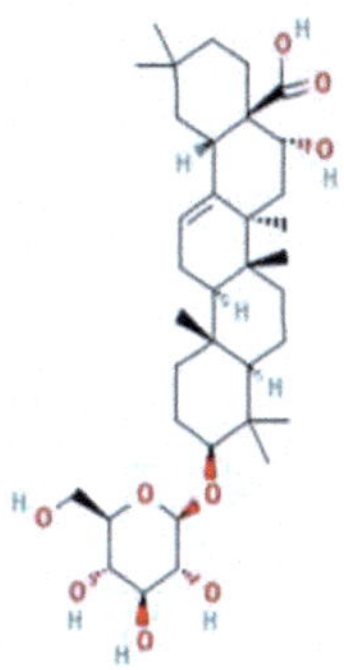

ecliptasaponin –A

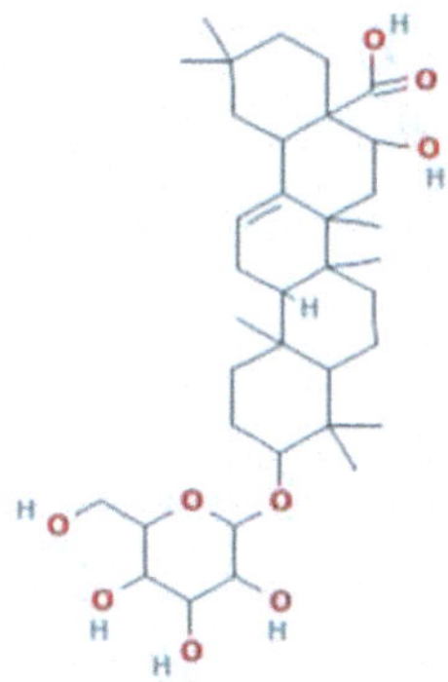

ecliptasaponin –D

Fig. 11.2 (continued)

Heptadecane

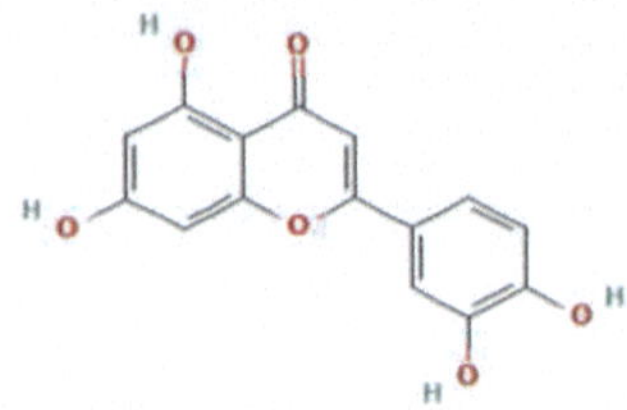

Luteolin

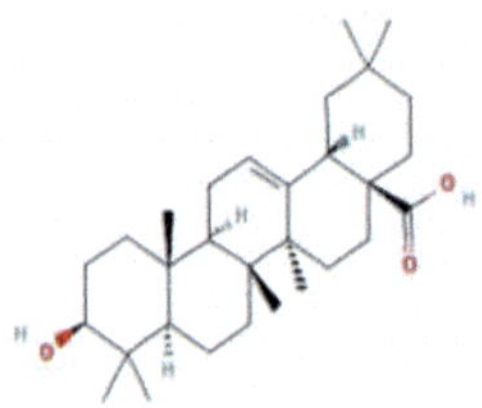

Oleanolic acid

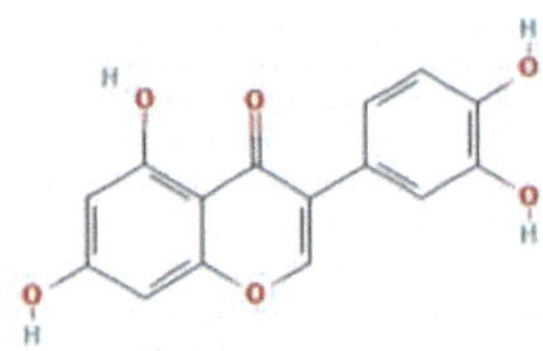

Orobol

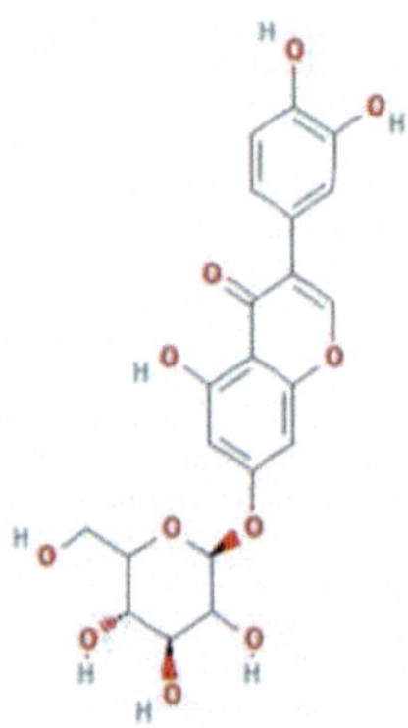

Oroboside

Pentadecane

Fig. 11.2 (continued)

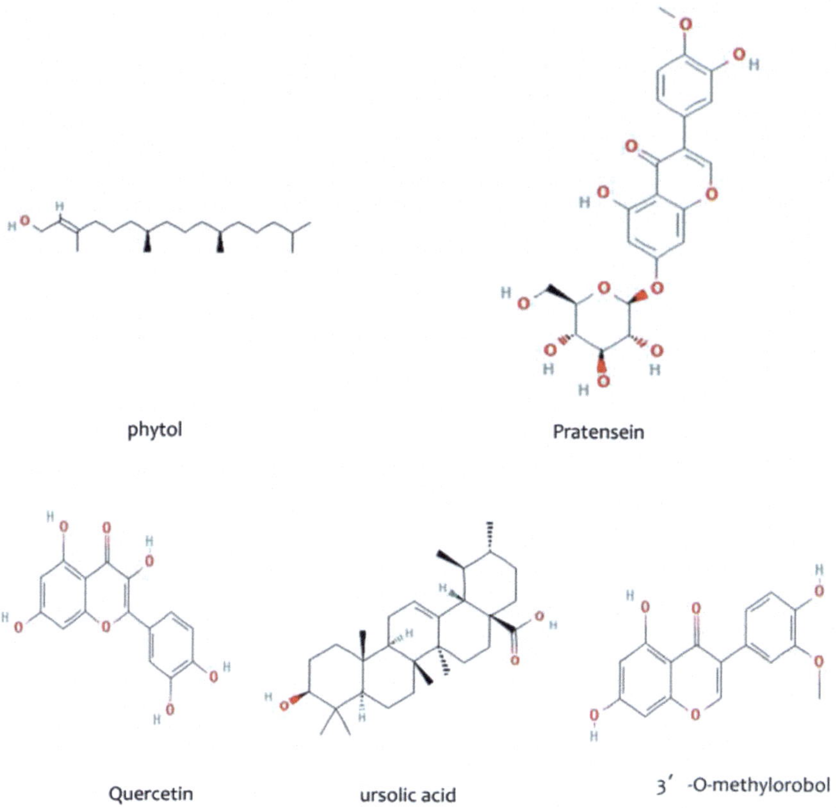

Fig. 11.2 (continued)

11.3.2 Alkaloids

The major alkaloids found in the plant is [(20S)(25S)-22,26-imino-cholesta-5,22(N)-dien-3β-ol] known as verazine. The additional alkaloids found includes [(20R)-20-pyridyl-cholesta-5-ene-3β,23-diol] (ecliptalbine), [20-epi-3-dehydroxy-3-oxo-5,6-dihydro-4,5-dehydroverazine], [(20R)-4β-hydroxyverazine], [(20R)-25 β-hydroxyverazine], [4 β-hydroxyverazine], and [25 β-hydroxyverazine]. Phytochemical studies also unveil the occurrence of alkaloids such as nicotine and ecliptine and bioactive steroidal alkaloids such as ecliptalbine, verazine, and dehydroverazine (Abdel-Kader et al. 1998).

11.3.3 Terpenoids and Their Glycosides

E. prostrata possesses a wide range of triterpenoids namely triterpenoid saponins which typically prevails in the form of glucosides. Until now, about 37 triterpenoids had been purified from this plant, consisting of eclalbasaponins I–XIII, echinocystic acid, ecliptasaponin A–D, ursolic acid, oleanolic acid, β-amyrin, α-amyrin, and their derivatives. All of them contain a pentacyclic ring and could be categorized into four primary types which include α-amyrane, β-amyrane, lupine, and taraxerane. Amidst, the most diverse kind is the oleanane-type triterpenoids (Feng et al. 2019).

11.3.4 Volatile Oils

The aerial portion of the herb contains several varieties of volatile compounds, which had been isolated using the hydrodistillation method and evaluated using the GC–MS method. The chief compounds includes pentadecane, 6,10,14-trimethyl-2-pentadecanone, heptadecane, n-hexadecanoic acid, eudesma-4(14), octadec-9-enoic acid, phytol, 1,2-benzenedicarboxylic acid di isooctyl ester, (Z,Z,Z)-1,5,9,9-tetramethyl-1,4,7-cycloundecatriene, and (Z)-7,11-dimethyl-3-methylene-1,6,10-dodecatriene (Jahan et al. 2014).

11.3.5 Flavonoids

A large number of flavonoids had been identified and isolated from the herb as flavones, isoflavones, and flavonols. Zhao et al. in 2002 isolated flavonol quercetin using column chromatography and HPLC method using the aerial portion of the plant. *E. prostrata* contains several types of flavonoids including luteolin-7-o-glucoside, orobol (isoluteolin), orobol-5-O-β-D-glucopyranoside, quercetin, luteolin, apigenin-7-O-glucoside, diosmetin, buddleosid, 7-O-methylorobol-4′-O-β-D-glucopyranoside, apigenin, pratensein-7-O-β-D-glucopyranoside, pratensein, 3′-O-methylorobol, and 3′-O-methyl orobol-7-O-β-D-glucopyranoside (Jahan et al. 2014; HAN et al. 2013; Timalsina and Devkota 2021).

11.4 Pharmacological Activities

11.4.1 Antihepatotoxic Effect

E. prostrata is considered a powerful tonic for the liver and generally utilized in the treatment of jaundice and hepatic disease. Wedelolactone, a chief coumestan constituent, was found to exhibit antihepatotoxic activity. It brings down liver inflammation and hepatocytes apoptosis in hepatitis mice induced by concanavalin A (Luo et al. 2018). An investigation of the hepatoprotective effect of the *E. prostrata* has been made on rats at subcellular level. The investigation showed that the hepatoprotective effect of *E. prostrata* works by modulating the degree of hepatic microsomal drug metabolizing enzymes known as glucose 6-phosphatase and aminopyrine N-demethylase (Saxena et al. 1993). The hepatoprotective effect was also examined on acute hepatitis in mice and rats. It was found that the severe upgradation of serum transaminases induced by carbon tetrachloride and β-D-galactosamine in mice and rats, respectively, were significantly inhibited by *E. prostrata* extracts. However, though an inhibiting tendency was noticed in acetaminophen-induced experimental models, no statistical significance was observed (Lin et al. 1996). Antioxidant as well as hepatoprotective activity of the plant was estimated against carbon tetrachloride (CCl_4) prompted hepatotoxicity in rats by Dheeba et al. (2012). Oral dose of 200 mg/kg of *E. prostrata* extract was given for 15 days, once daily. Due to CCl_4 treatment, the degree of marker enzymes of serum including aspartate transaminase, alkaline phosphatase, alanine transaminase, acid phosphatase, and antioxidant enzymes like superoxide dismutase, glutathione peroxidase, and catalase considerably got elevated. They were restored to regularity after giving the extract. The study concluded that *E. prostrata* possesses a potent antihepatotoxic activity against CCl_4 prompted hepatotoxicity.

11.4.2 Antidiabetic Effect

Diabetes mellitus alludes to a group of metabolic diseases marked by hyperglycemia caused by a defect in production and secretion of insulin. Chronic hyperglycemia is related to a number of end organ failure, harm, and dysfunction which includes the kidney, eye retina, heart, nervous system, and blood vessels (Alam et al. 2014). The extract of *E. prostrata* along with eclalbasaponin II were given to diabetic rats. A strong antidiabetic activity was detected during the experiment. Blood sugar level was remarkably lowered in contrast to diabetic untreated rats (Rahman et al. 2011). An investigation on antidiabetic efficacy of the plant against diabetic nephropathy and hyperglycemia was made in diabetic rats. A single dose reduced the level of blood sugar by 17.6% at 250 mg/kg dose after 5 h of oral administration. The extract manifested a remarkable inhibitory effect on alpha-glucosidase and to eye lens aldose reductase. The outcome indicated that *E. prostrata* extract has an antidiabetic effect related with inhibition of aldose reductase and alpha-glucosidase (Jaiswal et al. 2012). From the experiment conducted, it was observed by Ananthi et al. (2003) that oral intake of leaf extract of the plant for 60 days caused a significant lowering of glucose level in blood, glycosylated hemoglobin HbA_1c. It reduced the actions of glucose-6-phosphatase and fructose-1,6-bisphosphatase and elevated the liver hexokinase activity. Hence, it was evident that the plant possesses potent anti-hyperglycemic activity.

11.4.3 Alopecia and Hair Growth

E. prostrata is a notable medicinal herb aiding in hair development. As it advances hair growth and nourishes black hairs, it is used for preparation of hair oil. 10% w/v of *E. prostrata* is a chief element for preparing such herbal formulation. Alopecia is a dermatological disorder which results in hair loss of the patient. In the work reported by Roy et al. (2008), attempts were taken to analyze the effect of plant extract on hair development in denuded albino rats. Oleaginous cream was mixed with the extract and administered on the surface of shaved skin of albino rats. On treatment with the extract, the initiation time of the hair development and complete hair growth was significantly decreased in contrast to control rats. The observation certainly indicated that the extract has potential to successfully reduce the amount of time essential for initiation of hair growth. Datta et al. (2009) confirmed that methanol extract of *E. prostrata* surely advances growth of hair by inciting anagen in telogen phase hair follicles. It was a dose-dependent activity, 3.2 mg/15 cm^2 of the methanol extract exhibited greater efficiency in comparison to low doses. From the observation it was evident that this fraction of *E. prostrata* has considerable capability of promoting hair growth.

11.4.4 Analgesic Activity

In rats and mice, with the help of tail flick, hot plate, and writhing methods, the analgesic activity of alcoholic extract of *E. prostrata* had been ascertained. A dose of 200 mg/kg dose of the extract exhibited a notable analgesic effect (Pandey et al. 1997). The data obtained from the experiments of Sawant et al. (2004) suggested that an orally administered dose of 250 mg/kg and 500 mg/kg of *E. alba* possesses a beneficial analgesic activity. The information acquired also showed that the total alkaloid content at doses of 150 mg/kg, which is regarded as the plant extract's chloroform soluble part, demonstrated a satisfactory analgesic activity. It is concluded from the experiments that a number of compounds might be present in the ethanolic extract and the total alkaloid content which causes the analgesic activity.

11.4.5 Anti-Inflammatory Activity

It was assessed with the help of carrageenin, histamine, and serotonin prompted paw oedema and cotton pellet prompted granuloma. The highest inhibition of about 55.85% was found by a dose of 500 mg/kg of the plant extract in paw edema formed by the carrageenin drug treatment; on the other hand, the standard drug, indomethacin, exhibited inhibition of 61.30%. In the case of granuloma produced by cotton pellets, the plant extract and indomethacin exhibited a reduced development of granuloma tissue. The potential anti-inflammatory activity and pharmacological efficiency of *E. prostrata* extract has been indicated by the results on animal models (Kumar et al. 2005). The methanol extract of the plant leaves was analyzed for anti-inflammatory activity on albino rats. It was orally given to the animals at a concentration of 100 and 200 mg/kg which resulted in a remarkable dose dependent anti-inflammatory activity against hind paw oedema in rats induced by carrageenin and egg white. This result favored the traditional utilization of *E. prostrata* as an inflammatory agent (Arunachalam et al. 2009).

11.4.6 Antioxidant Properties

Multiple disorders like neurodegenerative diseases, atherosclerosis, cancer, and angina pectoris involve free radicals from the oxidative stress. Due to the scavenging activity of antioxidants, they are useful in the management of such diseases. The presence of a high quantity of ascorbic acid (9.83 mg/100 g) in *E. prostrata* performs the primary function of manifesting scavenging activity of free radicals of the herb. *E. alba* extract contains a total phenol of about 30.4 mg/1 g. Plants phenolic constitute is a crucial category of components which behaves as primary antioxidants. The phenols consist of hydroxyl, mainly due to its redox properties; they are responsible for the scavenging effect of radicals. The stronger free radical scavenging property of *E. prostrata* can be explained due to the occurrence of a greater amount of phenolic content and ascorbic acid (Uddin et al. 2010). The antioxidant activity of *E. alba* hydroalcoholic extract was evaluated using standard procedure. It was evident from the result that the plant extract comprises a good amount of tannins, phenolic, flavonoids, and ascorbic acid. The hydroalcoholic extract of the plant was successful in scavenging free radicals at various concentrations and manifested satisfactory antioxidant property (Patel et al. 2016). Ferric thiocyanate (FTC) was utilized to ascertain the quantity of peroxide produced and that reacted with $FeCl_2$ to produce pigment of red color. In the process, the peroxide concentration lowers with an increase in the antioxidant activity. Concentration-dependent antioxidant activities were exhibited at 500 μg/mL concentration; ethanol extract portrayed 77.62% antioxidant property which is nearer to 500 μg/mL concentration of reference compound tocopherol (80.06%). It was found from the experiment that the extract manifested a good antioxidant effect with an elevated polarity which indicates that polyphenols or flavonoids or flavanone might perform a crucial part in the activity (Karthikumar et al. 2007).

11.4.7 Antimicrobial Activity

The results of an antibacterial assay of *E. prostrata* demonstrated that the gram-positive bacteria were more susceptible to the extract when compared to gram negative one. This may be because of the presence of an outer membrane in gram negative species which acts as a successful barrier. *Bacillus subtilis* and *Staphylococcus aureus* were found to be the most vulnerable Gram-positive bacteria, a consideration which might be credited to the occurrence of one membrane in the microbes that makes it more open to penetration by the active constituents of *E. prostrata*. On the other hand, gram-negative bacteria *Escherichia coli* and *Salmonella typhi* exhibited the minimum susceptibility to the plant extract (Uddin et al. 2010). This may be due to the presence of an extra outer membrane that acts as a barrier and trans envelope multidrug resistance pumps (MDRs) in the microorganisms that provides them an intrinsic resistance from the extract (Girish and Satish 2008). For in vitro antimicrobial studies, the methanolic extract and active principle compound of *E. alba* was tested. The extract demonstrated antimicrobial property against each of the six strains examined. The occurrence of tannins, coumestans, flavonoids, alkaloids, saponins, etc. were reported. Ethyl acetate fraction of the herb and isolated pure wedelolactone exhibited intensified antimicrobial property. *Staphylococcus epidermidis, S. aureus*, and *Salmonella typhimurium* were found to be most susceptible. Among all the strains *Shigella flexneri* was reported to be the most resistant. Hence the results suggested that wedelolactone has been a positive antimicrobial phytochemical (Dalal et al. 2010). A marked antifungal activity had also been studied from the plant extract against *Microsporum canis, Trichophyton rubrum*, and *Microsporum gypseum* (Nagabhushan et al. 2013).

11.4.8 Immunomodulatory Activity

A principal concept of Ayurvedic medicine is to elevate the body's natural resistance to stress and diseases. Among the various plant used in ayurveda, *E. prostrata* had been identified as one of the promising immunostimulant agents (Karthikumar et al. 2011). At five dose levels (100–500 mg/kg body wt), the immunomodulatory effect of the whole plant extract had been evaluated using parameters such as antibody titer, carbon clearance, and cyclophosphamide immunosuppression. The extract remarkably raised the phagocytic index and antibody titer, and the F ratios of the phagocytic index and WBC count was also notable (Jayathirtha and Mishra 2004). A study was carried out on the immunostimulatory property of oral intake of the plant leaf extract in tilapia, *Oreochromis mossambicus*. The fishes were fed with diets containing the leaf aqueous extract at 0, 0.01, 0.1, or 1% levels for 1, 2, or 3 weeks. At the end of every week, nonspecific humoral, cellular responses, and disease resistance against *Aeromonas hydrophila* were estimated. The observation confirmed that aqueous extract of *E. prostrata* notably boosted many of the immune responses and also showed the resistance of *O. mossambicus* against *A. hydrophila* (Christybapita et al. 2007).

11.4.9 Anticancer and Antitumor Properties

Along with the expanding utility of complementary anticancer medications, there is a necessity to analyze the commercially accessible natural products under standard experimental conditions for their relative tumoricidal properties. In the experiment performed by Ali et al. (2014), it was evident that *E. alba* leaf extract exhibited an antitumor activity in skin cancer mice. Administration of the plant extract lowered the number, and formation of tumors by hindering the procedure related to development of carcinoma and promoting p53-mediated apoptosis. The data obtained from the experiments conducted by Liu et al. (2012) demonstrated the possible antitumor property of the 30% fraction and the compound eclalbasaponin I extracted from this fraction. The hydroalcoholic extract of the plant was found to exhibit apoptosis, antiproliferation, and anti-invasion activities as reported by Chaudhary et al. (2011). The MTT assay and phase contrast study revealed that in a dose-dependent manner, the extract was successful to stop the proliferation of Kidney (A498), liver (HepG2), and brain (C6 glioma) cell line. It could prompt apoptosis through DNA fragmentation in all the three tested cell lines. It was capable of reducing the matrix metalloproteinase MMP2 and MMP9 activity to a remarkable degree in the HepG2 and A498 cells. Hence, acting as a potent antimetastasis agent. Metastasis refers to a complicated multiple step procedure which includes the loosening of tumor and cancer cells from the original tumor, extravasation, traveling through blood or lymph, tumor cell proliferation, and angiogenesis at other organs or tissues of the body from the original sites (Steeg 2006).

A schematic representation of some of these pharmacological activities is shown in Fig. 11.3.

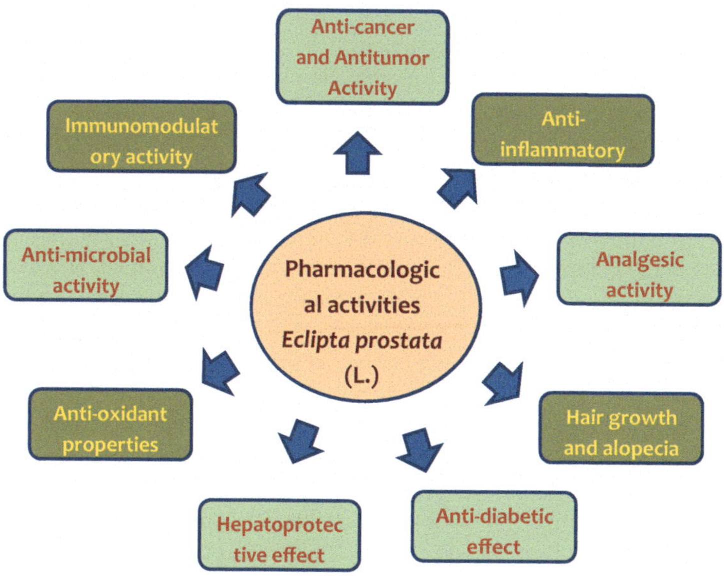

Fig. 11.3 Diagrammatic presentation of pharmacological activity of *Eclipta prostata* (L.) L

11.5 Safety and Toxicity

The natural products derived from plants are known to be safe and effective and have ample use as biologically active compounds for treatment of various ailments particularly in the area of infectious diseases (Cragg et al. 1997). It is also offered as a safer alternative to synthetic drugs and can be achieved easily by individuals at a very minimal price. However, to confirm their safety in administration, toxicity tests of the active plants need to be done. The use of *E. prostrata* as a source of medication requires persuading evidences regarding absence of toxic or harmful effects. Hence, it is mandatory to conduct toxicity analysis on laboratory animals to designate the plant as a secure promising medicinal plant (Baqar 1989). Acute toxicological studies of *E. prostrata* were executed on albino mice along three routes: oral, parenteral, and systemic. The extremity and intensity of toxicities manifested by each route were dependent on dose. The calculated LD_{50} (lethal dose 50) for oral, intravenous, and intraperitoneal routes were 7.841 g/kg, 302.8 mg/kg, and 328.3 mg/kg, respectively. A dose of 2.0 g/kg was found nontoxic and safe for oral route and 200 mg/kg for intravenous and intraperitoneal routes (Qadri et al. 2001). Acute toxicity of the plant was evaluated by Singh et al. (1993) during their experiment, the plant extract exhibited no signs of toxicity and when administered orally, and intraperitoneally the minimum lethal dose was observed above 2.0 gm/kg. Lin et al. (1996) from their study also found that the LD_{50} of *E. prostrata* was more than 2.0 gm/kg in mice and rats, so it can be regarded as a relatively safe drug. On that account it may be possible to isolate the hepatoprotective principles from the plant and make use of them in clinical studies in the future. The LD_{50} of the test extract was also calculated by Shaikh et al. (2012) using AOT 425 software supplied by the US Environmental Protection Agency which showed the same result. The value was found greater than 2000 mg/kg signifying that the drug is safe up to 2000 mg/kg. A similar result was obtained by Tanuja et al. 2011 during their experiment to estimate the impact of aqueous extract on the liver of male Swiss albino mice. Higher doses above 2000 mg/kg body weight manifested loss in weight of body and organ and behavioral change in the experimental groups IV, V, and VI which were statistically significant ($p < 0.05$) in contrast to control group I. In an experiment conducted by Kumari et al. (2006), it was found that administration of *E. prostrata* extract in different doses (50–400 mg/kg) didn't cause any kind of alterations in the behavioral and autonomic responses of the animal, and even after 7 days, no death was witnessed. It signified that the extract in the doses studied were safe and nontoxic.

Regarding toxicological studies of many common herbal medicines, a limited amount of information is available. Therefore, the risk associated with potential toxicity of herbal therapies utilized throughout a significant stretch of time requires a close follow-up to reported cases of renal and hepatic toxicity, due to the consumption of these medicine herbs. Abnormal laboratory test results can be confronted in healthy patients who are consuming herbal products as a reflection of the toxicity of herbs in many cases (Tanuja et al. 2011). It is evident from the experiments that

E. prostrata manifested a decent safety margin and had a good potency in all the doses analyzed on different parameters. It validates future investigations to explore its probable pharmaceutical role from every aspect in modern medicinal studies.

11.6 Conclusion

E. prostrata furnishes multiple beneficial possibilities for traditional as well as modern medicine. For a number of disorders, the herb provides promising herbal therapies. The plant has been considered as a potent cosmetic as well as therapeutic means. The herb is well known for hair growth, often called keshraj. It has been extensively used in preparation of hair oil to promote hair development and maintain black hair. For curing different ailments, the plant had been long utilized in traditional medicinal systems. Various chemical compounds had been obtained and identified from the herb which plays a significant role in providing evidence for its utility in herbal medicines. The variety of bioactive components found in the herb including alkaloids, coumestans, triterpenoids, glycosides, and flavonoids are responsible for the therapeutic activities manifested by the plant which includes anti-inflammatory activity, antimicrobial activity, hepatoprotective activity, antitumor activity, antioxidant activity, and anticancer and immunomodulatory activities. Due to the diverse medicinal value manifested by the herb, it has huge commercial demand in the market which promotes further studies on the herb at the molecular level. Hence, the plant requires foremost attention for conservation. The book chapter grants insight into the taxonomy, distribution, chemical constituents, traditional uses, pharmacological studies, clinical studies, and safety and toxicity related to the studies of *E. prostrata*. On the basis of pharmacological activities exhibited by the herb, several in vitro and in vivo works have been discussed in the chapter. The popularity and utility of *E. prostrata* may get restrained because of insufficient research and understanding. The details and information discussed here is expected to enlighten people about the utility of the herb and may be fruitful for future investigations. Clinical studies have been performed on different pharmacological activities like antidiabetic, immunomodulatory, anticancer, antioxidant, antitumor, hepatoprotective, anti-inflammatory, and antimicrobial. Further investigation of the herb can open new doors for modern medicines. Isolation and identification of new bioactive molecules from the plant will help in the study of the plant pharmacology against incurable human diseases, promote their future use in modern medicine, and will save the world from environmental and economic losses. In future, in vivo animal experiments should be executed to confirm the pharmacological activity and to explicate their parallel fundamental mechanism. The acute toxicology of the herb should be studied briefly so that safety and toxicological limits could be established which will provide confirmation for medical applications of the herb.

References

Abdel-Kader MS, Bahler BD, Malone S, Werkhoven MC, van Troon F, David WJH, Bursuker I, Neddermann KM, Mamber SW, Kingston DG (1998) DNA-damaging steroidal alkaloids from *Eclipta alba* from the Suriname rainforest. J Nat Prod 61(10):1202–1208. https://doi.org/10.1021/np000312f

Agnihotri N, Gupta AK (2013) Folklore medicines for cuts and wounds in Kalyanpur block of Kanpur District, Uttar Pradesh, India. Ph Tech Med 2(5):381–386

Alam U, Asghar O, Azmi S, Malik RA (2014) General aspects of diabetes mellitus. Handb Clin Neurol 126:211–222. https://doi.org/10.1016/B978-0-444-53480-4.00015-1

Ali F, Khan R, Khan AQ, Lateef MD, Maqbool T, Sultana S (2014) Assessment ofaugmented immune surveillance and tumor cell death by cytoplasmic stabilization ofp53 as a chemopreventive strategy of 3 promising medicinal herbs in murine 2-stageskin carcinogenesis. Integr Cancer Ther 13(4):351–367. https://doi.org/10.1177/1534735413513831

Ananthi J, Prakasam A, Pugalendi KV (2003) Antihyperglycemic activity of *Eclipta alba* leaf on alloxan-induced diabetic rats. Yale J Biol Med 76(3):97–102

Arunachalam G, Subramanian N, Pazhani GP, Ravich V (2009) Anti-inflammatory activity of methanolic extract of *Eclipta prostrata* L (Asteraceae). Afr J Pharm Pharmacol 3(3):097–100

Baqar SR (1989) Medicinal and poisonous plants of Pakistan. Karachi, Printas, p 179

Bhalerao SA, Verma DR, Teli NC, Murukate VR (2013) *Eclipta alba* (L): an overview. Int J Bioassays 2(11):1443–1447

Chan CF, Huang WY, Guo HY, Wang BR (2014) Potent antioxidative and UVB protective effect of water extract of *Eclipta prostrata* L. Sci World J 2014:759039. https://doi.org/10.1155/2014/759039

Chaudhary H, Dhuna V, Singh J, Kamboj SS, Seshadri S (2011) Evaluation of hydro-alcoholic extract of *Ecliptaalba*for its anticancer potential: an *in vitro* study. J Ethnopharmacol 136(2):363–367. https://doi.org/10.1016/j.jep.2011.04.066

Chauhan N, Singh D, Painuli RM (2012) Screening of bioprotective properties and phytochemical analysis of various extracts of *Eclipta alba* whole plant. Int J Pharm Pharm Sci 4(2):554–560

Chinese Pharmacopoeia Commission (2015) Pharmacopoeia of the People's Republic of China, vol 1. China Medical Science, Beijing, pp 374–375

Chopra RN, Nayar SL, Chopra IC (1955) Glossary of Indian medicinal plants. C.S.I.R, New Delhi

Chopra RN et al (1969) Supplement to glossary of Indian medicinal plants. C.S.I.R, New Delhi

Christybapita D, Divyagnaneswari M, Michael RD (2007) Oral administration of *Eclipta alba* leaf aqueous extract enhances the non-specific immune responses and disease resistance of *Oreochromismossambicus*. Fish Shellfish Immunol 23(4):840–842. https://doi.org/10.1016/j.fsi.2007.03.010

Chung IM, Rajakumar G, Lee JH, Kim SH, Thiruvengadam M (2017) Ethnopharmacological uses, phytochemistry, biological activities, and biotechnological applications of *Eclipta prostrata*. Appl Microbiol Biotechnol 101(13):5247–5257. https://doi.org/10.1007/s00253-017-8363-9

Cragg GM, Newman DJ, Snader KM (1997) Natural products in drug discovery and development. J Nat Prod 60(1):52–60. https://doi.org/10.1021/np9604893

Dalal S, Kataria SK, Sastry KV, Rana SVS (2010) Phytochemical screening of methanolic extract and antibacterial activity of active principles of hepatoprotective herb, *Eclipta alba*. Ethnobot Leaflets 14:248–258

Datta K, Singh AT, Mukherjee A, Bhat B, Ramesh B, Burman AC (2009) *Eclipta alba* extract with potential for hair growth promoting activity. J Ethnopharmacol 124(3):450–456. https://doi.org/10.1016/j.jep.2009.05.023

Dheeba B, Vaishnavi E, Sampathkumar P, Kannan M (2012) Hepatoprotective and curative effect of *Eclipta prostrata* on CCl4 induced hepatotoxicity in albino rats. Biosci Biotechnol Res Asia 9(1):309–314. https://doi.org/10.13005/bbra/1001

Feng L, Zhai YY, Xu J, Yao WF, Cao YD, Chenga FF, Bao BH, Zhang L (2019) A review on traditional uses, phytochemistry and pharmacology of *Eclipta prostrata* (L.) L. J Ethnopharmacol 245:112109. https://doi.org/10.1016/j.jep.2019.112109

Gani AMS, Devi ND (2015) Antioxidant activity of methanolic extract of *Eclipta prostrata* L. Int J Phytopharm 5(2):21–24. https://doi.org/10.7439/ijpp.v5i2.1754

Girish HV, Satish S (2008) Antibacterial activity of important medicinal plants on human pathogenic bacteria-a comparative analysis. World Appl Sci J 5(3):267–271

HAN L-f, Jing Z, Zhang Y, Agyemang K, Liu E-w, Wang T (2013) Chemical constituents from dried aerial parts of *Eclipta prostrata*. Chin Herb Med 5(4):313–316. https://doi.org/10.1016/S1674-6384(13)60047-7

Hussain K, Nisar MF, Majeed A, Nawaz K, Bhatti KH (2010) Ethnomedicinal survey for important plants of Jalalpur Jattan, district Gujrat, Punjab, Pakistan. Ethnobot Leafl 14:807–825

Indian Herbal Pharmacopoeia (1998) A joint publication of IDMA and RRL Jammu-Tavi, vol I, pp 81–85

Jahan R, Al-Nahain A, Majumder S, Rahmatullah M (2014) Ethnopharmacological significance of *Eclipta alba* (L.) Hassk. (Asteraceae). Int Sch Res Notices 8:1–22. https://doi.org/10.1155/2014/385969

Jaiswal N, Bhatia V, Srivastava SP, Srivastava AK, Tamrakar AK (2012) Antidiabetic effect of *Eclipta alba* associated with the inhibition of alpha-glucosidase and aldose reductase. Nat Prod Res 26(24):2363–2367. https://doi.org/10.1080/14786419.2012.662648

Jayathirtha MG, Mishra SH (2004) Preliminary immunomodulatory activities of methanol extracts of *Ecliptaalba* and *Centellaasiatica*. Phytomedicine 11(4):361–365. https://doi.org/10.1078/0944711041495236

Karthikumar S, Vigneswari K, Jegatheesan K (2007) Screening of antibacterial and antioxidant activities of leaves of *Eclipta prostrata* (L). Sci Res Essays 2(4):101–104

Karthikumar S, Jegatheesan K, Thangaraja A, Banupriya K, Dhivya T, Malarvizhi JM (2011) Immunomodulatory activity of *Eclipta prostrata* in SRBC immunized mice. J Pharmacogn Phytother 3(4):52–55. https://doi.org/10.5897/JPP.9000031

Kaushik-Basu N, Bopda-Waffo A, Talele TT, Basu A, Costa PR, da Silva AJ, Sarafianos SG, Noel F (2008) Identification and characterization of coumestans as novel HCV NS5B polymerase inhibitors. Nucleic Acids Res 36(5):1482–1496. https://doi.org/10.1093/nar/gkm1178

Khan AV, Khan AA (2008) Ethnomedicinal uses of *Ecliptaprostrta* Linn. Indian J Tradit Knowl 7(2):316–320

Khan AV, Ahmed QU, Khan MW, Khan AA (2014) Herbal cure for poisons and poisonous bites from Western Uttar Pradesh, India. Asian Pac J Trop Dis 4(1):S116–S120. https://doi.org/10.1016/S2222-1808(14)60425-4

Khanna VG, Kannabiran K (2008) Antimicrobial activity of saponin fractions of the leaves of Gymnemasylvestre and *Ecliptaprostrata*. World J Microbiol Biotechnol 24(11):2737–2740. https://doi.org/10.1007/s11274-008-9758-7

Kirtikar KR, Basu BD (2005) Indian medicinal plants, vol I. International Book Distributors, Dehradun, pp 478–479

Kumar SS, Sivakumar T, Chandrasekar MJ, Suresh B (2005) Evaluation of anti-inflammatory activity of *Eclipta alba* in rats. Anc Sci Life 24(3):112–118

Kumar A, Agarwal S, Singh A, Deepak D (2012) Medicobotanical study of some weeds growing in moradabad district of Western Uttar Pradesh in India. Indian J Sci Res 3(1):107–111

Kumari CS, Govindasamy S, Sukumar E (2006) Lipid lowering activity of *Eclipta prostrata* in experimental hyperlipidemia. J Ethnopharmacol 105(3):332–335. https://doi.org/10.1016/j.jep.2005.10.031

Leal LKAM, Ferreira AAG, Bezerra GA, Matos FJA, Viana GSB (2000) Antinociceptive, anti-inflammatory and bronchodilator activities of Brazilian medicinal plants containing coumarin: a comparative study. J Ethnopharmacol 70(2):151–159. https://doi.org/10.1016/S0378-8741(99)00165-8

Lin SC, Yao CJ, Lin CC, Lin YH (1996) Hepatoprotective activity of Taiwan folk medicine: *Eclipta prostrata* linn. Against various hepatotoxins induced acute hepatotoxicity. Phytother Res 10(6): 483–490. https://doi.org/10.1002/(SICI)1099-1573(199609)10:6%3C483::AID-R884%3E3.0.CO;2-2

Liu QM, Zhao HY, Zhong XK, Jiang JG (2012) *Eclipta prostrata* L. phytochemicals: isolation, structure elucidation, and their antitumor activity. Food Chem Toxicol 50(11):4016–4022. https://doi.org/10.1016/j.fct.2012.08.007

Luo QQ, Ding JY, Zhu LP, Chen FX, Xu LL (2018) Hepatoprotective effect of wedelolactone against concanavalin A-induced liver injury in mice. Am J Chin Med 46(4):1–15. https://doi.org/10.1142/S0192415X1850043X

Malan DF, Neuba DF (2011) Traditional practices and medicinal plants use during pregnancy by Anyi-Ndenye women (eastern Côte dIvoire). Afr J Reprod Health 15(1):85–93. https://doi.org/10.4314/ajrh.v15i1.67861

Marble C, Steed S, Boyd NS (2015) Biology and management of Eclipta (*Eclipta prostrata*) in ornamental crop production. UF/IFAS Extension ENH 1251:1–4. http://edis.ifas.ufl.edu

Mithun NM, Shashidhara S, Kumar V (2011) *Eclipta alba* (L.) a review on its phytochemical and pharmacological profile. Pharmacology 1:345–357

Mukhopadhyay G, Kundu S, Sarkar A, Sarkar P, Sengupta R, Kumar C (2018) A review on physicochemical & pharmacological activity of *Eclipta alba*. J Pharm Innov 7(9):78–83

Nagabhushan, Raveesha K, Shrisha DL (2013) Antidermatophytic activity of *Eclipta prostrata* L. against human infective trichophyton and microsporum spp. Int J Chem Anal Sci 4(2): 136–138. https://doi.org/10.1016/j.ijcas.2013.05.003

Neeraja P, Margaret E (2012) *Eclipta alba* (L.) Hassk: a valuable medicinal herb. Int J Curr Pharm Rev Res 2(4):188–197

Pandey PS, Upadhyay KK, Pandey DN (1997) Experimental evaluation of the analgesic property of *Eclipta alba* (L) hassk. Anc Sci Life 17(1):36–40

Panghal M, Arya V, Yadav S, Kumar S, Yadav JP (2010) Indigenous knowledge of medicinal plants used by Saperas community of Khetawas, Jhajjar District, Haryana, India. J Ethnobiol Ethnomed 6:4. https://doi.org/10.1186/1746-4269-6-4

Panthi MP, Singh AG (2013) Ethnobotany of Arghakhanchi district, Nepal: plants used in dermatological and cosmetic disorders. Int J Appl Sci Biotechnol 1(2):27–32. https://doi.org/10.3126/ijasbt.v1i2.8199

Patel M, Verma R, Srivastav P (2016) Antioxidant activity of *Eclipta alba* extract. J Med Plants Stud 4(5):92–98

Puri HS (2003) Rasayana: ayurvedic herbs for longevity and rejuvenation. Taylor & Francis, London, pp 80–85

Qadri NM, Ahmad S, Qureshi S, Badar Y (2001) Acute toxicological evaluation of the aqueous extract of *Ecliptaalba* hassk. Pak J Sci Ind Res 44(1):38–41

Rahman MS, Rahman MZ, Begum B, Chowdhury R, Islam SN, Rashid MA (2011) Antidiabetic principle from *Eclipta prostrata*. Lat Am J Pharm 30(8):1656–1660

Rahmatullah M, Mollik MAH, Azam ATMA et al (2009) Ethnobotanical survey of the Santal tribe residing in thakurgaon District, Bangladesh. Am Eurasian J Sustain Agric 3(4):889–898

Rahmatullah M, Hasan ME, Islam MA, Islam MT, Jahan FI, Seraj S, Chowdhury AR, Jamal F, Islam MS, Miajee ZUMEU, Jahan R, Chowdhury MH (2010) A survey on medicinal plants used by the folk medicinal practitioners in three villages of Panchagarh and Thakurgaon district, Bangladesh. Am Eurasian J Sustain Agric 4(3):291–301

Roy RK, Thakur M, Dixit VK (2008) Hair growth promoting activity of *Eclipta alba* in male albino rats. Arch Dermatol Res 300(7):357–364. https://doi.org/10.1007/s00403-008-0860-3

Sahu CR, Nayak RK, Dhal NK (2013) Traditional herbal remedies for various diseases used by tribals of Boudh district, Odisha, India for sustainable development. Int J Herb Med 1(1):12–20

Sawant M, Isaac JC, Narayanan S (2004) Analgesic studies on total alkaloids and alcohol extracts of *Eclipta alba* (Linn.) Hassk. Phytother Res 18(2):111–113. https://doi.org/10.1002/ptr.1165

Saxena AK, Singh B, Anand KK (1993) Hepatoprotective effects of *Eclipta alba* on subcellular levels in rats. J Ethnopharmacol 40(3):155–161. https://doi.org/10.1016/0378-8741(93)90063-B

Shaikh MF, Sancheti J, Sathaye S (2012) Phytochemical and pharmacological investigations of *Ecliptaalba* (Linn.) Hassak leaves for antiepileptic activity. Int J Pharm Pharm Sci 4(4):319–323

Shekokar S, Nayak SU (2017) A phytopharmacological review of prospective of bhrungaraj (*Eclipta alba* Hassk.). Int J Ayurved Med 8:1–7. https://doi.org/10.47552/ijam.v8i1.892

Singh B, Saxena AK, Chandan BK, Agarwal SG, Bhatia SMS, Anand KK (1993) Hepatoprotective effect of ethanolic extract of *Eclipta alba* on experimental liver damage in rats and mice. Phytother Res 7(2):154–158. https://doi.org/10.1002/ptr.2650070212

State Administration of Traditional Chinese Medicine of the People's Republic of China (1999) Zhong Hua Ben Cao. Shanghai Scientific & Technical Publishers, Shanghai

Steeg PS (2006) Tumor metastasis: mechanistic insights and clinical challenges. Nat Med 12(8): 895–904. https://doi.org/10.1038/nm1469

Tanuja SA, Sinha N, Kumar R (2011) Evaluation of acute toxicity of aqueous extract of *Eclipta alba* and it's effects on liver of male Swiss albino mice. J Herb Med Toxicol 5(2):89–95

Tewari RC, Kotecha M, Sharma AK, Sharma P (2013) Ethno-medicinal heritage of Chandi Devi hills of Haridwar, Uttarakhand. Int J Innov Res Dev 2(7):233–241

Tewtrakul S, Subhadhirasakul S, Cheenpracha S, Karalai C (2007) HIV-1 protease and HIV 1 integrase inhibitory substances from *Eclipta prostrata*. Phytother Res 21(11):1092–1095. https://doi.org/10.1002/ptr.2252

Tewtrakul S, Subhadhirasakul S, Tansakul P, Cheenpracha S, Karalai C (2011) Antiinflammatory constituents from *Eclipta prostrata* using RAW 264.7 macrophage cells. Phytother Res 25(9): 1313–1316. https://doi.org/10.1002/ptr.3383

The Indian Pharmacopoeia (2010) The Indian Pharmacopoeia 2010, vol 3. Govt of India Publication, Ghaziabad

Timalsina D, Devkota HP (2021) *Eclipta prostrata* (L.) L.(Asteraceae): ethnomedicinal uses, chemical constituents, and biological activities. Biomol Ther 11(11):1738

Uddin MN, Atiar Rahman A, Ahmed N, Sohel Rana MS, Akter R, Chowdhury AMMA (2010) Antioxidant, cytotoxic and antimicrobial properties of ethanol extract. Int J Biol Med Res 1(4): 341–346

Yadav P, Harisha CR, Prajapati PK (2011) Validation of pharmacopoeial characters of Bhringaraja (*Eclipta alba* LinnHassk). J Curr Pharm Res 8(1):17–24

Yesodharan K, Sujana KA (2007) Status of ethnomedicinal plants in the Parambikulam wildlife sanctuary, Kerala, South India. Ann For 15(2):322–334

Zhang JS, Guo QM (2001) Studies on the chemical constituents of *Eclipta prostrata* (L). Acta Pharm Sin 36(1):34–37

Chapter 12
Helianthus annuus L.: Traditional Uses, Phytochemistry, and Pharmacological Activities

Arpit Singh, Shivaraman Pandey, Ravi Prakash Srivastava, Hari Prasad Devkota, Lav Singh, and Gauri Saxena

Abstract The Asteraceae (formerly Compositae) family contains approximately 1600 genera, many of which are known for their phytochemistry and pharmacological properties. *Helianthus annuus* is one of the most important plants in this family, known for producing oil and playing an important role in the economies of countries. This plant's traditional uses date back more than 3000 years. The plants are ethnomedicinally significant and contain a variety of alkaloids, terpenoids, carbohydrates, fixed oils, steroids, amino acids, and other compounds. An important point to emphasize in this book chapter is that plants are highly adaptable to various environmental conditions, making them easier to cultivate and yielding higher yields. There is a substantial literature on the plant, but as research-based knowledge grows, it must be updated, so we will attempt to update this plant's information in terms of its traditional uses, phytochemistry, and pharmacology properties; the pharmacological properties of *H. annuus* were investigated using a variety of sources, including medicinal plant databases, ethnobotanical and

A. Singh · R. P. Srivastava · G. Saxena (✉)
Department of Botany, University of Lucknow, Lucknow, Uttar Pradesh, India

S. Pandey
Department of Botany, University of Lucknow, Lucknow, Uttar Pradesh, India

Plant Systematic, Biodiversity and Herbarium Division, CSIR-NBRI, Lucknow, Uttar Pradesh, India

Department of Botany, Government PG College, Datia, Madhya Pradesh, India

H. P. Devkota
Headquarters for Admissions and Education, Kumamoto University, Kumamato, Japan
e-mail: devkotah@kumamoto-u.ac.jp

L. Singh
Post Graduate Department of Botany, RD and DJ College, Munger University, Munger, Bihar, India

Central Academy for State Forest Services, Burnihat, Assam, India
e-mail: lavupc@gmail.com

H. P. Devkota, T. Aftab (eds.), *Medicinal Plants of the Asteraceae Family*,
https://doi.org/10.1007/978-981-19-6080-2_12

ethnopharmacological books, and peer-reviewed papers. This book chapter delves deeply into the chemical, nutritional, and pharmacological properties of *H. annuus*.

Keywords *Helianthus annuus* · Sunflower · Asteraceae · Fatty acids · Phenolic acids

12.1 Introduction

Asteraceae, also known as the sunflower family, is one of the largest family of flowering plants, with 1600 genera and 25,000 species that are found all over the world except the Arctic and Antarctic regions. The family is a well-known source of ornamental flowers, medicinal properties, food, and cosmetics (Rolnik and Olas 2021). *Helianthus annuus* L., commonly known as sunflower (Fig. 12.1), is distributed worldwide, and its domestication can be traced back to Mexico and North America around 2600 BC (Lentz et al. 2008). The plant is annual in nature and is 1–3 m in height with a taproot system that later change to fibrous and lateral roots. The stem of the plant is unbranched and round with internodes. The shape of leaves is ovate, 4–20 cm in length with a width of 3–15 cm. Upper leaves alternate along the stem, while lower leaves are opposite. Inflorescence of the plant is capitulum comprising of composite heads of 1000 to 2000 individual ray and disk florets and fruit that is an achene (Mashwani et al. 2015). The plant has important bioactive compounds that include alkaloids, phenols, saponins, terpenes, steroids, carbohydrates, fixed oil, flavonoids, active proteins, and vitamin B, D, E, and K which make it of high economic value (Mashwani et al. 2015; Subashini and Rakshitha 2016). According to a report (agricultural marketing resource center) in 2020, the sunflower crop was valued at nearly $587 million and the total harvest accounted for 2.98 billion pounds. Sunflower is the world's third most produced oilseed, trailing only soybean and rapeseed. In the vegetable oil market, it holds the fourth position behind palm oil, soybean oil, and rapeseed oil. Seeds of the plant (Fig. 12.2) contain both, oil (44%), and protein (16%) making it valuable in the vegetable oils market as well as vegetable protein-rich products. The crop is mainly traded after processing. As in 2018/2019, 55% of the oil was exported, 38% of the meals, and only 5% of the seeds were exported globally (Pilorge 2020). Russia, Ukraine, European Union, and Argentina are the leading producers of sunflower with a 70% share alone in the global market. Food and agriculture organization (FAO) is expecting the total world output of *Helianthus* to reach 60 million tons in 2050 (Fernández-Luqueño et al. 2014).

The uses of *H. annuus* are innumerable as it possesses a high content of polyunsaturated and monounsaturated oil that makes it an excellent food while leftovers are used in animal feed. It is also used for the production of biofuel, dyes, cosmetics, paints, and surfactants (Adeleke and Babalola 2020). The plant evidenced efficiency in the reclamation of heavy metals contaminated industrial soil and accumulated a diverse range of Pb, Cd, Zn, Cu, and Fe and thus can be used for phytoremediation in industries such as plastic, paper, dye, and textile (Chauhan and Mathur 2020). The

Fig. 12.1 Photographs of the plant, flower, and seeds bearing flower pod of *Helianthus annuus*

chapter on *Helianthus annuus* is written with the objective to gain insight into various aspects of this plant such as its botany, economics, traditional uses, phytochemistry, nutritional benefits, etc. It also aims to explore the potential of this crop as functional food being used as dietary supplements.

12.2 Nutritional Benefits

Sunflower has various biological properties like antioxidant, anti-inflammatory, antidiabetic, antimicrobial, and antihypertensive effects (Guo et al. 2017). Plant-derived lipids are considered healthier than animal-derived fats due to their high polyunsaturated fatty acid content. Sunflower oil contains high levels of omega

Fig. 12.2 Photograph of the seeds of *Helianthus annuus*

Table 12.1 *Helianthus annuus* and its nutritional benefits

Antioxidant benefits	Protect body cells from ROS, reduces risk of diseases like carcinoma, atherosclerosis, chronic inflammation and cataracts, biological constituents are enzymes (glutathione reductase, catalase, glutathione dehydrogenase) and phenolic compounds, L-ascorbic acid, peptides, carotenoids	Jiraungkoorskul (2016) and Guo et al. (2017)
Anti-inflammatory benefits	High magnesium and vitamin E reduce symptoms of osteoarthritis, rheumatoid and asthma	Bashir et al. (2015)
Cardiovascular benefits	Vitamin E lowers the risk of heart attack and reduces blood pressure, soreness, and fatigue	Vijayakumar et al. (2016)
Anti-cholesterol benefits	Phytosterol and tocopherol reduce LDL cholesterol and plasma cholesterol, and sterols inhibit the absorption of cholesterol in intestine	Rani et al. (2017) and Jesch and Carr (2017)
Anti-cancer benefits	Selenium found in sunflower oil improves immunity against cancerous cells and help in DNA repair	Roy et al. (2015) and Pisoschi and Pop (2015)

3 and omega 6, which adds to its nutritional value (Van Nieuwenhove et al. 2019). Sunflower seeds contain flavonoids such as heliannone, kaempferol, quercetin, apigenin, and luteolin, as well as phenolic acids such as sinapic acid, gallic acid, chlorogenic acid, caffeic acids, and coumaric acid, among other things (Guo et al. 2017) (Table 12.1, Fig. 12.3).

12.3 Ethnobotanical Uses

Traditional uses of *H. annuus* date back 3000 years. It has been used to treat the common cold, heart disease, cough, respiratory infections, whooping cough, and other ailments. Tannins, flavonoids, saponins, phenolic acids, steroids, and terpenoids are found in the plant's leaves, stems, and roots (Table 12.2).

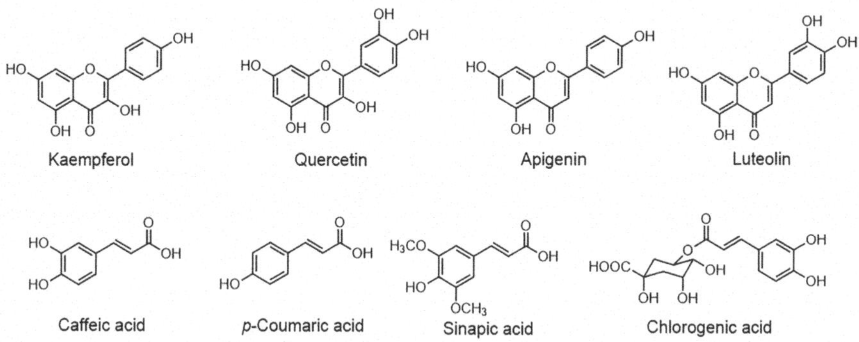

Fig. 12.3 Structures of common flavonoids and phenolic acids of *Helianthus annuus*

Table 12.2 *Helianthus annuus* and its traditional uses

Country	Plant part used	Effects	References
Iraq	Seeds	Diuretic and expectorant	Al-douri (2000)
India	Flower, leaves, seed, and roots	Bronchiectasis and skin allergy	Balkan et al. (2020) and Xavier et al. (2015)
Europe	Seeds	Pulmonary infection	Balkan et al. (2020)
Russia	Leaves	Fever	Balkan et al. (2020)
America	Leaves	Kidney diseases	Balkan et al. (2020)
Mexico	Roots	Wounds and cuts	Bashir et al. (2015)
Morocco	Leaves	Hypoglycemic, gastrointestinal stimulant, and anthelmintic	Bnouham et al. (2002)
Africa	Leaves	Malaria	Oladeji et al. (2020)
Nigeria	Leaves	Treatment of diabetes mellitus	Onoja and Anaga (2013)
South Africa	Seeds	Antiretroviral therapy	Malangu (2007)

12.4 Phytochemistry

Helianthus annuus plant is widely studied for its unique chemical constituents and their uses. Flavonoids, alkaloids, tannins, saponins, steroids, and carbohydrates have been found in earlier studies in various extracts (Düsterhöft et al. 1992; Subashini and Rakshitha 2012). Sunflower seed contains proteins, carbohydrates, acids, vitamins, minerals, and other micronutrients. Entkaurane glycoside, dubbed "helikauranoside A," was isolated from the aerial parts of *H. annuus*, along with three other compounds: grandifloric acid, paniculoside, and entkaurane-type diterpenoids: (−) kaur-16-en-19-oic acid (Spring and Hager 1982). In ethanolic extract, a new germacranolide with an amethylene-y-lactone moiety, heliangolideniveusin B, and its ethoxy derivatives were isolated, and the structures of two other sesquiterpenes were determined using spectroscopic methods (Spring and Hager 1982). *H. annuus* L. methanolic seed extract contained a significant amount of carbohydrates (Düsterhöft et al. 1992). Glenn and Chapman (1987) isolated a proteinaceous competitive inhibitor of lipase from the seeds of *H. annus*

(Glenn and Chapman 1987). Giudici also isolated a 16-kDa protein called SAP16 (Giudici et al. 2000). Five flavonoids were isolated from *H. annuus* L, including chalconeskukulcanin B, heliannone A, the flavanonesheliannones B and C, and flavonoltambulin (Macías et al. 2008; Rao et al. 2001). Ukiya et al. (2007) isolated eight triterpene alcohol fatty acid esters, four free triterpene alcohols, four diterpene acids, four estolides, three syn-alkane-4,6-diols, two tocopherol-related compounds, one 1,3-dioxoalkanoic acid, and one aliphatic ketone from the pollen grains of *H. annuus* L. (Ukiya et al. 2007). Nevadensin, a bioflavonoid isolated from *H. annuus* L., exhibits a wide range of significant biological activities including hypotensive, antitubercular, antimicrobial, anti-inflammatory, antitumor, and anti-cancer properties (Brahmachari 2010). Using thin-layer chromatography, allelochemicals were discovered and analyzed in sunflower leaves, roots, and stems. Spectrophotometry, on the other hand, has been used to examine alkaloids, flavonoids, and phenols (Kamal 2011). *H. annuus* has produced some light-colored proteins, and four tocopherol isomers ((α, β, γ, and δ) have also been discovered in sunflower seed oil that contain helianthinin as a globulin molecule (Pickardt et al. 2011). Tannins, as well as oleic acid, alkaloid, fixed oils, and simple phenolic compounds, have been found in *H. annuus* (Aziz et al. 2014). Phenols were isolated from sunflower florets and found to be a good source of dietary fiber, iron, and phenols (Aziz et al. 2014).

Caffeic acid, chlorogenic acid, and dicaffeoylquinic acid were extracted from seeds using an aqueous methanolic extract (Ibrahim et al. 2014). Acetone extract of *H. annuus* L. flower disk revealed four compounds: (2R)-2-hydroxyl-N-[(2S, 3S, 4R, 10E)-1,3,4-trihydroxyicos10-en-2-y1] docosahexanoic acid (2R,3R) [(2S,3S,4R,10E)-1,3,4-trihydroxyicos-10-en-2-yl]-2,3-dihydroxy-N [(2S,3S,4R,10E)-1,3,4-trihydroxyicos-10-en-2-N-(2-phenylethyl) tetracosanamide, docosanamide, and a known ceramide, (2R)-N-[(2S,3S,4R,8E)-(−D-Glucopyranosyloxy)-1 3–4-dihydroxyoctadec-8-en-2-yl]-2,2-Hexadecanamide (Suo and Yang 2014). *H. annuus* L. contains chemical constituents such as heliannone, quercetin, kaempferol, luteolin, and apigenin; phenolic acids such as caffeoylquinic acid, protocatechuic, coumaric, ferulic acid, and sinapic acids; and fatty acids like lauric, palmitic, oleic, linoleic, stearic, and linolenic (Arruda et al. 2018). Flavones undergo 5,7,4-trioxygenation (apigenin type) and 5,7,3,4-tetraoxygenation (luteolin type), whereas flavonols undergo 3,5,7,4-tetraoxygenation (kaempferol type) and 3,5,7,3,4-pentaoxygenation (quercetin type) (Arruda et al. 2018). Recently, GC-MS revealed the presence of -pinene, verbenone, terpinolene, and -terpineol in sunflower receptacles (Liu et al. 2020) (Tables 12.3 and 12.4].

Table 12.3 Phytochemicals found in *Helianthus annuus*

Compound name	Plant part	References
Alkaloids	Seed	Saini and Sharma (2013)
Allelochemicals	Whole plant	Düsterhöft et al. (1992)
Carbohydrates and phenolic compounds	Seed and florets	Liang et al. (2013)
Flavonoids	Seed	Kamal (2011)
Saponins and steroids	Seeds	Saini and Sharma (2013)
Tannins	Seed	Düsterhöft et al. (1992)

Table 12.4 List of phenolic compounds and peptides compounds present in *H. annuus*

Phenolic compounds	Trans-ferulic acid Cis-ferulic acid 3-O- Caffeoylquinic acid	Seed	Karamać et al. (2012), Aziz et al. (2014) and Guo et al. (2017)
	Flavanone	Seed	Karamać et al. (2012), Guo et al. (2017) and Gai et al. (2020)
	Quercetin diglycoside	Seed	Aziz et al. (2014) and Guo et al. (2017)
	Caffeoyl feruloylquinic acid	Seed	Zeb (2020)
	Quercetin rutinoside (Rutin)	Seed	Weisz et al. (2009), Karamać et al. (2012) and Gai et al. (2020)
	Ferulic acid dehydrotrimer	Seed	Saftić-Panković et al. (2006), Weisz et al. (2009) and Karamać et al. (2012)
	Dicaffeoylquinic acid	Seed	Karamać et al. (2012) and Gai et al. (2020)
Peptides	Albumin	Seed	Zilic et al. (2010)
	Helianthinin	Seed	González-Pérez et al. (2004)
	ACE inhibitory peptide	Seed	Megías et al. (2004)
	Total acid subunit of helianthinin ($\alpha + \alpha'$)	Seed	Guo et al. (2017)
	Total basic subunit of helianthin (β)	Seed	Lawson et al. (2019)

12.5 Pharmacological Studies

12.5.1 As Anti-Inflammatory and Analgesic

Sunflower (*Helianthus annuus*) is well known for its anti-inflammatory and analgesic properties due to its high concentration of vitamin E and other microelements such as magnesium, which serves as an excellent source of fat-soluble antioxidants to the body (Muratspahić et al. 2021; Kumar et al. 2017). Triterpene glycosides derived from the methanol extract of sunflower petals were found to have anti-inflammatory and inhibitory effects in mice with ear edema (Bashir et al. 2015). The analgesic activity of a methanolic extract of sunflower seeds was tested using the acetic acid-induced writhing and hot plate methods. The extract exhibited significant analgesic potential (even greater than aspirin) (Islam et al. 2016a, b). The anti-inflammatory effect of helianthoside compounds isolated from an n-butanol-soluble fraction of a methanol extract against 12-O-tetradecanoylphorbol-13-acetate [TPA]-induced inflammation in mice was studied in another study. Díaz-Viciedo et al. (2008) were given an ethanol extract of sunflower leaves to rats. When compared to control animals, the treatment effectively inhibited paw edema caused by egg albumin and significantly increased the mean tolerance time of rats to thermal noxious stimuli (Emamuzo et al. 2010).

12.5.2 Cardiovascular Benefits

Sunflowers contain vitamin E and magnesium, which have anti-inflammatory properties, protect biomolecular components and reduce symptoms of osteoarthritis, rheumatoid arthritis, and asthma by scavenging free radicals. Consumption of vitamin E and magnesium-rich foods lowers the risk of atherosclerosis, lower blood pressure, asthma, osteoarthritis, sudden heat sensation in menopausal women, high blood pressure, stroke, cardiovascular disease, and migraine headaches (Bashir et al. 2015).

12.5.3 As Antimicrobial, Antiasthmatic, and Antidiabetic

Methanolic extracts of sunflower seed were tested for antimicrobial activity against Gram-positive and Gram-negative bacteria such as *Staphylococcus aureus*, *S. epidermidis*, *Escherichia coli*, *Proteus vulgaris*, and *Pseudomonas aeruginosa*. Another study found that the extract had high activity against *Salmonella typhi*, *Aspergillus fumigates*, and *Rhizopus stolonifer*; moderate activity against *S. aureus*, *Candida albicans*, and *Vibrio cholera*; less activity against *Bacillus subtilis*; and resistance to *Fusarium oxysporum* (Subashini and Rakshitha 2012; Menzel et al. 2019). As a result, sunflower seeds can be used as natural food preservatives (Thielmann et al. 2017). Sunflower ethanolic stem extract was tested against *S. aureus*, *E. coli*, *A. Niger*, and *Candida albicans*. Except for *E. coli*, the extract exhibited antimicrobial activity against all bacteria tested (Adetunji et al. 2014). Sunflower extracts have also been shown to have antiasthmatic and antidiabetic properties (Gad and El-Ahmady 2018). In addition, an in vivo antiasthmatic assay of aqueous extract on ovalbumin-induced mice, as well as hematoxylin and eosin staining of their lungs, revealed the extract's potency in reducing asthma effect on mice (Kim et al. 2020).

12.5.4 As Anti-Ulcer and Antidiarrheal

Hydroalcoholic extracts of *A. indicum*, *H. annuus*, and a combination of both were tested for anti-ulcer activity in Albino Wistar rats against ethanol-induced gastric ulcer and pyloric ligation-induced gastric ulcer (Venkateswarlu et al. 2015). In the castor oil-induced diarrhea and gastrointestinal transit model in mice, an ethanolic extract of the leaves of *H. annuus* demonstrated antidiarrheal activity (Dwivedi and Kaushik 2015).

12.5.5 As Antihistaminic

The antihistaminic activity of an ethanolic extract of sunflower leaves was tested in guinea pigs with histamine-induced bronchoconstriction and rabbits with a microshock model. Sunflower had a high antihistaminic potential (Dwivedi and Kaushik 2015).

12.5.6 As Anticancerous

The antiproliferative effect of chloroform root extract of sunflower was studied against HeLa, MCF-7, and A-431 cell lines. MIC50 of the extract was 3.51, 3.36, and 4.19 μg/mL against the three cell lines respectively (Csupor-Löffler 2012). Selenium in sunflower has induced DNA repair and production in degenerated cells, inhibition of cancer cell growth, induction of apoptosis, and self-destruction of unwanted or worn-out cells. Furthermore, the addition of selenium to protein active sites like glutathione peroxidase protects body cells from cancer (Khan et al. 2018; Pisoschi and Pop 2015; Cardoso et al. 2017). The antiproliferative effect of chloroform root extract of sunflower on HeLa, MCF-7, and A-431 cell lines was investigated. The extract's MIC50 against the three cell lines was 3.51, 3.36, and 4.19 g/mL, respectively (Csupor-Löffler 2012). The presence of selenium in sunflower-induced DNA repair and production in degenerated cells, inhibited the growth of cancer cells, induced apoptosis, and caused unwanted or worn-out cells to self-destruct. Furthermore, the addition of selenium to protein active sites like glutathione peroxidase protects body cells from cancer (Khan et al. 2018; Pisoschi and Pop 2015; Cardoso et al. 2017; Zoumpoulakis et al. 2017).

12.5.7 As an Antioxidant

Antioxidants protect body cells from reactive oxygen species damage and lower the risk of cataracts, carcinoma, chronic inflammation, atherosclerosis, cardiovascular disease, neurodegenerative diseases, and premature ageing (Jiraungkoorskul 2016). Toxicology limits the use of synthetic antioxidants. Thus, the discovery of novel antioxidants from plants may be promising as a source of dietary antioxidants (Singh et al. 2016). Sunflower natural antioxidants can inhibit or prevent oxidation, scavenge free radicals, and thus prevent disease proliferation within cells (Guo et al. 2017; Rauf et al. 2020).

12.5.8 Anti-Obesity Activity

Sunflower seeds were also tested for anti-obesity activity in mice fed a cafeteria diet, atorvastatin [10 mg/kg], and methanolic extract at 200 mg/kg daily for 6 weeks. The methanolic extract increased locomotor activity [rearing, grooming, and ambulation] with HDL while decreasing food consumption, body weight, BMI, LIO, total cholesterol, triglyceride, LDL, and glucose (Islam 2016).

12.5.9 Others

In rats, the effects of an ethanol extract of sunflower leaves [0.5 g/kg orally for 2 weeks] on fecundity were investigated. The findings revealed that coital frequency was unaffected, but the pregnancy rate and number of pups per rat and per group were significantly reduced. The ethanol extract-induced histo-degenerative changes in the gonads may be responsible for the reduced fecundity observed in treated adult rats (Emamuzo et al. 2010). The study on the histology of the testes, blood levels of some reproductive hormones, and epididymal sperm in Wistar rats after treatment with the extract revealed that the extract had some antifertility effects (Ejebe et al. 2008).

12.6 Clinical Trials

A number of clinical trials have been conducted to test the clinical efficacy and allergy potential of *Helianthus annuus*. Topical application of sunflower seed oil three times per day significantly improved skin condition in preterm infants and is effective for improving neonatal outcomes, particularly in low birth-weight infants whose skin barrier is temporarily but critically compromised due to immaturity and who are frequently susceptible to infection due to their underdeveloped skin. According to the study, infants who received a daily skin treatment with sunflower oil were 41% less likely to develop infections in the hospital (Moroń et al. 2009). In randomized forearm-controlled mechanistic studies, 19 volunteers with and without a history of atopic dermatitis were recruited. In 19 volunteers, topical application of olive oil resulted in a significant decrease in stratum, corneum integrity, and the induction of mild erythema (Danby et al. 2013).

Ozonoid sunflower oil was evaluated on a patient with bedsores who was treated with Oz.Or.Oil 30 and showed faster ulcer healing (Serio et al. 2017). Ozonized sunflower oil (OSO) was studied for its efficacy as a treatment for generalized demodicosis in dogs. Since the seventh day of application, there has been a significant reduction in mite counts, clinical signs, and pruritus (Rodriguez et al. 2021). In a population-based, cluster-randomized, controlled trial in 276 clusters in rural Uttar Pradesh, India, hospitalized preterm infants with compromised skin barrier function treated topically with sunflower seed oil (SSO) showed reductions in sepsis and neonatal mortality rate (NMR) (Kumar et al. 2021). A randomized, controlled, unblinded clinical trial on 212 children aged 2 to 24 months with severe acute malnutrition was conducted at Dhaka Hospital in Bangladesh. For 10 days, children in the emollient group received three times daily whole-body treatments of 3 g/kg sunflower seed oil (SSO). Emollient therapy resulted in systematically greater increases in 26 of 29 fatty acids (FAs) over time when compared to the control. These effects were significantly stronger in the younger age groups, with only minor differences in the older children (Shahunja et al. 2021).

12.7 Studies Related to Safety and Toxicity

On male Wistar rats, an ethanolic extract of sunflower leaves was tested, and the LD50 was determined to be 14 g/kg (Ejebe et al. 2008). Although allergies to sunflower seeds are uncommon, several cases of occupational allergies to sunflowers have been reported. A case study of a 40-year-old male patient admitted to the hospital after experiencing shortness of breath and urticaria shortly after eating sunflower seeds. According to the study, a patient who is allergic to mugwort pollen could have an anaphylactic reaction after eating sunflower seeds (Ukleja-Sokołowska et al. 2016). Data from 40 Wistar rats (male and female) revealed a decrease in fecundity (Emamuzo et al. 2010). The antidiabetic effects of methanolic extracts of *H. annuus* leaves at various doses were studied in alloxan-induced diabetic rats, and it was discovered that the extract showed a significant dose- and time-dependent decrease in blood glucose levels in alloxan-induced diabetic rats (Onoja and Anaga 2014). On 56 Wistar rats, the effects of various concentrations of hydromethanolic extract of *H. annuus* on the leukocyte profile and spleen histology after prolonged exposure were studied. The results indicate that a hydromethanolic extract of *H. annuus* have no effect on the leucocyte profile or the histology of the spleen (Ezeh et al. 2021).

12.8 Conclusion

The plants are ethnomedicinally important and are a rich source of various alkaloids, terpenoids, carbohydrates, fixed oils, steroids, amino acids, etc. There are many reports of this plant having the potential of substituting soybean as a source for proteins. Studies have shown the importance of protein challenge (quantitative and qualitative) to feed the growing population of the world and sunflower is an excellent vegetal protein demanding limited resources. The plant can be used locally as a cropping system to improve soil quality showing its potential in the field of phytoremediation. The oil of the plant produces biodegradable and sustainable products making it important for petroleum chemistry and bioplastics. The cellulosic fraction and fibers from pith and stem can also be used for making different biomaterials and contributing to renewable sourcing. The important point to be highlighted is that plants have robust adaptability to various environmental conditions and hence are easier to cultivate with higher yields. Horticulturally too they are extremely in demand since they can be used for ornamental purposes. There is a substantial literature on the plant, but as research-based knowledge grows, it must be updated regularly as plant faces many challenges on several platforms such as proper adaptation of cropping practices, function being used as food, CO_2 fertilization, etc. Investment in sunflower being used as an important oil crop is slowly making progress, especially over last decade key developments in the research area have been done and the crop holds great potential as a source of oil, food, fiber, medicines, and ornamentals.

References

Adeleke BS, Babalola OO (2020) Oilseed crop sunflower (*Helianthus annuus*) as a source of food: nutritional and health benefits. Food Sci Nutr 8:4666–4684

Adetunji CO, Olatunji OM, Ogunkunle ATJ, Ogundare MO (2014) Antimicrobial activity of ethanolic extract of Helianthus annuus stem. SMU Med J 1(1):79–88

Al-douri NA (2000) A survey of medicinal plants and their traditional uses in Iraq. Pharm Biol 38(1):74–79

Arruda HS, Pereira GA, de Morais DR, Eberlin MN, Pastore GM (2018) Determination of free, esterified, glycosylated and insoluble-bound phenolics composition in the edible part of araticum fruit (Annonacrassiflora Mart.) and its by-products by HPLC-ESI-MS/MS. Food Chem 245:738–749. https://doi.org/10.1016/j.foodchem.2017.11.1

Aziz FM, Darweesh MJ, Rahi FA, Saeed RT (2014) In vivo and in vitro studies of a polar extract of Helianthus annuus (sunflower) seeds in treatment of napkin dermatitis. Int J Pharm Sci Rev Res 24(2):1–3

Balkan AA, Yildiz A, Berber D, Gokalsin B, Sesal NC (2020) Infusion or decoction extracts of *Helianthus annuus* leaves: potential inhibitors for QS system and biofilm formation in *Pseudomonas aeruginosa*. Int J Adv Eng Pure Sci 32(4):499–506

Bashir T, Zahara K, Haider S, Tabassum S (2015) Chemistry, pharmacology and ethnomedicinal uses of Helianthus annuus (sunflower): a review. Pure Appl Biol 4(2):226. https://doi.org/10.19045/bspab.2015.42011

Bnouham M, Mekhfi H, Legssyer A, Ziyyat A (2002) Ethnopharmacology forum medicinal plants used in the treatment of diabetes in Morocco. Int J Diabetes Metab 10:33–50

Brahmachari G (2010) Nevadensin: isolation, chemistry and bioactivity. Int J Green Pharmacy 4: 213–219

Cardoso BR, Duarte GBS, Reis BZ, Cozzolino SM (2017) Brazil nuts: Nutritional composition, health benefits and safety aspects. Food Res Int 100:9–18. https://doi.org/10.1016/j.foodres.2017.08.036

Chauhan P, Mathur J (2020) Phytoremediation efficiency of *helianthus annuus* L. For reclamation of heavy metals-contaminated industrial soil. Environ Sci Pollut Res 27:29954–29966

Csupor-Löffler B (2012) Activity-guided investigation of antiproliferative secondary metabolites of Asteraceae species. University of Szeged

Danby SG, AlEnezi T, Sultan A, Lavender T, Chittock J, Brown K et al (2013) Effect of olive and sunflower seed oil on the adult skin barrier: implications for neonatal skin care. Pediatr Dermatol 30(1):42–50

Díaz-Viciedo R, Hortelano S, Girón N, Massó JM, Rodriguez B, Villar A, de Las HB (2008) Modulation of inflammatory responses by diterpene acids from *Helianthus annuus* L. Biochem Biophys Res Commun 369(2):761–766

Düsterhöft EM, Posthumus MA, Voragen GJ (1992) Non-starch polysaccharides from sunflower (Helianthus annuus) meal and palmkernel (Elaeisguineensis) meal investigation of the structure of major polysaccharides. J Sci Food Agr 59:151–160

Dwivedi A, Kaushik AY (2015) Evaluation of Helianthus annuus L. leaves extract for the anti-diarrheal and anti-histaminic activity. Int J Res Ayur Pharmacy 6(1):118–123

Ejebe DE, Emudainohwo JOT, Akonoghrere R, Olise CC, Amadi CN, Siminialayi IM (2008) Effects of ethanol extract of leaves of Helianthus annuus on the reproductive system of male wistar rats: testicular histology, epididymal sperm properties and blood levels of reproductive hormones. Biomed Pharmacol J 1(1):65

Emamuzo ED, Miniakiri SI, Tedwin EJO, Delesi KH, Precious A (2010) Effects of ethanol extract of leaves of *Helianthus annus* on the fecundity of Wistar rats. Asian Pac J Trop Med 3(6): 435–438

Ezeh BR, Heleno SA, Oliveira MBPP, Barros L, Ferreira ICFR (2021) Phenolic compounds: current industrial applications, limitations and future challenges. Food Funct 12(1):14–29

Fernández-Luqueño F, López-Valdez F, Miranda-Arámbula M, Rosas-Morales M, Pariona N, Espinoza-Zapata R (2014) An introduction to the sunflower crop. In: Sunflowers: growth and development, environmental influences and pests/diseases. Nova Science, New York

Gad HA, El-Ahmady SH (2018) Prediction of thymoquinone content in black seed oil using multivariate analysis: an efficient model for its quality assessment. Ind Crop Prod 124:626–632. https://doi.org/10.1016/j.indcrop.2018.08.037

Gai F, Karamać M, Janiak MA, Amarowicz R, Peiretti PG (2020) Sunflower (Helianthus annuus L.) plants at various growth stages subjected to extraction—comparison of the antioxidant activity and phenolic profile. Antioxidants 9(6):535

Giudici AM, Regente MC, Canal L (2000) A potent antifungal protein from Helianthus annuus flower is a trypsin inhibitor. Plant Physiol Biochem 38(11):881–888

Glenn W, Chapman J (1987) A proteinaceous competitive inhibitor of lipase isolated from Helianthus annuus seeds. Phytochemistry 26(12):3127–3131

González-Pérez S, Vereijken JM, Merck KB, Koningsveld GA, Gruppen H, Voragen AG (2004) Conformational states of sunflower (Helianthus annuus) helianthinin: effect of heat and pH. J Agric Food Chem 52(22):6770–6778

Guo S, Ge Y, Jom KN (2017) A review of phytochemistry, metabolite changes, and medicinal uses of the common sunflower seed and sprouts (Helianthus annuus L.). Chem Cent J 11(1):95. https://doi.org/10.1186/s13065-017-0328-7

Ibrahim TA, Ajongbolo KF, Aladekoyi G (2014) Phytochemical screening and antimicrobial activity of crude extracts of Basella alba and Helianthus annuus on selected food pathogens. Res Rev J Microbiol Biotechnol 3(2):27–29

Islam AT (2016) In vivo anti-obesity activity of methanolic extract of Helianthus annuus seeds. Available at SSRN 2862458
Islam R, Islam AT, Mazumder K (2016a) In vivo antidepressant and anxiolytic activity of the methanol extract of Helianthus annuus seeds. Int J Innov Appl Res 4(2):5–10
Islam R, Islam AT, Hossain M, Mazumder K (2016b) In vivo analgesic activity of methanolic extract of Helianthus annuus seeds. Int Curr Pharmaceut J 5(4):38–40
Jesch ED, Carr TP (2017) Food ingredients that inhibit cholesterol absorption. Prev Nutr Food Sci 22(2):67–80
Jiraungkoorskul W (2016) Review of nutraceutical uses of an antioxidant sunflower sprout, Helianthus annuus. Asian J Pharm Clin Res 9(6):21–23. https://doi.org/10.22159/ajpcr.2016.v9i6.12874
Kamal J (2011) Quantification of alkaloids, phenols and flavonoids in sunflower (Helianthus annuus L.). Afr J Biotechnol 10(16):3149–3151
Karamać M, Kosińska A, Estrella I, Hernández T, Duenas M (2012) Antioxidant activity of phenolic compounds identified in sunflower seeds. Eur Food Res Technol 235(2):221–230
Khan N, Zandi P, Ali S, Mehmood A, Adnan Shahid M, Yang J (2018) Impact of Salicylic acid and PGPR on the drought tolerance and phytoremediation potential of Helianthus annus. Front Microbiol 9:2507. https://doi.org/10.3389/fmicb.2018.02507
Kim K, Yoo HJ, Jung JH, Lee R, Hyun JK, Park JH, Yeon JH et al (2020) Cytotoxic effects of plant sap-derived extracellular vesicles on various tumor cell types. J Funct Biomater 11(2):22. https://doi.org/10.3390/jfb11020022
Kumar S, Sharma S, Vasudeva N (2017) Review on antioxidants and evaluation procedures. Chin J Integr Med 6:1–12. https://doi.org/10.1007/s11655-017-2414-z
Kumar A, Mishra S, Singh S, Ashraf S, Kan P, Ghosh AK, Shivgarh Emollient Research Group (2021) Effect of sunflower seed oil emollient therapy on newborn infant survival in Uttar Pradesh, India: a community-based, cluster randomized, open-label controlled trial. PLoS Med 18(9):e1003680
Lawson SK, Sharp LG, Powers CN, McFeeters RL, Satyal P, Setzer WN (2019) Essential oil compositions and antifungal activity of sunflower (helianthus) species growing in North Alabama. Appl Sci 9(15):3179
Lentz DL, Phol MD, Alvarado JL, Taright S, Bye E (2008) Sunflower (Helianthus annuus L.) as a pre-Columbian domesticate in Mexico. Proc Natl Acad Sci U S A 105(17):6232–6237
Liang Q, Cui J, Zhao G (2013) Florets of sunflower (*Helianthus annuus* L.): potential new sources of dietary fiber and phenolic acids. J Agric Food Chem 61(14):3435–3442
Liu XS, Gao B, Li XL, Li WN, Qiao ZA, Han L (2020) Chemical composition and antimicrobial and antioxidant activities of essential oil of sunflower (Helianthus annuus L.) receptacle. Molecules 25(22):5244
Macías FA, Lopez A, Varela RM, Torres A, Molinillo JM (2008) Helikauranoside a new bioactive diterpene. J Chem Ecol 34(1):65–69
Malangu N (2007) Self-reported use of traditional, complementary and over-the-counter medicines by Hiv-infected patients on antiretroviral therapy in Pretoria, South Africa. Afr J Tradit Complement Altern Med 4(3):273–278
Mashwani ZR, Bashir T, Zahara K, Haider S, Tabassum S, Mudrikah M (2015) Chemistry, pharmacology and ethnomedicinal uses of *Helianthus annuus* (sunflower): a review. Pure Appl Biol 4(2):226–225
Megías C, Yust MDM, Pedroche J, Lquari H, Girón-Calle J, Alaiz M, Vioque J et al (2004) Purification of an ACE inhibitory peptide after hydrolysis of sunflower (Helianthus annuus L.) protein isolates. J Agric Food Chem 52(7):1928–1932
Menzel C, González-Martínez C, Chiralt A, Vilaplana F (2019) Antioxidant starch films containing sunflower hull extracts. Carbohydr Polym 214:142–151. https://doi.org/10.1016/j.carbpol.2019.03.022

Moroń D, Lenda M, Skórka P, Szentgyörgyi H, Settele J, Woyciechowski M (2009) Wild pollinator communities are negatively affected by invasion of alien goldenrods in grassland landscapes. Biol Conserv 142:1322–1332
Muratspahić E, Tomašević N, Koehbach J, Duerrauer L, Hadžić S, Castro J, .. Gruber CW (2021). Design of a stable cyclic peptide analgesic derived from sunflower seeds that targets the κ-opioid receptor for the treatment of chronic abdominal pain. J Med Chem, 64(13), 9042–9055
Oladeji SO, Oluyori AP, Bankole DT, Afolabi TY (2020) Natural products as sources of antimalarial drugs: ethnobotanical and ethnopharmacological studies. Scientifica 2020:7076139. 22 pages
Onoja SO, Anaga AO (2013) Evaluation of the antidiabetic and antioxidant potentials of methanolic leaf extract of Helianthus annuus L. on alloxan-induced hyperglycemic rats. Comp Clin Pathol 23(5):1565–1573
Onoja SO, Anaga AO (2014) Evaluation of the antidiabetic and antioxidant potentials of methanolic leaf extract of Helianthus annuus L. on alloxan-induced hyperglycemic rats. Comp Clin Pathol 23(5):1565–1573
Pickardt C, Weiszb GM, Eisnera P, Kammererb DR, Neidhartb S, Carle R (2011) Processing of low polyphenol protein isolates from residues of sunflower seed oil production. Proc Food Sci 1: 1417–1424
Pilorge E (2020) Sunflower in the global vegetable oil system: situation, specificities and perspectives. Oilseeds Fats Crops Lipids 27:34
Pisoschi AM, Pop A (2015) The role of antioxidants in the chemistry of oxidative stress: a review. Eur J Med Chem 97:55–74. https://doi.org/10.1016/j.ejmech.2015.04.040
Rani R, Sheoran R, Sharma B (2017) Perspectives of breeding for altering sunflower oil quality to obtain novel oils. Int J Curr Microbiol App Sci 6(8):949–962. https://doi.org/10.20546/ijcmas.2017.608.117
Rao YK, Rao CV, Kishore PH, Gunasekar D (2001) Total synthesis of Heliannone A and (R,S)-Heliannone B, two bioactive flavonoids from Helianthus annuus cultivars. J Nat Prod 64(3): 368–369
Rauf S, Ortiz R, Shehzad M, Haider W, Ahmed I (2020) The exploitation of sunflower (Helianthus annuus L.) seed and other parts for human nutrition, medicine and the industry. Helia 43(73): 167–184. https://doi.org/10.1515/helia-2020-001960
Rodriguez ZZ, Lemus M, Fleitas E (2021) Efficacy of ozonized sunflower oil as treatment of canine generalized demodicosis. Insights Vet Sci 5:015–021
Rolnik A, Olas B (2021) The plants of the Asteraceae family as agents in the protection of human health. Int J Mol Sci 22(6):3009
Roy RD, Hossan MS, Rahmatullah M (2015) A review of anticancer potential of Elephantopus scaber and its phytoconstituents. World J Pharmacol Pharmaceut Sci 4(10):86–94
Saftić-Panković D, Veljović-Jovanović S, Pucarević M, Radovanović N, Mijić A (2006) Phenolic compounds and peroxidases in sunflower near-isogenic lines after downy mildew infection/ papel de compuestosfenólicos y peroxidasis en la respuesta de diferente-resistenteslíneasisogénicas de girasol a la infecciónpor el causante de tizó/rôle des composésphénoliques et des peroxydasesdans les lignées quasi-isogéniques du tournesolrésistantes à l'infection du mildiou. Helia 29(45):33–42
Saini SNI, Sharma S (2013) Antidiabetic effect of Helianthus annuus L., seeds ethanolicextract instreptozotocinnicotinamide induced type 2 diabetes mellitus. Int J Pharm Pharm Sci 25(2): 382–387
Serio F, Alba M, Cozzolino G, Idolo A, Grassi T (2017) Efficacy of a dermatological gel based on ozonized sunflower seed oil (Oz. or. Oil 30) on bedsores: a pilot study. Clin Derm Res J 2:2
Shahunja KM, Sévin DC, Kendall L, Ahmed T, Hossain MI, Mahfuz M, Darmstadt GL et al (2021) Effect of topical applications of sunflower seed oil on systemic fatty acid levels in under-two children under rehabilitation for severe acute malnutrition in Bangladesh: a randomized controlled trial. Nutr J 20(1):1–17

Singh JP, Kaur A, Singh N, Nim L, Shevkani K, Kaur H, Arora DS (2016) In vitro antioxidant and antimicrobial properties of jambolan (Syzygium cumini) fruit polyphenols. LWT Food Sci Technol 65:1025–1030

Spring O, Hager A (1982) Inhibition of elongation growth by two sesquiterpene lactones isolated from Helianthus annuus L. Planta 156(5):433–440

Subashini R, Rakshitha SU (2012) Phytochemical screening, antimicrobial activity and in vitro antioxidant investigation of methanolic extract of seeds from Helianthus annuus L. Chem Sci Rev Lett 1(1):30–34

Subashini R, Rakshitha SU (2016) Phytochemical screening, antimicrobial activity and in vitro antioxidant investigation of Methanolic extract of seeds from *Helianthus annuus* L. Chem Sci Rev Lett 1:30–34

Suo M, Yang J (2014) Ceramides isolated from Helianthus annuus L. Helvetica Chimica Acta 97 (3):355–360

Thielmann J, Kohnen S, Hauser C (2017) Antimicrobial activity of OleaeuropaeaLinné extracts and their applicability as natural food preservative agents. Int J Food Microbiol 251:48–66. https://doi.org/10.1016/j.ijfoodmicro.2017.03.019

Ukiya M, Akihisa T, Yasukawa K, Koike K, Takahashi A, Suzuki T, Kimura Y (2007) Triterpene glycosides from the flower petals of sunflower (Helianthus annuus) and their anti-inflammatory activity. J Nat Prod 70(5):813–816

Ukleja-Sokołowska N, Gawrońska-Ukleja E, Żbikowska-Gotz M, Bartuzi Z, Sokołowski L (2016) Sunflower seed allergy. Int J Immunopathol Pharmacol 29(3):498–503

Van Nieuwenhove E, Lagou V, Van Eyck L, Dooley J, Bodenhofer U, Roca C, Liston A (2019) Machine learning identifies an immunological pattern associated with multiple juvenile idiopathic arthritis subtypes. Ann Rheum Dis, annrheumdis–214354. https://doi.org/10.1136/annrheumdis-2018-214354

Venkateswarlu K, Vijayabhaskar K, Krishna OS, Chandra Sekhar KB (2015) Evaluation of anti-ulcer activity of hydro alcoholic extracts of Abutilon indicum, Helianthus annuus and combination of both against ethanol and pyloric ligation induced gastric ulcer in albino wistar rats. Br J Pharmaceut Res 5(1):42–51

Vijayakumar M, Vasudevan DM, Sundaram KR, Krishnan S, Vaidyanathan K, Nandakumar S, Mathew N et al (2016) A randomized study of coconut oil versus sunflower oil on cardiovascular risk factors in patients with stable coronary heart disease. Indian Heart J 68(4):498–506. https://doi.org/10.1016/j.ihj.2015.10.384

Weisz GM, Kammerer DR, Carle R (2009) Identification and quantification of phenolic compounds from sunflower (Helianthus annuus L.) kernels and shells by HPLC-DAD/ESI-MSn. Food Chem 115(2):758–765

Xavier TF, Kannam M, Auxilia A (2015) Traditional medicinal plants used in the treatment of different skin diseases. Int J Curr Microbiol Appl Sci 4(5):1043–1053

Zeb A (2020) Concept, mechanism, and applications of phenolic antioxidants in foods. J Food Biochem 44(9):e13394

Zilic S, Maksimovic-Dragisic J, Maksimovic V, Maksimovic M, Basic Z, Crevar M, Stankovic G (2010) The content of antioxidants in sunflower seed and kernel. Helia 33(52):75–84. https://doi.org/10.2298/hel1052075z

Zoumpoulakis P, Sinanoglou VJ, Siapi E, Heropoulos G, Proestos C (2017) Evaluating modern techniques for the extraction and characterisation of sunflower (Hellianthus annus L.) seeds phenolics. Antioxidants 6(3):46

Chapter 13
Silybum marianum (L.) Gaertn.: Traditional Uses, Phytochemistry, and Pharmacological Activities

Deepa Khatri, Sumit Bahadur Baruwal Chhetri, and Hari Prasad Devkota

Abstract *Silybum marianum* (L.) Gaertn. (family: Asteraceae), commonly known as milk thistle, has been widely used from ancient period of times for the treatment of various ailments related to liver, kidneys, gallbladder, etc. It is an extensively used and studied herb for the treatment of hepatobiliary diseases. Silymarin is the major active fraction of seeds/fruits and is a mixture of taxifolin and several flavonolignans such as silibinin A (silybin A), silibinin B (silybin B), isosilibinin A (isosilybin A), isosilibinin B (isosilybin B), silychristin, isosilychristin, and silydianin. Silibinin (a mixture of diastereoisomers silibinin A and B) is the major component of silymarin and possesses the greatest degree of biological activity. Various pharmacological activities such as antioxidant, hepatoprotective, renoprotective, neuroprotective, anti-inflammatory, anticancer, cardioprotective, and antidiabetic activities have been reported from milk thistle. Several clinical trials have been performed to evaluate the safety and efficacy profile of this plant. However, these researches/clinical trials are inadequate, so well-designed clinical studies should be conducted and an extensive research on pharmacokinetics profile should be carried out to launch *S. marianum* as a clinically proven drug for different health disorders in the market.

Keywords Silymarin · Milk thistle · Hepatoprotective · Silibinin

D. Khatri (✉)
Department of Pharmacy, Novel Academy, Purbanchal University, Pokhara, Nepal

S. B. B. Chhetri
Department of Pharmaceutical Sciences, School of Health and Allied Sciences, Pokhara University, Pokhara, Nepal

H. P. Devkota
Headquarters for Admissions and Education, Kumamoto University, Kumamoto, Japan
e-mail: devkotah@kumamoto-u.ac.jp

H. P. Devkota, T. Aftab (eds.), *Medicinal Plants of the Asteraceae Family*,
https://doi.org/10.1007/978-981-19-6080-2_13

Fig. 13.1 Photographs of whole plant, flowers, and seeds of *Silybum marianum*

13.1 Introduction

Silybum marianum (L.) Gaertn. (family: Asteraceae) (Fig. 13.1) is an annual or biennial herb, which has been used as a traditional medicinal plant for almost 2000 years. It is commonly known as milk thistle, and its name was derived from its prickle leaves with white veins, which were believed to be carrying Virgin Mother Mary's milk (Flora et al. 1998). It is also known as Doodh patra (India), Cardo di Maria (Italy), *da ji* (China), silybon (Greece), and Ostropestro (Russia) (Porwal et al. 2019). It is native to Mediterranean regions of Europe and North America, but nowadays it is widely distributed throughout the world (Abenavoli et al. 2018; Flora et al. 1996). It is also grown as an ornamental flower. According to The Plant List, there are three accepted species of the genus *Silybum* known as *S. marianum* (L.) Gaertn., *S. eburneum* Coss. & Durieu, and *S. morianum* Gaertn. (http://www.theplantlist.org/browse/A/Compositae/Silybum/).

The plant is characterized by lanceolate leaves and red-purple flower heads which are 4 to 12 cm wide. The leaves are alternate, glossy, green, milky-white veined, and glabrous with strongly spiny margins. The inflorescences are large and round capitula, solitary at the apex of the stem or its branches, and surrounded by triangular, hairless, thorny bracts (Benhouhou 2005). The fruits are black achenes about 5–8 mm long, 2–3 mm thick, with a white, silky pappus, surrounded by a yellow basal ring. Fruits possess cocoa-like odor and an oily bitter taste (Bijak 2017).

It grows well in a warm atmosphere and blooms in mid-May. The plant has the ability to thrive on light soils with periodic water deficit (Andrzejewska et al. 2011).

In 2020, *S. marianum* was one of the top-selling herbal supplements and ranked 11th in US natural channel with the total sales of $9,152,946 and was ranked 23rd in US mainstream multioutlet channel with 8.4% rise in sales from 2019. (https://www.herbalgram.org/resources/herbalgram/issues/131/table-of-contents/hg131-mkrpt/).

The aim of this chapter is to provide the detailed insight on the traditional uses, phytochemistry, pharmacological activities, and clinical studies of *S. marianum*.

13.2 Traditional Uses

Milk thistle was used from the ancient period of time for the treatment of various disorders related to liver, kidney, spleen, and gallbladder (Flora et al. 1998; Schadewaldt 1969). Pedanius Dioscorides, a Greek physician, pharmacologist, botanist, and author of pharmacopeia of medicinal plants "De material medica," has mentioned the use of tea of milk thistle in the treatment of poisonous snake bites. Pliny the Elder, a Roman naturalist, has noted that milk thistle juice together with honey helps to remove the excess bile from the body. In the late 16 centuries, the infusion of roots and seeds were used for the treatment of jaundice and gallstones. In the seventeenth century, the famous English botanist and herbalist, Nicholas Culpeper, wrote the use of milk thistle in opening the obstructions of the liver and spleen (Karimi et al. 2011). Traditionally, milk thistle was used in the treatment of depression, varicose veins, menstrual problems, diabetes, amenorrhea, hemorrhage, stimulation of breast milk production and bile secretion, and protection against mushroom poisoning (Jaggi and Singh 2016).

In Harighal, Pakistan, the leaves of *S. marianum* are used in liver problems, scanty urination, and respiratory tract infections (Amjad et al. 2020). In Frontier Region, Bannu located in the south of Khyber Pakhtunkhwa Province, Pakistan, herbal tea prepared from the root of *S. marianum* and vinegar is used for the treatment of liver disease and for carminative purposes (Adnan et al. 2014).

In Italy, leaves were used for the treatment of high fever, sores, and hemorrhoid, whereas the aerial parts of plants were used for the treatment of liver disorders and stomach pain. In Navarra, Spain, the boiled leaves and fruits were used as antihypertensive agents (Calvo and Cavero 2014). In France, decoction and tincture of fruits are used for the treatment of hemorrhoids, varicose ulcers, hay fever, asthma, nettle rash, and various hepatic disorders such as jaundice, bile stones, hepatitis, and steatosis (Benhouhou 2005). In Sinai Peninsula, Egypt, fruits are used in the treatment of CNS disorders such as depression and migraine, whereas tinctures from seeds are used for liver disorders and mushroom *(Amanita phalloides)* poisoning (Eissa et al. 2014). The flower and seed decoction were used by Arabic indigenous herbal practitioners for liver disease, poisoning, infertility, and sexual weakness (Said et al. 2002). The decoction of seeds was used as herbal remedy by breast cancer women in the West Bank of Palestine (Jaradat et al. 2016). In

Kohghiluyehva Boyer Ahmad province of Iran, powdered seed is used as sedative and galactagogue (Mosaddegh et al. 2012).

13.3 Chemical Constituents

The major active constituents obtained from the seeds are bioactive flavonoid and flavonolignans. Silymarin is the complex mixture of taxifolin and several flavonolignans, namely, silibinin A, silibinin B, isosilibinin A, isosilibinin B, silychristin, isosilychristin, and silydianin. Other flavonolignans isolated are 2,3-dehydrosilychristin, 2,3-dehydrosilymarin, 2,3-dehydrosilibinin (Takemoto et al. 1975; Kurkin et al. 2001). Structures of some of these compounds are represented in Fig. 13.2. Similarly, polyphenolic compounds such as hydroxycinnamic acids (caffeic, chlorogenic, ferulic, and cynarine acid) and flavonoids (taxifolin, apigenin, catechin, luteolin, luteolin-7-*O*-glucoside, quercetin, miricetin, 3′-*O*-methyltaxifolin and dihydrokaempferol), fixed oil (60% linoleic acid, 30% oleic acid, 9% palmitic acid), sterols (cholesterol, campesterol, stigmasterol, sitosterol), sugars (arabinose, rhamnose, xylose, glucose), tocopherol, protein, and mucilage were isolated from the plant seeds (Abenavoli et al. 2010; Lucini et al. 2016).

The aerial parts contain carbohydrates such as water-soluble polysaccharides (WSPS), pectinic substances (PS), hemicellulose (HMC-A and HMC-B), and arabino-4-galactan (Zhauynbaeva et al. 2017). The aerial parts of *S. marianum* subsp. Anatolicum contains flavonoids, namely, kaempferol 3-sulphate, apigenin

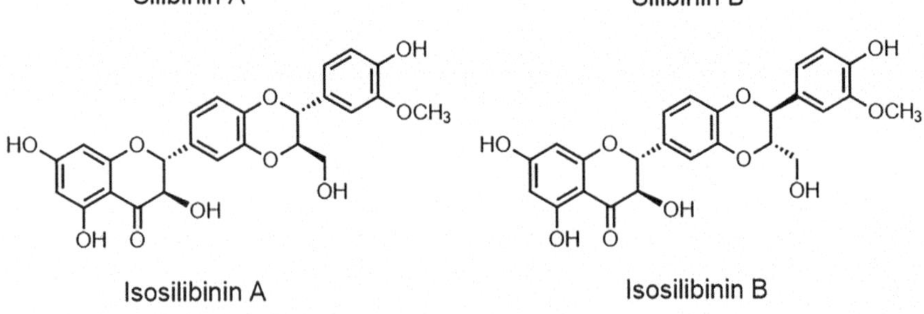

Fig. 13.2 Structure of major compounds from *Silybum marianum*

4′-7-diglucoside, kaempferol 7-glucoside, luteolin 7-glucoside, apigenin 7-glucoside, luteolin, and apigenin (Meriçli 1988).

The difference in flavonolignans constituents was reported in the two variants (white flowered and purple flowered) of *S. marianum*. Purple flowered variants were found rich in silibinin and silydianin, whereas these flavonolignans were not reported in white flower variants (Stieber et al. 1977). Various phytoconstituents reported from white flower variants were 3-deoxyflavonolignans, silymonin, silandrin (A and B), isosilandrin (A and B) silyhermin, neosilyhermine A, neosilyhermine B, naringenin, eriodictyol, apigenin, and chrysoeriol (Szilági et al. 1981; Fiebig and Wagner 1984; Samu et al. 2004),

13.4 Pharmacological Activities

S. marianum has been used from ancient times for the treatment of various medical ailments, which has led to the discovery of various pharmacological potentials such as antioxidant, antidiabetic, anti-inflammatory, analgesic, hepatoprotective, anticancer, and antiviral activities. Numerous studies have been conducted to explore the pharmacological activity, but this chapter provides the insight of recent studies conducted in *S. marianum*.

13.4.1 Antioxidant Activity

The in vitro and in vivo antioxidant potential of *S. marianum* extract and its active constituents have been reported by several researchers. Silymarin has shown effective DPPH scavenging, ABTS scavenging, superoxide anion radical scavenging, hydrogen peroxide scavenging, ferric ions (Fe^{3+}) reducing power by Fe^{2+}-Fe^{3+} transformation, cupric ions (Cu^{2+}) reducing ability by Cuprac method, and ferrous ion (Fe^{2+}) chelating activities (Serçe et al. 2016; Sulas et al. 2016; Köksal et al. 2009).

Silymarin exerts its antioxidant properties by increasing the expression of hepatic antioxidant enzymes such as superoxide dismutase (SOD), catalase (CAT), and glutathione peroxidase (GPx) (Ramadan et al. 2011; Surai 2015).

13.4.2 Anti-Inflammatory Activity

Several studies have shown that silymarin exerts a potent anti-inflammatory activity through the suppression of the release of cytokines, adhesion molecules such as VCAM-1, ICAM-1 and E-selectin, and nuclear factor kappa-light-chain-enhancer of activated B cells (NF-κB) signaling pathway, nitric oxide and 5-lipooxygenase pathways (Kang et al. 2003).

The anti-inflammatory activity of methanolic extract of dried leaf callus (100 mg/kg, b.w.) and leaf extract (100 mg/kg, b.w.) was evaluated using carrageenan and formalin-induced rat paw edema models. The leaf extract and leaf callus extract of *S. marianum* showed significant inhibition of paw edema formation in both carrageenan and formalin induced models. In both models, leaf callus extract showed a greater percentage of inhibition of paw oedema in comparison to that of standard aspirin (150 mg/kg) (Balian et al. 2006).

13.4.3 Antidiabetic Activity

Several studies have demonstrated the antidiabetic potential of silimarin/silibinin obtained from *S. marianum*. In alloxan and streptozotocin-induced diabetic rats, silibinin, and silibinin nanoparticles were able to reduce the serum glucose level, increase the level of antioxidant enzymes, and protect the pancreas from damage from alloxan and streptozotocin-induced damage (Soto et al. 2003, 2004; Das et al. 2014). The ZnO nanoparticles prepared from *S. marianum* seed extract at a dose of 10 mg/kg showed superior antidiabetic activity in alloxan-induced diabetic rats than the ZnO-treated group and extract treated group (Arvanag et al. 2019).

The new amide, namely, mariamides A ($IC_{50} = 1.50 \pm 0.01$ μM), along with other known compounds such as naringenin 7-O-β-D-glucopyranoside ($IC_{50} = 1.44 \pm 0.01$ μM) and dehydrodiconiferyl alcohol-4-β-D-glucoside ($IC_{50} = 0.38 \pm 0.01$ μM) isolated from the seeds of milk thistle showed more potent α-glucosidase inhibitory activity than the standard acarbose ($IC_{50} = 2.68 \pm 0.07$ μM) (Qin et al. 2017a).

Qin et al. (2017b) studied the in vitro α-glucosidase inhibitory activity and in vivo hypoglycemic activity of silycristin A isolated from the fruit of *S. marianum*. Silychristin A showed the potent α-glucosidase inhibitory activity by reducing the digestion rate of carbohydrates. Similarly, it was able to lower the glucose level, increased insulin secretion, and improved the structure of β cells in STZ-induced rats.

13.4.4 Anti-Atherosclerotic Activity

Radjabian and Fallah (2010) studied the anti-atherosclerotic activity of silymarin obtained from cultivated and wild plants of *S. marianum* in a rabbit fed a high cholesterol diet. At a dose of 200 mg/kg/day, silymarin showed significant inhibition of atherosclerotic plaque formation in the aortas of rabbits.

Kang et al. (2003) demonstrated the possible mechanism behind the anti-atherosclerotic activity of Silymarin. Silymarin inhibits TNF-alpha-induced expression of adhesion molecules such as VCAM-1, ICAM-1, and E-selectin in human umbilical vein endothelial cells and also inhibits the THP-1cell adhesion to human umbilical vein endothelial cells (HU-VECs).

13.4.5 Hepatoprotective Activity

Various studies reported that silymarin exhibited a protective effect against hepatic toxicity induced by a wide variety of agents (Ramadan et al. 2011; Freitag et al. 2015). The hepatoprotective effect of ethanolic extract of *S. marianum*/silymarin on hepatotoxic liver induced by carbon tetrachloride/acetaminophen was evaluated by measuring the liver function, tissue antioxidant enzymes, and histological examination of liver. The levels of ALT, AST, and ALP were found to decrease significantly in the group pretreated with ethanol extract/silymarin than the control group. Similarly, the levels of antioxidant enzymes (CAT, SOD, GSH-px) were found to be increased in the ethanol extract/silymarin pretreated group in comparison to the control group. The histopathological examination also revealed almost normal hepatocytes and sinusoids in the extract/silymarin pretreated liver samples, which indicates that the ethanolic extract/silymarin was able to offer protection to the rat liver against the CCL_4/acetaminophen intoxication.

13.4.6 Anticancer Activity

Many studies have reported the anticarcinogenic effects of silymarin for cancers of the colon, breast, prostate, bladder, oral, skin, and hepatocellular carcinoma (Zi et al. 1998; Vinh et al. 2002; Lahiri-Chatterjee et al. 1999; Tyagi et al. 2002; Kohno et al. 2002; Won et al. 2018; Varghese et al. 2005). Silymarin may have exhibited anticancer activities by regulating cell cycle, inducing apoptosis, inhibiting angiogenesis, invasion and metastasis, and inhibiting the growth of cancer cells (Ramasamy and Agarwal 2008).

13.4.7 Antiviral Activity

Several evidences have suggested that silymarin and its derivatives possess potent antiviral activity against numerous viruses such as flaviviruses (hepatitis C virus and dengue virus), togaviruses (Chikungunya virus and Mayaro virus), influenza virus, human immunodeficiency virus, and hepatitis B virus (Liu et al. 2019). Silymarin appears to block virus entry and transmission, possibly by targeting the host cell, and exhibit the antiviral action (Wagoner et al. 2010; Liu et al. 2019; Lalani et al. 2020).

13.4.8 Cardioprotective Activity

Tai Ping Sheng Hui Fang, the first medical formulary of the Song Dynasty of China, has mentioned the use of *S. marianum* in the prevention and treatment of cardiovascular diseases (Zhao et al. 2019).

S. marianum exerts cardioprotective effects against ischemia-reperfusion-induced myocardial infarction (MI) in rats in a dose-dependent manner with the greatest protective effects at a dose of 500 mg/kg by alleviating the oxidative stress, protecting the endogenous antioxidants, and suppressing the neutrophil infiltration (Rao and Viswanath 2007).

Vilahur et al. (2018) have reported the cardioprotective effect of *S. marianum* in pigs during the acute phase of MI and remodeling period post-MI, which may be attributed to the antioxidant and antifibrotic properties of *S. marianum*.

13.4.9 Neuroprotective Effects

Several studies in animal and cell model systems have demonstrated the neuroprotective effect of silymarin against CNS disorders such as Alzheimer's disease (Murata et al. 2010; Yaghmaei et al. 2014; Aboelwafa et al. 2020), Parkinson's disease (Haddadi et al. 2014; Perez-H et al. 2014), cerebral ischemia (Hou et al. 2010; Moghaddam et al. 2020), aging (Galhardi et al. 2009), and cognitive impairment (Yön et al. 2019; Shokouhi et al. 2020). Although the ability of silymarin to inhibit oxidative stress in the brain is considered the underlying mechanism behind its neuroprotective effects, there are several other possible molecular mechanisms supporting the neuroprotective effect of silymarin. The neuroprotective effect of silymarin may be attributed to its anti-inflammatory properties, inhibition of the mitochondrial apoptotic pathway, inhibition of β-amyloid aggregation, inhibition of microglia activation, and modulation of the estrogenic receptor-mediated pathway of neuronal death (Borah et al. 2013; Ullah and Khan 2018).

13.4.10 Renoprotective Effects

Several studies have demonstrated the renoprotective effect of silymarin against different drugs (vancomycin, gentamicin, cisplastin, thioacetamide, cyclophosphamide)-induced renal injury (Guzel et al. 2020; Mohammad 2012; Turgut et al. 2008; Cengiz 2018; Amien et al. 2015). In different studies, animals pretreated with silymarin showed a significant reduction in renal, apoptotic (caspase-3, caspase-8, and caspase-9 enzymes), and oxidative stress (NO and MDA levels) markers; serum blood urea nitrogen (BUN); and creatinine level. The renoprotective effect of silymarin may be attributed to its antioxidant, anti-inflammatory, and antiapoptotic properties.

13.5 Clinical Studies

Several numbers of clinical trials of milk thistle have been conducted in the last 40 years. Clinical studies have suggested that milk thistle extracts possess anticancer, antidiabetic, hepatoprotective, renoprotective effects, etc.

13.5.1 Hepatoprotective Effect

The hepatoprotective effect of silymarin is well researched and has undergone several clinical trials. As a herbal formulation, silymarin is available in various dosage forms (tablets and capsules) and dosages ranging from 210 to 1200 mg daily. Clinical studies were performed in patients with different liver diseases such as alcoholic liver disease, acute viral hepatitis, chronic viral hepatitis, chronic hepatitis C, alcoholic cirrhosis, nonalcoholic steato-hepatitis (NASH), and nonalcoholic fatty liver disease (NAFLD) (Federico et al. 2006; Polyak et al. 2013; Feher et al. 1989; Ferenci et al. 1989; Huber et al. 2005; Loguercio et al. 2012).The commercially available silymarin that has undergone clinical studies are Realsil, Legalon, Silipide, and Eurosil 85.

The outcomes of some clinical studies (Trinchet et al. 1989; Bunout et al. 1992; Parés et al. 1998; Tanamly et al. 2004; El-Zayadi et al. 2005; Gordon et al. 2006; Hawke et al. 2010; Fried et al. 2012) have demonstrated no effect of silymarin on biochemical parameters such as total bilirubin content, and alanine transaminase (ALT) level, whereas others (Salmi and Sarna 1982; Feher et al. 1989; Marcelli et al. 1992; Buzzelli et al. 1993; Melhem et al. 2005) have shown improvement in liver enzymes levels.

In a double-blind, prospective, randomized controlled clinical trial of silymarin (420 mg) treatment in 170 cirrhosis patients for 2 years, silymarin was able to increase the four- year survival rate in comparison to the placebo group with no side effects reported during the treatment period (Ferenci et al. 1989). Similarly, in another double-blind, randomized placebo-controlled trial, which was conducted in 55 patients with tuberculosis to evaluate the efficacy of silymarin in antituberculosis drug-induced liver injury, the silymarin-treated group showed a 28% reduction in the risk of liver injury in comparison to placebo (Luangchosiri et al. 2015).

Recently, Derakhshandeh-Rishehri et al. (2020) systematically reviewed the effects of Realsil (silibinin-phospholipid-vitamin E complex) on liver enzymes in patients with NASH and NAFLD. The intake of Realsil was able to significantly decrease the gamma-glutamyl transpeptidase (GGT) level without any effect on AST and ALT levels.

13.5.2 Anticancer Activity

Milk thistle has been used as an anticarcinogenic agent and as an agent to attenuate the side effects associated with anticancer treatment. The number of clinical trials on cancer patients has been increasing day by day.

In colorectal cancer patients, the blood level of insulin-like growth factor (IGF-1) is increased, while IGF- binding protein 3 (IGFBP-3) is decreased. Silibinin was able to demonstrate anticancer activity both in in vivo (Singh et al. 2002) and in vitro (Zi et al. 2000) studies by increasing the level of IGFBP-3. However, in the pilot study of silipide (silibinin-phosphatidylcholine), which was conducted in colorectal cancer patient at the dosage of 360, 720, and 1440 mg daily for 7 days, no significant differences in the serum level of IGF-1 and IGFBP-3 was observed before and after the treatment (Hoh et al. 2006). This study suggested that although the daily dose of 1.4 g of silibinin for 7 days was found to be safe for human but was unable to exhibit human colorectal cancer chemotherapeutic agents.

Flaig et al. (2007) performed phase I clinical trial to measure the toxic effect of high doses of silibinin-phytosome in prostate cancer patients and concluded that 13 g daily, in three divided doses, was found to be well tolerated by prostate cancer patients. In phase II trial, silibinin content was found high in blood levels and low in prostate/tissue level, and no significant differences were observed in IGF-1 and IGFBP-3 in baseline and posttreatment blood levels (Flaig et al. 2010).

13.5.3 Renoprotective Effects

Cisplastin, an anticancer drug, is usually associated with nephrotoxicity. In a clinical trial (Momeni et al. 2015), the renoprotective effect of silymarin (140 mg/b.i.d, 7 days) was evaluated in 60 patients with malignancy and who are the candidate of cisplastin treatment. In the silymarin pretreated group, serum creatinine and BUN levels were found to decrease after 14 days of cisplastin administration in comparison to the control group. Possible mechanisms behind its renoprotective effects may be due to its antioxidant and anti-inflammatory activities.

13.5.4 Obsessive Compulsive Disorder (OCD)

In 2010, the first double-blind randomized clinical trial on the effect of *S. marianum* leaf extract for the treatment of OCD was reported. The results revealed that *S. marianum* (capsule) when administered at a dose of 600 mg/day for 8 weeks was able to reduce the symptoms of OCD. Furthermore, there was no significant difference observed between the extract and fluoxetine in the treatment of OCD (Sayyah et al. 2010).

13.5.5 *Antidiabetic Effect*

Huseini et al. (2006) conducted a 4-month randomized, double-blind clinical trial in 51 type II diabetic patients. The results revealed that silymarin (200 mg, t.i.d)-treated group showed a significant decrease in HbA1c, FBS, total cholesterol, LDL, triglyceride SGOT, and SGPT levels compared with placebo.

Ebrahimpour-Koujan et al. (2018) performed a paralleled, randomized, triple-blinded, placebo-controlled clinical trial to evaluate the effect of silymarin supplementation on glycemic index and serum lipid profile in 40 type II diabetes patients. The results revealed that the patient who received 140 mg of silymarin thrice daily for 45 days showed significant improvement in glycemic index and lipid profile in comparison to the placebo group.

13.5.6 *Effect on Menopausal Symptoms*

A randomized, double-blind, placebo-controlled clinical trial was performed to evaluate the effect of *S. marianum* extract on 80 postmenopausal women (40–60 years) with symptoms of hot flashes. The results showed that *S. marianum* extract was able to improve the hot flash frequency and severity as well as total Greene Climacteric Scale and Hot Flash Related Daily Interference Score without any considerable side effects (Saberi et al. 2020).

13.5.7 Amanita phalloides *Poisoning*

About 1491 documented cases have reported that there was a survival rate of 93% in Amanita-poisoned patients treated with intravenous Legalon® SIL (silibinin-C-2,3-dihydrogen succinate) alone or in combination with the drug penicillin or N-acetylcysteine or thioctic acid (Mengs et al. 2012).

13.5.8 *COVID-19*

The antiviral, antioxidant, and immune-modulatory properties of silymarin make it one of the potential candidates in the treatment of SARS-CoV-2. Several clinical trials are being carried out to study the effect of silymarin in COVID-19 patients.

In a clinical trial SIL-COVID19 (EudraCT number: 2020–001794-77), the effects of oral Legalon® (nutraceutical containing STAT3 inhibitor silibinin) in two hospitalized patients with cancer and COVID-19 were studied. In patients treated with Legalon® (1050 mg/day, 7 days), only minimal oxygen support (2–4 L/min) was required, and there was a slight but significant reduction in inflammatory biomarkers, LDH, and STAT3-regulated C-reactive protein (CRP) (Bosch-Barrera et al. 2021).

13.6 Studies Related to Safety and Toxicity

Several acute and subacute toxicity tests of orally or intravenously administered silymarin in different animal models revealed no toxicity signs in general behavior and mortality rate (Desplaces et al. 1975; Hahn et al. 1968; Singh and Agarwal 2006). However, one study has demonstrated the teratogenic effect of silymarin on mouse fetus. At the higher dose (200 mg/kg/day), silymarin leads to embryo resorption, intrauterine growth retardation, and craniofacial, vertebral, and limb abnormalities (Gholami et al. 2016). More precise studies should be conducted to know the detailed mechanism behind this tetratogenic effect, and caution should be taken by pregnant women in the administration of silymarin.

In clinical trials of silymarin and its products, a low rate of adverse effects with no serious adverse effects and fatalities has been reported so far. Some side effects reported are mild laxative effect, sweating, GI disturbance (nausea, abdominal pain, diarrhea), headache, abdominal colic, fatigue, and pruritus (Saller et al. 2007; Tanamly et al. 2004;Fallahzadeh et al. 2012).

13.7 Conclusions and Future Recommendations

The worldwide ethnomedicinal relevance, safety profile, and diverse chemical constituents have made *S. marianum* an interesting and widely used plant for the exploration of its pharmacological properties and discovery of new plant-derived drugs. An immense number of in vivo and in vitro pharmacological tests have been carried out. Furthermore, in the last 40 years, several clinical studies have been conducted to explore the effect of *S. marianum* in patients with hepatic disorders, diabetes, cancer, renal disorders, etc. Many of these clinical trials are conducted in a small number of patients, so a well-designed, placebo-controlled clinical study in a larger number of populations should be conducted to get clinical approval for use against different diseases.

Although several researches exploring the traditional uses and pharmacological properties have been conducted, there are few researches exploring the pharmacokinetic profile. Therefore, researchers should also focus on this area of research. *S. marianum* is one of the top selling nutraceutical products in US markets, so cultivation and commercialization of *S. marianum* should be encouraged in regions where geographical and climatic conditions favor their cultivation. The massive production and commercialization can help to boom the economic growth of an individual and country.

References

Abenavoli L, Capasso R, Milic N, Capasso F (2010) Milk thistle in liver diseases: past, present, future. Phytother Res 24(10):1423–1432

Abenavoli L, Izzo AA, Milić N, Cicala C, Santini A, Capasso R (2018) Milk thistle (Silybum marianum): a concise overview on its chemistry, pharmacological, and nutraceutical uses in liver diseases. Phytother Res 32(11):2202–2213

Aboelwafa HR, El-Kott AF, Abd-Ella EM, Yousef HN (2020) The possible neuroprotective effect of silymarin against aluminum chloride-prompted Alzheimer's-like disease in rats. Brain Sci 10(9):628

Adnan M, Ullah I, Tariq A, Murad W, Azizullah A, Khan AL, Ali N (2014) Ethnomedicine use in the war affected region of Northwest Pakistan. J Ethnobiol Ethnomed 10(1):1–16

Amien AI, Fahmy SR, Abd-Elgleel FM, Elaskalany SM (2015) Renoprotective effect of *Mangifera indica* polysaccharides and silymarin against cyclophosphamide toxicity in rats. J Basic Appl Zool 72:154–162

Amjad MS, Zahoor U, Bussmann RW, Altaf M, Gardazi SMH, Abbasi AM (2020) Ethnobotanical survey of the medicinal flora of Harighal, Azad Jammu & Kashmir. Pak J Ethnobiol Ethnomed 16(1):1–28

Andrzejewska J, Sadowska K, Mielcarek S (2011) Effect of sowing date and rate on the yield and flavonolignan content of the fruits of milk thistle (*Silybum marianum* L. Gaertn.) grown on light soil in a moderate climate. Ind Crop Prod 33(2):462–468

Arvanag FM, Bayrami A, Habibi-Yangjeh A, Pouran SR (2019) A comprehensive study on antidiabetic and antibacterial activities of ZnO nanoparticles biosynthesized using *Silybum marianum* L seed extract. Mater Sci Eng C 97:397–405

Balian S, Ahmad S, Zafar R (2006) Anti-inflammatory activity of leaf and leaf callus of *Silybum marianum* (L.) Gaertn. In albino rats. Indian J Pharm 38(3):213

Benhouhou S (2005) A guide to medicinal plants in North Africa. IUCN, Malaga

Bijak M (2017) Silibinin, a major bioactive component of milk thistle (*Silybum marianum* L. Gaernt.)—chemistry, bioavailability, and metabolism. Molecules 22(11):1942

Borah A, Paul R, Choudhury S, Choudhury A, Bhuyan B, Das Talukdar A, Dutta Choudhury M, Mohanakumar KP (2013) Neuroprotective potential of silymarin against CNS disorders: insight into the pathways and molecular mechanisms of action. CNS Neurosci Ther 19(11):847–853

Bosch-Barrera J, Roqué A, Teixidor E, Carmona-Garcia MC, Arbusà A, Brunet J, Martin-Castillo-B, Cuyàs E, Verdura S, Menendez JA (2021) Clinical management of COVID-19 in cancer patients with the STAT3 inhibitor Silibinin. Pharmaceuticals 15(1):19

Bunout D, Hirsch S, Petermann M, De La Maza MP, Silva G, Kelly M, Ugarte G, Iturriaga H (1992) Controlled study of the effect of silymarin on alcoholic liver disease. Rev Med Chil 120(12): 1370–1375

Buzzelli G, Moscarella S, Giusti A, Duchini A, Marena C, Lampertico M (1993) A pilot study on the liver protective effect of silibinin-phosphatidylcholine complex (IdB1016) in chronic active hepatitis. Int J Clin Pharmacol Ther Toxicol 31(9):456–460

Calvo MI, Cavero RY (2014) Medicinal plants used for cardiovascular diseases in Navarra and their validation from official sources. J Ethnopharmacol 157:268–273

Cengiz M (2018) Renoprotective effects of Silybum marianum (L.) Gaertn (Silymarin) on thioacetamide-induced renal injury: biochemical and histopathological approach. Pak J Pharm Sci 31(5):2137–2141

Das S, Roy P, Pal R, Auddy RG, Chakraborti AS, Mukherjee A (2014) Engineered silibinin nanoparticles educe efficient control in experimental diabetes. PLoS One 9(7):e101818

Derakhshandeh-Rishehri SM, Heidari-Beni M, Eftekhari MH (2020) The effects of Realsil (silibinin-phospholipid-vitamin E complex) on liver enzymes in patients with non-alcoholic fatty liver disease (NAFD) or non-alcoholic steato-hepatitis (NASH): a systematic review and meta-analysis of RCTS. Acta Endocrinol 16(2):223

Desplaces A, Choppin J, Vogel G, Trost W (1975) The effects of silymarin on experimental phalloidine poisoning. Arzneimittelforschung 25(1):89–96

Ebrahimpour-Koujan S, Gargari BP, Mobasseri M, Valizadeh H, Asghari-Jafarabadi M (2018) Lower glycemic indices and lipid profile among type 2 diabetes mellitus patients who received novel dose of *Silybum marianum* (L.) Gaertn. (silymarin) extract supplement: a triple-blinded randomized controlled clinical trial. Phytomedicine 44:39–44

Eissa TAF, Palomino OM, Carretero ME, Gómez-Serranillos MP (2014) Ethnopharmacological study of medicinal plants used in the treatment of CNS disorders in Sinai peninsula, Egypt. J Ethnopharmacol 151(1):317–332

El-Zayadi AR, Attia M, Badran HM, El-Tawil A, Zalata K, Barakat E, Selim O, El-Nakeeb A, Saied A (2005) Non-interferon-based therapy: an option for amelioration of necro-inflammation in hepatitis C patients who cannot afford interferon therapy. Liver Int 25(4):746–751

Fallahzadeh MK, Dormanesh B, Sagheb MM, Roozbeh J, Vessal G, Pakfetrat M, Daneshbod Y, Kamali-Sarvestani E, Lankarani KB (2012) Effect of addition of silymarin to renin-angiotensin system inhibitors on proteinuria in type 2 diabetic patients with overt nephropathy: a randomized, double-blind, placebo-controlled trial. Am J Kidney Dis 60(6):896–903

Federico A, Trappoliere M, Tuccillo C, De Sio I, Di Leva A, Blanco CDV, Loguercio C (2006) A new silibinin-vitamin E-phospholipid complex improves insulin resistance and liver damage in patients with non-alcoholic fatty liver disease: preliminary observations. Gut 55(6):901–902

Feher J, Deák G, Müzes G, Lang I, Niederland V, Nekam K, Karteszi M (1989) Liver-protective action of silymarin therapy in chronic alcoholic liver diseases. Orvosihetilap 130(51): 2723–2727

Ferenci P, Dragosics B, Dittrich H, Frank H, Benda L, Lochs H, Meryn S, Base W, Schneider B (1989) Randomized controlled trial of silymarin treatment in patients with cirrhosis of the liver. J Hepatol 9(1):105–113

Fiebig M, Wagner H (1984) Neue antihepatotoxisch wirksame Flavonolignane aus einer weißblühenden Silybum-Varietät. Planta Med 50(04):310–313

Flaig TW, Gustafson DL, Su LJ, Zirrolli JA, Crighton F, Harrison GS, Pierson AS, Agarwal R, Glodé LM (2007) A phase I and pharmacokinetic study of silibinin-phytosome in prostate cancer patients. Invest New Drugs 25(2):139–146

Flaig TW, Glodé M, Gustafson D, van Bokhoven A, Tao Y, Wilson S, Su LJ, Li Y, Harrison G, Agarwal R, Crawford ED (2010) A study of high-dose oral silibinin-phytosome followed by prostatectomy in patients with localized prostate cancer. Prostate 70(8):848–855

Flora KD, Rosen HR, Benner KG (1996) The use of naturopathic remedies for chronic liver disease. Am J Gastroenterol 91(12):2654–2655

Flora K, Hahn M, Rosen H, Benner K (1998) Milk thistle (*Silybum marianum*) for the therapy of liver disease. Am J Gastroenterol 93(2):139–143

Freitag AF, Cardia GFE, da Rocha BA, Aguiar RP, Silva-Comar FMDS, Spironello RA, Grespan R, Caparroz-Assef SM, Bersani-Amado CA, Cuman RKN (2015) Hepatoprotective effect of silymarin (*Silybum marianum*) on hepatotoxicity induced by acetaminophen in spontaneously hypertensive rats. Evid Based Complement Alternat Med 2015:538317. https://doi.org/10.1155/2015/538317

Fried MW, Navarro VJ, Afdhal N, Belle SH, Wahed AS, Hawke RL, Doo E, Meyers CM, Reddy KR (2012) Effect of silymarin (milk thistle) on liver disease in patients with chronic hepatitis C unsuccessfully treated with interferon therapy: a randomized controlled trial. J Am Med Assoc 308(3):274–282

Galhardi F, Mesquita K, Monserrat JM, Barros DM (2009) Effect of silymarin on biochemical parameters of oxidative stress in aged and young rat brain. Food Chem Toxicol 47(10): 2655–2660

Gholami M, Moallem SA, Afshar M, Amoueian S, Etemad L, Karimi G (2016) Teratogenic effects of silymarin on mouse fetuses. Avicenna J Phytomed 6(5):542

Gordon A, Hobbs DA, Bowden DS, Bailey MJ, Mitchell J, Francis AJ, Roberts SK (2006) Effects of *Silybum marianum* on serum hepatitis C virus RNA, alanine aminotransferase levels and well-being in patients with chronic hepatitis C. J Gastroenterol Hepatol 21(1):275–280
Guzel S, Sahinogullari ZU, Canacankatan N, Antmen SE, Kibar D, Coskun Yilmaz B (2020) Potential renoprotective effects of silymarin against vancomycin-induced nephrotoxicity in rats. Drug Chem Toxicol 43(6):630–636
Haddadi R, Nayebi AM, Farajniya S, Brooshghalan SE, Sharifi H (2014) Silymarin improved 6-OHDA-induced motor impairment in hemi-parkisonian rats: behavioral and molecular study. DARU J Pharmaceut Sci 22(1):1–9
Hahn G, Lehmann HD, Kürten M, Uebel H, Vogel G (1968) On the pharmacology and toxicology of silymarin, an antihepatotoxic active principle from *Silybum marianum* (L.) Gaertn. Arzneimittelforschung 18(6):698–704
Hawke RL, Schrieber SJ, Soule TA, Wen Z, Smith PC, Reddy KR, Wahed AS, Belle SH, Afdhal NH, Navarro VJ, Berman J (2010) Silymarin ascending multiple oral dosing phase I study in noncirrhotic patients with chronic hepatitis C. J Clin Pharmacol 50(4):434–449
Hoh C, Boocock D, Marczylo T, Singh R, Berry DP, Dennison AR, Gescher AJ (2006) Pilot study of oral silibinin, a putative chemopreventive agent, in colorectal cancer patients: silibinin levels in plasma, colorectum, and liver and their pharmacodynamic consequences. Clin Cancer Res 12(9):2944–2950
Hou YC, Liou KT, Chern CM, Wang YH, Liao JF, Chang S, Chou YH, Shen YC (2010) Preventive effect of silymarin in cerebral ischemia–reperfusion-induced brain injury in rats possibly through impairing NF-κB and STAT-1 activation. Phytomedicine 17(12):963–973
Huber R, Futter I, Ludtke R (2005) Oral silymarin for chronic hepatitis C—a retrospective analysis comparing three dose regimens. Eur J Med Res 10(2):68–70
Huseini HF, Larijani B, Heshmat R, Fakhrzadeh H, Radjabipour B, Toliat T, Raza M (2006) The efficacy of *Silybum marianum* (L.) Gaertn. (silymarin) in the treatment of type II diabetes: a randomized, double-blind, placebo-controlled, clinical trial. Phytother Res 20(12):1036–1039
Jaggi AS, Singh N (2016) Silymarin and its role in chronic diseases. Drug Discov Mother Nat 929: 25–44. https://doi.org/10.1007/978-3-319-41342-6_2
Jaradat NA, Shawahna R, Eid AM, Al-Ramahi R, Asma MK, Zaid AN (2016) Herbal remedies use by breast cancer patients in the West Bank of Palestine. J Ethnopharmacol 178:1–8
Kang JS, Park SK, Yang KH, Kim HM (2003) Silymarin inhibits TNF-α-induced expression of adhesion molecules in human umbilical vein endothelial cells. FEBS Lett 550(1–3):89–93
Karimi G, Vahabzadeh M, Lari P, Rashedinia M, Moshiri M (2011) "Silymarin", a promising pharmacological agent for treatment of diseases. Iran J Basic Med Sci 14(4):308
Kohno H, Tanaka T, Kawabata K, Hirose Y, Sugie S, Tsuda H, Mori H (2002) Silymarin, a naturally occurring polyphenolic antioxidant flavonoid, inhibits azoxymethane-induced colon carcinogenesis in male F344 rats. Int J Cancer 101(5):461–468
Köksal E, Gülçin I, Beyza S, Sarikaya O, Bursal E (2009) *In vitro* antioxidant activity of silymarin. J Enzyme Inhib Med Chem 24(2):395–405
Kurkin VA, Zapesochnaya GG, Volotsueva AV, Avdeeva EV, Pimenov KS (2001) Flavolignans of *Silybum marianum* fruit. Chem Nat Compd 37(4):315–317
Lahiri-Chatterjee M, Katiyar SK, Mohan RR, Agarwal R (1999) A flavonoid antioxidant, silymarin, affords exceptionally high protection against tumor promotion in the SENCAR mouse skin tumorigenesis model. Cancer Res 59(3):622–632
Lalani SS, Anasir MI, Poh CL (2020) Antiviral activity of silymarin in comparison with baicalein against EV-A71. BMC Complement Med Therapies 20(1):1–12
Liu CH, Jassey A, Hsu HY, Lin LT (2019) Antiviral activities of silymarin and derivatives. Molecules 24(8):1552
Loguercio C, Andreone P, Brisc C, Brisc MC, Bugianesi E, Chiaramonte M, Cursaro C, Danila M, de Sio I, Floreani A, Freni MA (2012) Silibinin combined with phosphatidylcholine and vitamin E in patients with nonalcoholic fatty liver disease: a randomized controlled trial. Free Radic Biol Med 52(9):1658–1665

Luangchosiri C, Thakkinstian A, Chitphuk S, Stitchantrakul W, Petraksa S, Sobhonslidsuk A (2015) A double-blinded randomized controlled trial of silymarin for the prevention of antituberculosis drug-induced liver injury. BMC Complement Altern Med 15(1):1–7

Lucini L, Kane D, Pellizzoni M, Ferrari A, Trevisi E, Ruzickova G, Arslan D (2016) Phenolic profile and in vitro antioxidant power of different milk thistle [*Silybummarianum* (L.) Gaertn.] cultivars. Ind Crop Prod 83:11–16

Marcelli R, Bizzoni P, Conte D, Lisena MO, Lampertico M, Marena C, De Marco MF, Del Ninno E (1992) Randomized controlled study of the efficacy and tolerability of a short course of IdB 1016 in the treatment of chronic persistent hepatitis. Eur Bull Drug Res 1(3):131–135

Melhem A, Stern M, Shibolet O, Israeli E, Ackerman Z, Pappo O, Hemed N, Rowe M, Ohana H, Zabrecky G, Cohen R (2005) Treatment of chronic hepatitis C virus infection via antioxidants: results of a phase I clinical trial. J Clin Gastroenterol 39(8):737–742

Mengs U, Pohl T, Mitchell T (2012) Legalon® SIL: the antidote of choice in patients with acute hepatotoxicity from amatoxin poisoning. Curr Pharm Biotechnol 13(10):1964–1970

Meriçli AH (1988) Flavonolignans, kaempferol 3-sulphate and other flavonoids from *Silybum marianum* subsp. anatolicum. Planta Med 54(01):44–45

Moghaddam AH, Sangdehi SRM, Ranjbar M, Hasantabar V (2020) Preventive effect of silymarin-loaded chitosan nanoparticles against global cerebral ischemia/reperfusion injury in rats. Eur J Pharmacol 877:173066

Mohammad M (2012) Renoprotective effect of silymarin on gentamicin-induced nephropathy. Afr J Pharm Pharmacol 6(29):2241–2246

Momeni A, Hajigholami A, Geshnizjani S, Soleiman K (2015) Effect of silymarin in the prevention of cisplatin nephrotoxicity, a clinical trial study. J Clin Diagn Res 9(4):OC11–OC13. https://doi.org/10.7860/JCDR/2015/12776.5789

Mosaddegh M, Naghibi F, Moazzeni H, Pirani A, Esmaeili S (2012) Ethnobotanical survey of herbal remedies traditionally used in Kohghiluyehva Boyer Ahmad province of Iran. J Ethnopharmacol 141(1):80–95

Murata N, Murakami K, Ozawa Y, Kinoshita N, Irie K, Shirasawa T, Shimizu T (2010) Silymarin attenuated the amyloid β plaque burden and improved behavioral abnormalities in an Alzheimer's disease mouse model. Biosci Biotechnol Biochem 74(11):2299–2306

Parés A, Planas R, Torres M, Caballería J, Viver JM, Acero D, Panés J, Rigau J, Santos J, Rodés J (1998) Effects of silymarin in alcoholic patients with cirrhosis of the liver: results of a controlled, double-blind, randomized and multicenter trial. J Hepatol 28(4):615–621

Perez-H J, Carrillo-S C, Garcia E, Ruiz-Mar G, Perez-Tamayo R, Chavarrıa A (2014) Neuroprotective effect of silymarin in a MPTP mouse model of Parkinson's disease. Toxicology 7:38–43

Polyak SJ, Ferenci P, Pawlotsky JM (2013) Hepatoprotective and antiviral functions of silymarin components in HCV infection. Hepatology 57(3):1262

Porwal O, Ameen MSM, Anwer ET, Uthirapathy S, Ahamad J, Tahsin A (2019) *Silybum marianum* (Milk thistle): review on its chemistry, morphology, ethno medical uses, phytochemistry and pharmacological activities. J Drug Deliv Therapeut 9(5):199–206

Qin NB, Jia CC, Xu J, Li DH, Xu FX, Bai J, Li ZL, Hua HM (2017a) New amides from seeds of *Silybum marianum* with potential antioxidant and antidiabetic activities. Fitoterapia 119:83–89

Qin N, Hu X, Li S, Wang J, Li Z, Li D, Xu F, Gao M, Hua H (2017b) Hypoglycemic effect of silychristin a from *Silybum marianum* fruit via protecting pancreatic islet β cells from oxidative damage and inhibiting α-glucosidase activity in vitro and in rats with type 1 diabetes. J Funct Foods 38:68–179

Radjabian T, Fallah HH (2010) Anti-hyperlipidemic and anti-atherosclerotic activities of silymarins from cultivated and wild plants of *Silybum marianum* L. with different content of flavonolignans. Iran J Pharmacol Therapeut 9(2):63–67

Ramadan SI, Shalaby MA, Afifi N, El-Banna HA (2011) Hepatoprotective and antioxidant effects of *Silybum marianum* plant in rats. Int J Agro Vet Med Sci 5:541–547

Ramasamy K, Agarwal R (2008) Multitargeted therapy of cancer by silymarin. Cancer Lett 269(2): 352–362

Rao PR, Viswanath RK (2007) Cardioprotective activity of silymarin in ischemia-reperfusion-induced myocardial infarction in albino rats. Exp Clin Cardiol 12(4):179

Saberi Z, Gorji N, Memariani Z, Moeini R, Shirafkan H, Amiri M (2020) Evaluation of the effect of *Silybum marianum* extract on menopausal symptoms: a randomized, double-blind placebo-controlled trial. Phytother Res 34(12):3359–3366

Said O, Khalil K, Fulder S, Azaizeh H (2002) Ethnopharmacological survey of medicinal herbs in Israel, the Golan Heights and the West Bank region. J Ethnopharmacol 83(3):251–265

Saller R, Melzer J, Reichling J, Brignoli R, Meier R (2007) An updated systematic review of the pharmacology of silymarin. Complement Med Res 14(2):70–80

Salmi HA, Sarna S (1982) Effect of silymarin on chemical, functional, and morphological alterations of the liver: a double-blind controlled study. Scand J Gastroenterol 17(4):517–521

Samu Z, Nyiredy S, Baitz-Gács E, Varga Z, Kurtán T, Dinya Z, Antus S (2004) Structure elucidation and antioxidant activity of (−)-Isosilandrin isolated from *Silybummarianum* l. Chem Biodivers 1(11):1668–1677

Sayyah M, Boostani H, Pakseresht S, Malayeri A (2010) Comparison of *Silybum marianum* (L.) Gaertn. With fluoxetine in the treatment of obsessive− compulsive disorder. Prog Neuropsychopharmacol Biol Psychiatry 34(2):362–365

Schadewaldt H (1969) The history of Silymarin. Contribution to the history of liver therapy. Med Welt 20(15):902–914

Serçe A, Toptancı BÇ, Tanrıkut SE, Altaş S, Kızıl G, Kızıl S, Kızıl M (2016) Assessment of the antioxidant activity of *Silybum marianum* seed extract and its protective effect against DNA oxidation, protein damage and lipid peroxidation. Food Technol Biotechnol 54(4):455–461

Shokouhi G, Kosari-Nasab M, Salari AA (2020) Silymarin sex-dependently improves cognitive functions and alters TNF-α, BDNF, and glutamate in the hippocampus of mice with mild traumatic brain injury. Life Sci 257:118049

Singh RP, Agarwal R (2006) Prostate cancer chemoprevention by silibinin: bench to bedside. Mol Carcinog 45(6):436–442

Singh RP, Dhanalakshmi S, Tyagi AK, Chan DC, Agarwal C, Agarwal R (2002) Dietary feeding of silibinin inhibits advance human prostate carcinoma growth in athymic nude mice and increases plasma insulin-like growth factor-binding protein-3 levels. Cancer Res 62(11):3063–3069

Soto C, Recoba R, Barron H, Alvarez C, Favari L (2003) Silymarin increases antioxidant enzymes in alloxan-induced diabetes in rat pancreas. Comp Biochem Physiol Pt C Toxicol Pharmacol 136(3):205–212

Soto C, Mena R, Luna J, Cerbon M, Larrieta E, Vital P, Uria E, Sanchez M, Recoba R, Barron H, Favari L (2004) Silymarin induces recovery of pancreatic function after alloxan damage in rats. Life Sci 75(18):2167–2180

Stieber GG, Szilagyi I, Tetenyi P (1977) Hatoanyagtartalmi es osszetetelbelikulonbsegek a Silybumtgenusketfajaban (*Silybum marianum* (L.) Gaertn. es *Silybum eburneum* Coss. et Durr.). Herba Hungarica:55–75

Sulas L, Re GA, Bullitta S, Piluzza G (2016) Chemical and productive properties of two Sardinian milk thistle (*Silybum marianum* (L.) Gaertn.) populations as sources of nutrients and antioxidants. Genet Resour Crop Evol 63(2):315–326

Surai PF (2015) Silymarin as a natural antioxidant: an overview of the current evidence and perspectives. Antioxidants 4(1):204–247

Szilági I, Tétényi P, Antus S, Seligmann O, Chari VM, Seitz M, Wagner H (1981) Structure of silandrin and silymonin, two new flavanolignans from a white blooming *Silybum marianum* variety. Planta med 43(2):121–127

Takemoto T, Ikegawa S, Nomoto K (1975) Studies on constituents of Silybummarianum (L.) GAERTN. I. New flavonolignans named 2, 3-dehydrosilymarin and 2, 3-dehydrosilychristin. Yakugaku Zasshi 95(8):1017–1021

Tanamly MD, Tadros F, Labeeb S, Makld H, Shehata M, Mikhail N, Abdel-Hamid M, Abu-Baki L, Medhat A, Magder LS, Afdhal NH (2004) Randomised double-blinded trial evaluating silymarin for chronic hepatitis C in an Egyptian village: study description and 12-month results. Dig Liver Dis 36(11):752–759

Trinchet JC, Coste T, Levy VG, Vivet F, Duchatelle V, Legendre C, Gotheil C, Beaugrand M (1989) Treatment of alcoholic hepatitis with silymarin. A double-blind comparative study in 116 patients. Gastroenterol Clin Biol 13(2):120–124

Turgut F, Bayrak O, Catal F, Bayrak R, Atmaca AF, Koc A, Akbas A, Akcay A, Unal D (2008) Antioxidant and protective effects of silymarin on ischemia and reperfusion injury in the kidney tissues of rats. Int Urol Nephrol 40(2):453–460

Tyagi AK, Singh RP, Agarwal C, Chan DC, Agarwal R (2002) Silibinin strongly synergizes human prostate carcinoma DU145 cells to doxorubicin-induced growth inhibition, G2-M arrest, and apoptosis. Clin Cancer Res 8(11):3512–3519

Ullah H, Khan H (2018) Anti-Parkinson potential of silymarin: mechanistic insight and therapeutic standing. Front Pharmacol 9:422

Varghese L, Agarwal C, Tyagi A, Singh RP, Agarwal R (2005) Silibinin efficacy against human hepatocellular carcinoma. Clin Cancer Res 11(23):8441–8448

Vilahur G, Casaní L, Peña E, Crespo J, Juan-Babot O, Ben-Aicha S, Mendieta G, Béjar MT, Borrell M, Badimon L (2018) *Silybum marianum* provides cardioprotection and limits adverse remodeling post-myocardial infarction by mitigating oxidative stress and reactive fibrosis. Int J Cardiol 270:28–35

Vinh PQ, Sugie S, Tanaka T, Hara A, Yamada Y, Katayama M, Deguchi T, Mori H (2002) Chemopreventive effects of a flavonoid antioxidant silymarin on N-butyl-N-(4-hydroxybutyl) nitrosamine-induced urinary bladder carcinogenesis in male ICR mice. Jpn J Cancer Res 93(1): 42–49

Wagoner J, Negash A, Kane OJ, Martinez LE, Nahmias Y, Bourne N, Owen DM, Grove J, Brimacombe C, McKeating JA, Pécheur EI (2010) Multiple effects of silymarin on the hepatitis C virus lifecycle. Hepatology 51(6):1912–1921

Won DH, Kim LH, Jang B, Yang IH, Kwon HJ, Jin B, Oh SH, Kang JH, Hong SD, Shin JA, Cho SD (2018) *In vitro* and *in vivo* anti-cancer activity of silymarin on oral cancer. Tumor Biol 40(5): 1010428318776170

Yaghmaei P, Azarfar K, Dezfulian M, Ebrahim-Habibi A (2014) Silymarin effect on amyloid-β plaque accumulation and gene expression of APP in an Alzheimer's disease rat model. DARU J Pharmaceut Sci 22(1):1–7

Yön B, Belviranlı M, Okudan N (2019) The effect of silymarin supplementation on cognitive impairment induced by diabetes in rats. J Basic Clin Physiol Pharmacol 30(4). https://doi.org/10.1515/jbcpp-2018-0109

Zhao Z, Hodge J, Wang D, Liu Q (2019) New light shed on the old herb–*Silybum marianum*. Int J Cardiol 288:123

Zhauynbaeva KS, Rakhmanberdyeva RK, Abdurakhmanov BA (2017) Polysaccharides from *Silybum marianum*. Chem Nat Compd 53(5):820–822

Zi X, Feyes DK, Agarwal R (1998) Anticarcinogenic effect of a flavonoid antioxidant, silymarin, in human breast cancer cells MDA-MB 468: induction of G1 arrest through an increase in Cip1/p21 concomitant with a decrease in kinase activity of cyclin-dependent kinases and associated cyclins. Clin Cancer Res 4(4):1055–1064

Zi X, Zhang J, Agarwal R, Pollak M (2000) Silibinin up-regulates insulin-like growth factor-binding protein 3 expression and inhibits proliferation of androgen-independent prostate cancer cells. Cancer Res 60(20):5617–5620

The manufacturer's authorised representative in the EU is Springer Nature Customer Service Centre GmbH, Europaplatz 3, 69115 Heidelberg, Germany. If you have any concerns regarding our products, please contact ProductSafety@springernature.com

Printed and bound by CPI Group (UK) Ltd, Croydon, CR0 4YY
30/09/2026
02225299-0002